International Conference of Applied Sciences and Technology for Infrastructure Engineering (ICASIE 2017)

Material for Infrastructure Engineering

IOP Conference Series: Materials Science and Engineering
Volume 267

Surabaya, Indonesia
5 August 2017

Editors:

Shelvy Elvina Santoso
Ekky Hardianto

ISBN: 978-1-5108-5266-2
ISSN: 1757-8981

Printed from e-media with permission by:

Curran Associates, Inc.
57 Morehouse Lane
Red Hook, NY 12571

proceedings.com

Some format issues inherent in the e-media version may also appear in this print version.

Printed by Curran Associates, Inc. (2017)

For permission requests, please contact the Institute of Physics
at the address below.

Institute of Physics
Dirac House, Temple Back
Bristol BS1 6BE UK

Phone: 44 1 17 929 7481
Fax: 44 1 17 920 0979

techtracking@iop.org

Additional copies of this publication are available from:

Curran Associates, Inc.
57 Morehouse Lane
Red Hook, NY 12571 USA
Phone: 845-758-0400
Fax: 845-758-2633
Email: curran@proceedings.com
Web: www.proceedings.com

Table of contents

Volume 267

International Conference of Applied Science and Technology for Infrastructure Engineering

5 August 2017, Surabaya, East Java, Indonesia

Accepted papers received: 6 November 2017
Published online: 23 November 2017

Preface

Geopolymer Concrete and Materials Technology

Structural Dynamics, Engineering, and Sustainability

Seismic Design and Control of Structural Vibrations

Project Management

Transportation and Highway Engineering

Geotechnical Engineering

Water Engineering and Resources Management

Energy

Data Mining

IOP Conf. Series: Materials Science and Engineering **267** (2017) 011001 doi:10.1088/1757-899X/267/1/011001

Preface: International Conference of Applied Science and Technology for Infrastructure Engineering (ICASIE) 2017.

Authors: Shelvy Elvina Santoso, Ekky Hardianto.

The International Conference of Applied Science and Technology for Infrastructure Engineering (ICASIE) 2017 has been scheduled and successfully taken place at Swiss-Bell Inn Hotel, Surabaya, Indonesia, on August 5th 2017 organized by Department of Civil Infrastructure Engineering, Faculty of Vocation, Institut Teknologi Sepuluh Nopember (ITS).

This annual event aims to create synergies between government, private sectors; employers; practitioners; and academics. This conference has different theme each year and "MATERIAL FOR INFRASTUCTURE ENGINEERING" will be taken for this year's main theme. In addition, we also provide a platform for various other sub-theme topic including but not limited to Geopolymer Concrete and Materials Technology, Structural Dynamics, Engineering, and Sustainability, Seismic Design and Control of Structural Vibrations, Innovative and Green Buildings, Project Management, Transportation and Highway Engineering, Geotechnical Engineering, Water Engineering and Resources Management, Surveying and Geospatial Engineering, Coastal Engineering, Geophysics, Energy, Electronic and Mechatronic, Industrial Process, and Data Mining.

As the conference covered both engineering and science aspects. The main purpose of this seminar is to provide insights on building materials research to support the development of building infrastructure, develop the civil engineering knowledge and the related disciplines, by sharing, in order to contribute into the infrastructure problems-solving that affect the structures engineering, build and strengthen the networking among academics and practitioners of civil engineering both in the scope national and international. ICASIE 2017 was designated to provide professionals by giving them the exploratory experiences to develop and expand their technical/professional and educational skills which are necessary in the construction sectors.

Finally, we would like to express our greatest gratitude to Minister of Public Work and Housing – Republic of Indonesia DR. Ir. H. Moch. Basoeki Hadimoeljono, M.Sc, Minister of Research, Technology, and Higher Education – Republic of Indonesia Prof. H. Mohamad Nasir, Ph.D., Ak. Lecturer from Faculty of Engineering Technology (FETech) Universiti Malaysia Perlis, Head of Department Vocational University of Indonesia Prof. Dr. Ir. Sigit Pranowo Hadiwardoyo, DEA as the guest speaker of the conference. In addition, we would like to thank you the rector of ITS Ir. Joni Hermana, MSc.ES, Ph.D. and the dean of vocational faculty Prof. Ir. M. Sigit Darmawan M.Eng Ph.D.

We also would like to thank:

- The entire steering, scientific, and organizing committee members of ICASIE 2017 for their help and support.
- Faculty of Vocation, Institut Teknologi Sepuluh Nopember (ITS) for the support for ICASIE 2017.
- University partners and institutions: Universiti Teknologi Petronas, Malaysia; Universiti Malaysia Perlis, Malaysia;
- The distinguished keynote and invited speakers for their acceptance to give speeches and intellectual contribution on their respective fields of expertise.
- The participants of ICASIE 2017 for their contribution and sharing knowledge.
- The sponsors for cooperation and their consideration to support ICASIE 2017 including PT.Teknindo Geosistem Unggul, PT.Pelindo II, PT.Nindya Karya, PT.Cipta Mandiri Nusantara Abadi, PT.Marga Harjaya Infrastruktur, PT.Astra-Infra - Toll Road, PT. Pakuwon Jati, CV.Cipta Suramadu Consultant

IOP Conf. Series: Materials Science and Engineering **267** (2017) 011001 doi:10.1088/1757-899X/267/1/011001

List of Committees

Department of Civil Infrastructure Engineering Institut Teknologi Sepuluh Nopember 2017

Organized by:

Department of Civil Infrastructure Engineering, Faculty of Vocation,
Institut Teknologi Sepuluh Nopember, Indonesia

Journal Editor:

Ridho Bayuaji, ST, MT, Ph. D (Institut Teknologi Sepuluh Nopember, Indonesia)

Prof. Ir M Sigit Darmawan, M.Eng.Sc., PhD. (Institut Teknologi Sepuluh Nopember, Indonesia)

Assoc. Prof. Dr. Mohd Mustafa Al Bakri Abdullah, (Universiti Malaysia Perlis, Malaysia)

Steering Committee:

Ridho Bayuaji, ST, MT, Ph. D (Institut Teknologi Sepuluh Nopember, Indonesia)

Kamilia Aziz, ST, MT (Institut Teknologi Sepuluh Nopember, Indonesia)

Diana Wahyu Hayati S. T, M.Sc (Institut Teknologi Sepuluh Nopember, Indonesia)

International Scientific Committee:

Prof. Dr.Ir. Triwulan, DEA (Institut Teknologi Sepuluh Nopember, Indonesia)

Prof. Dr. Djwantoro Hardjito (Petra Christian University, Indonesia)

Prof. Dr. Nasir Shafiq, (Universiti Teknologi Petronas, Malaysia)

Dr. Januarti Jaya Ekaputri ST., MT. (Institut Teknologi Sepuluh Nopember, Indonesia)

Hamzah Fansuri , PhD (Institut Teknologi Sepuluh Nopember, Indonesia)

Assoc. Prof. Dr. Mohd Mustafa Al Bakri Abdullah, (Universiti Malaysia Perlis, Malaysia)

Dr. Subaer (Universitas Negeri Makasar, Indonesia)

Antoni, S.T, M.Eng., Ph.D. (Petra Christian University, Indonesia)

Dr. H. M. Jonkers (TU Delft, Netherland)

Dr. Benny Suryanto (Heriot-Watt University, UK)

Prof. Ir. Dr. Muhd Fadhil Nuruddin, MSc. (Universiti Teknologi Petronas, Malaysia)

Prof. Dr. rer. nat. Irmina Kris Murwani (Institut Teknologi Sepuluh Nopember, Indonesia)

Professor Dr. Aeslina Abdul Kadir, (University Tun Hussein Onn Malaysia, Malaysia)

Prof. Ir. Priyo Suprobo, MS., Ph.D (Institut Teknologi Sepuluh Nopember, Indonesia)

IOP Conf. Series: Materials Science and Engineering **267** (2017) 011001 doi:10.1088/1757-899X/267/1/011001

Prof. Ir M Sigit Darmawan, M.Eng.Sc., PhD. (Institut Teknologi Sepuluh Nopember, Indonesia)

Budi Suswanto, ST., MT., PhD (Institut Teknologi Sepuluh Nopember, Indonesia)

Prof. Dr. Ir. Han Ay Lie, M.Eng (Diponegoro University, Indonesia)

Ir. Agung Budipriyanto, MEng., PhD. (Institut Teknologi Sepuluh Nopember, Indonesia)

Ir. Christiono, MT., PhD (Institut Teknologi Sepuluh Nopember, Indonesia)

Dr. Machsus Fawzi, ST., MT. (Institut Teknologi Sepuluh Nopember, Indonesia)

Ir. Ervina, MSc., PhD (Institut Teknologi Sepuluh Nopember, Indonesia)

Prof. Buntara S. Gan (Nihon University, Japan)

Dr. Ir. Dicky Imam Wahyudi, MS. (Institut Teknologi Sepuluh Nopember, Indonesia)

Prof. Dr. Ir. Indarto, DEA (Institut Teknologi Sepuluh Nopember, Indonesia)

Mohamad Khoiri, ST., MT., PhD (Institut Teknologi Sepuluh Nopember, Indonesia)

Dr. Mohammad Muntaha, ST., MT. (Institut Teknologi Sepuluh Nopember, Indonesia)

Dr. Widya Utama DEA (Institut Teknologi Sepuluh Nopember, Indonesia)

Ir. Totok Ruki Biyanto, S.T., M.T., Ph.D (Institut Teknologi Sepuluh Nopember, Indonesia)

Prof. Dr. Ir. Soeprijanto, MSc (Institut Teknologi Sepuluh Nopember, Indonesia)

Dr. Ir. Kuntjoro, MT (Institut Teknologi Sepuluh Nopember, Indonesia)

Dr. Ir. Suharjoko, MT (Institut Teknologi Sepuluh Nopember, Indonesia)

Dr. Ir. Hendra Wahyudi, MS. (Institut Teknologi Sepuluh Nopember, Indonesia)

Prof. Dr. Ir. Bangun Muljo Sukojo DEA.DESS

Prof. Dr. Ir. H. Nadjadji Anwar, M.Sc (Institut Teknologi Sepuluh Nopember, Indonesia)

Silvianita, S.T., M.SC., Ph..D (Institut Teknologi Sepuluh Nopember, Indonesia)

Dr. Ir. Hitapriya Suprajitno M. Eng (Institut Teknologi Sepuluh Nopember, Indonesia)

Dr. Ir. Ria Asih Aryani Soemitro, M.Eng (Institut Teknologi Sepuluh Nopember, Indonesia)

Prof. Dr. Ir. Danawati Hari Prajitno (Institut Teknologi Sepuluh Nopember, Indonesia)

Chairman:

Ridho Bayuaji, ST, MT, Ph. D (Institut Teknologi Sepuluh Nopember, Indonesia)

IOP Conf. Series: Materials Science and Engineering **267** (2017) 011001 doi:10.1088/1757-899X/267/1/011001

Keynote Speakers:

Assoc. Prof. Dr. Mohd Mustafa Al-Bakri Abdullah (Universiti Malaysia Perlis, Malaysia)

Prof. Ir. Dr. Muhammad Fadhil Nuruddin (Universiti Teknologi Petronas, Malaysia)

Dr. Mohammad Anuar bin Kamaruddin (Universiti Sains Malaysia)

Ir. Wahyu P. Kuswanda (Marketing Director of PT. Teknindo Geosistem Unggul)

IOP Conf. Series: Materials Science and Engineering **267** (2017) 011002 doi:10.1088/1757-899X/267/1/011002

Peer review statement

All papers published in this volume of *IOP Conference Series: Materials Science and Engineering* have been peer reviewed through processes administered by the proceedings Editors. Reviews were conducted by expert referees to the professional and scientific standards expected of a proceedings journal published by IOP Publishing.

IOP Conf. Series: Materials Science and Engineering **267** (2017) 012001 doi:10.1088/1757-899X/267/1/012001

Review: Potential Strength of Fly Ash-Based Geopolymer Paste with Substitution of Local Waste Materials with High-Temperature Effect

S Subekti[1], R Bayuaji[1], M S Darmawan[1], N A Husin[1], B Wibowo[1], B Anugraha[1], S Irawan[2], D Dibiantara[1]

[1] Departement of Civil Infrastructure Engineering, Institut Teknologi Sepuluh Nopember, Surabaya Indonesia.
[2] Universiti Teknologi Petronas, Bandar Sri Iskandar, Malaysia

E-mail: bayuaji@ce.its.ac.id

Abstract. This research provided an overview of the potential fly ash based geopolymer paste for application in building construction. Geopolymer paste with various variations of fly ash substitution with local waste material and high-temperature influence exploited with the fresh and hardened condition. The local waste material which utilized for this study were sandblasting waste, carbide waste, shell powder, bagasse ash, rice husk and bottom ash. The findings of this study indicated that fly-based geopolymer paste with local waste material substitution which had high-temperature influence ash showed a similar nature of OPC binders potentially used in civil engineering applications.

Keywords: Geopolymer Paste, Fly Ash, Local Waste Material, High Temperature

1. Introduction

The growth of technology and current economic activity leads us to live in complex human activities and affect the growing demand for infrastructure business. Infrastructure and construction materials have a close connection; a component of building materials that become the primary requirement was cement. The demand for cement would continue to increase and was predicted to experience an imbalance compared to its production numbers until 2030, and this would be a global problem. On the other hand, the researchers' awareness towards this situation had been growing to comprehend and conceive more about the environmental sustainability system. The facts showed that cement industry had contributed to the accumulation of carbon dioxide (CO2) in the air close to 8% of other industrial waste contribution. Thus, it is suggested to conduct research in the field of building materials and construction focused on designing alternatives to cement usage. Therefore, this is the right time to promote alternative concrete with 100% non-cement hydration content, which is geopolymer concrete. Also, one of the advantages of geopolymer concrete is that it has an excellent resistance both in acid and salt environment [1] and shows a strength escalation as the curing time and temperature increase. Another factor that determines the strength of this concrete is the fineness of the material since it enhances the activation while the polymerization process held. But those aren't everything, some other

[1] Ridho Bayuaji, bayuaji@ce.its.ac.id, +62315938025

parameters, such as the molarity of alkaline, the ratio of alkaline activator composers, the addition of superplasticizer, and other stuff, play major rule for geopolymer concrete's performances.
The mechanism of the geopolymer binder was the synergy of the alumina-silicate material and the activator material. One of the most widely used activators of sodium silicate Na2SiO3 that served to speed up the polymerization reaction. Also, sodium hydroxide (NaOH) is used in alkaline solution to help bonding process between particles [2]. After those processes, usually, researchers cure the sample to harden the samples [3]. Fly Ash (FA), one of the alumina-silicate materials, is a waste from coal combustion process. The development of the FA-based geopolymer research in Indonesia was based on the physical and chemical characteristics of FA from each burning site that had highly variable characters [4-7]. It was due to the process and resources used in producing a different FA that became a separate issue to determine the overall FA based geopolymer quality[8]. Besides the ratio of alkaline activator composition, the molarity of alkaline and the usage of admixture will affect the compressive strength and workability of fresh geopolymer concrete. The utilized local waste materials were chosen to have the character of pozzolan or alumina-silicate to reduce the dominance of fly ash in geopolymer. This study reviews the possible usage of waste from local materials in geopolymer resistance of high temperatures.

2. Geopolymer Development with Local Waste Materials

Geopolymer is a new material which still developing the standard design procedure of its constituent material mixing up until this moment. However, trial and error method still used. Nevertheless, there are other approaches, for example, Taguchi method. One of the main reasons why there was no standard for mixing procedures is because fly ash's raw materials are quite varied based on the process of picking, transporting and burning coal. Below are the various compositions of local waste materials, with the objection to substituting FA with six local materials in East Java found by researchers.

2.1 Sand-blasting

The mixture of geopolymer paste, used by [13], with the composition of FA (50%) and Sand-blasting waste (50%) resulted in: time setting, compressive strength, porosity, UPV, and permeability. This result was better than Fly Ash composition (100%). Sand-blasting had the potential to increase the setting time of geopolymer. By using 100% sand-blasting composition to geopolymer, the setting time could be reached at about one week. Duan et al. [15]. Has varied the form of silica fume used in his study with 10%, 20%, and 30% respectively of its total binder (they compared with a reference sample of FA 100%). The curing method for the samples were made by heating exposures up to 200°C, 400°C, and 800°C after 28 days. Furthermore, the addition of silica fume could increase the compressive strength rapidly. The addition of silica fume resulted into a denser and more homogeneous sample, which supports the strong performance. The maximum silica used in the study was 30% of total binder. This lead into the creation of most impenetrable and homogeneous microstructure. In another word, it enhances the microstructure and improves its thermal resistance.

2.2 Shell Powder (SP)

The CaO's content on shell powder showed the interaction with FA. According to Freizna's study, it was noted that the composition of the geopolymer paste were FA (50%) and SLP (50%). This composition resulted in a better compressive strength than FA full replacement of 100%.
Olivia et al. [16] found that the partially substituted-OPC with seashell could decrease the compressive performance but supporting the tensile and flexural strength. Based on research by Lertwattanaruk et al. [17], the calcium content of seashell is less than the OPC's therefore it caused a slow disruption. They have studied that the more shell powder used to replace some part of the OPC the faster its setting time, but the particle remains unreacted and acts as a filler as its particle is smaller than the OPC's. Although, the unreactive act could cause a diminution of strength compared to the reference sample, the compressive strength still met the standard and requirement for plastering. Furthermore, its thermal insulation showed a slightly better value compared to 100% OPC's.

IOP Conf. Series: Materials Science and Engineering **267** (2017) 012001 doi:10.1088/1757-899X/267/1/012001

Restu et al. [12] concluded that higher calcium content could disrupted the polymerization process and changes the microstructure, so they have a worse compressive strength compared with the reference's sample (100% FA).

2.3 Sugar Cane Ash (SCA)

The SiO_2 content on SCA was above 70% making SCA a potential material for geopolymer paste mixture. Alfi proved that SCA and FA composition of 20%: 80% with additional of Na2SiO3 and NaOH within 1.5 ratio were resulted into a similar compressive strength with FA composition 100%. This result proved that FA dominance reduced by the same result as FA 100%. Castaldelli et al. [18] characterized the ash by its forming compound, as it composed by 46.3% of Silica Dioxides and 23.9% Calcium Oxide, both are from the inorganic fraction (which is 67.8% of the total composition). Beside, this ash composed by high percentage of crystallized matter (quartz mostly, calcite, etc.), which was reactive in a high alkaline environment [19]. The organic fraction which found inside the Sugar Cane Bagasse Ash could affect the setting time above heating temperature of 650°C during calcination process. The determination of mortar's mechanical properties samples for both compression and flexural test depend on the alkali activators used. Compressive mechanical strength was reported on the range of 30-55 Mpa. Bahurudeen et al. [20] did a deeper research of Raw Bagasse Ash preparation with 600, 700, and 800°C burning temperatures variations and used various fineness number of the ash from 210-45μm. They found that the optimum value of burning was at 700°C. A temperature higher than 700°C could produce trydimite and cristobalite crystallization which led into the degradation of pozzolanic activity[21]. Naturally, this material has less pozzolanic activity than the minimum requirement as per standard. Besides the burning process, we could enhance the rise of pozzolanic activity by grinding process. Sieving trough 300μm were suggested to achieve a maximum pozzolanic activity, lower the loss on ignition, and minimize the energy used in the course mixing. Finer particle gave a better chance of a denser sample. In this circumstances, the mix could fill the air and water cavity to enhance a better performance[3].

The existence of Sugar Cane Bagasse Ash triggered the reactivity of Blast Furnace Slag mixture at early ages [22]. The combination of mixed alkali activator solution and curing temperature of 20°C was resulted in a better compressive strength (55-65 Mpa). The different composition of BFS/SCBA mixture were 100/0, 85/15, 75/25, and 60/40. Pereira, et al.[19] Studied the same mix of BFS and SCBA. He replaced 25% of BFS with SCBA and got a similar/higher compressive strength than the reference (100% BFS) for a longer curing period. After seven hours of 65°C thermal activation, the samples cured for 3-90 days. Based on the results, found that the influence of alkali activator could determine its mechanical compressive strength. The usage of Natrium Hydroxide resulted in a high-early strength. However, when KOH were used as alkali activator, the final strength were two times greater than using NaOH.

2.4 Rice Husk Ash (RHA)

The use of rice husk ash in a geopolymer binder could decrease the compressive strength; it reinforced by the SEM RHA results described that RHA particles had "water absorbed" sponge surfaces and were slow to react. Therefore, the water which used for the polymerization process seemed to reduce due to the ASP absorption and resulted in longer binding time. This result was good for ASP combination with FA. This combination was proposed improvements in future research. Because particle refining with grinding and filtering could optimized the combustion temperatures and firing time. Hwang et al. [23] discussed the effect of substituting part of fly ash with RHA and found out that the optimum substitution proportion was 35% of total weight. Those two pozzolanic materials could be activated using Sodium Hydroxide and Sodium Silicate solution, with the concentration of 8; 10; 12; and 14 M of Sodium Hydroxide and the ratio of Silicate to Sodium Oxide 0.93; 0.81; 0.73; and 0.66. The research proved that the value of 10M Sodium Hydroxide concentrations in 35% RHA mixture at 56 days curing time was the optimum proportion to achieve an ideal compressive strength needed.

The samples were de-molded then cured at 35°C by placing it inside an automated temperature system with 50% relative humidity.
Nazari et al. [3] have varied the material based on its fineness mixture proportion. They have compared the fine fly ash (3μm) and coarse fly ash (75μm) with RHBA fine particle (7μm) and coarse particle (90μm). They used sodium hydroxide and sodium silicate solution with the molarity of 4; 8; or 12 M as alkali activator. The samples obtained were pre-cured at room temperature for 24 hours and soaked in water. There were three ways of curing method. First, the samples were placed in a 25°C ambient temperature and second, the samples were placed inside the oven for 36 hours at the elevated temperatures of 50-90°C; and the last method, samples were heated at 80°C for 36 hours. All samples were tested after 7 and 28 days after it was de-molded. Those processes have shown us that the highest compressive strength were obtained by using 12 M of sodium hydroxide with ratio of 2.99 silica to aluminum oxide and a proportion of 70:30 fine fly ash and fine rice husk ash which obtained 58.9 Mpa of compressive strength.
Sturm et al. [24] concluded some factors which affect the high compressive strength samples. Those factors including:
(a) No swelling found on the hardened paste,
(b) Better workability
(c) A good homogeneity for both for fresh and hardened geopolymer paste.
During geopolymerization, the formation of crystalline by product like Zeolite would enhance the bad strength, which fortunately did not find in the geopolymerization process of RHA one-part geopolymer based. Another advantage of RHA usage is its high reactivity towards the alkali activator. This circumstances will lead into a high SiO2/Al2O3 ratio, which affected the microstructure (ranged start from 3.5 mole/mole) density and homogeneity. RHA has a higher specific surface area compared to other silica-rich material, therefore, it has high reactivity.

2.5 Carbide Waste (CW)
The CaO compound in carbide waste was quite high. It can be seen from Phetchuay et al. research [25]. About 70% of Calcium Carbide chemical composition residue is CaO. It is an indication that this material is reactive and a potential cementitious material [26]. Some studies have done to observe its performances including strength, durability, and microstructure for pavement or subgrade material. Peng Du et al. [27] investigated two type of silicon carbide (silicon carbide particles and whiskers) into their metakaolin geopolymerization. The microstructure analysis concluded that a denser and a less porous geopolymer composite could be achieved by using spherical material particles and whisker's shape in reverse. A spherical carbide particle could induce a better compressive property. Although, both particles have a good interfacial bonding between the filler and geopolymer, its silica reactivity of silica is questioned.
Hanjutsiwan et al. [28] presented that the usage of bottom ash based geopolymer was varied by CCR. He figured out that it has high content calcium hydroxide (which is about the same as the OPC's). The Bottom Ash which induced with calcium rich-material is a good alternative for the geopolymer binder instead of using Fly Ash.
Previous research has provided a source that a large CaO would affect the acceleration of setting time. Putri's research showed that a good combination of CW and FA-based geopolymer paste at 1: 1 ratio was good portion to achieved compressive strength. While, NaOH 8M, Na2SiO3 ratio also NaOH 0.5 was still better than FA 100% composition.

2.6 Fly Ash and Bottom Ash
Apha Sathonsaowaphaka [29] utilized the bottom ash from power generation waste to develop geopolymer mortar. The conclusion of this study was NaOH addition could slightly increase geopolymer mortar's resilience while maintaining its strength.
A small ratio of silica to alumina compound, could reduce the compressive strength of bottom ash based geopolymer as reported by Hanjitsuwan et al. [28]. Geopolymer performance will increase after the

IOP Conf. Series: Materials Science and Engineering **267** (2017) 012001 doi:10.1088/1757-899X/267/1/012001

addition of SiO2/Al2O3 and calcium or silica promoter like rice husk ash, glass, sugar cane ash, carbide waste, and other materials. Through a microscopically scanning figured, the paste which made from bottom ash contains some unreacted particle or partially reacted particle in the matrix (less dense form compared to the mixture which involving a promoter.

According to Lee W et.al [30], aggregate-binder interfaces were stronger in geopolymer than it was in the OPC's, [8] conclude that this particular character could enhance the performance of geopolymer concrete. The use of bottom ash can't substitute the fly ash for improving geopolymer's strength. In other words, the existence of bottom ash could worsen the compressive strength's results instead of improve it [31]. In the same report, the researcher proved that the usage of gypsum gas desulfurization flue (FGDG) could increase the fly ash compressive strength up to 10%. According to SEM observation, a simple fly ash mixture of geopolymer was more homogenous and denser compared to a simple bottom ash mixture. In a high content of fly ash, the usage of FGDG could disrupt the matrices' density. It was in contrary with fly ash.

3 High Temperature Effect on Geopolymer Concrete

Geopolymer concrete has characteristics like ceramics. It shared the same strength characteristics and has a high temperature resistant. Daniel [32] undertook an experimental in geopolymer study under high temperatures that focused on metakaolin-based geopolymer. To compare the performance of metakaolin-based geopolymer on paste, mortar, and concrete, he used the FA-based geopolymer comparators. The highest temperature which used during the study was 800°C. Aggregate dimension factor was the most important parameter to show the performance of geopolymer concrete at high temperature, and the minimum aggregate dimension factor was 10.

The comparison of geopolymer mortar strength and OPC showed in Dao's research [33]. This study utilized F class FA from Phalai Power Plant and 14M NaOH activator with K2SiO3 modulus of 3. Complete result of compressive strength ratio with high-temperature influence on mortar presented in Table 1.

Table 1. Strong Press Mortar Geopolymer (GM) and Ordinary Portland Cement (OPC) with temperature variations

Temperature	**Geopolymer Mortar (GM)**	**OPC Mortar (OM)**
	(Strength)	
°C	MPa	MPa
25	41.98	40.71
200	71.01	38.05
400	68.53	35.21
600	47.42	29.35
800	31.77	12.79
1000	26.36	5.91

IOP Conf. Series: Materials Science and Engineering **267** (2017) 012001 doi:10.1088/1757-899X/267/1/012001

Table 1 explained that GM has excellent temperature resistance compared to OM at room temperature. The GM compressive strength sample was about 42 MPa, almost the same as the OM sample. As the temperature increased to 600°C, the compressive strength of OM gradually decreased from 40.71 MPa to 38, 35 and 29 MPa (7, 14 and 18%, respectively, compared to the value at room temperature). On the one hand, when the temperature rose to 600°C, the compressive strength of GM showed an increase of 70, 63 and 13%, compared to the value at room temperature. When the temperature was increased up to 1000°C, OM compressive strength dramatically reduced to 6 MPa, equivalent to only 14% of the strength at room temperature. Similarly, GM's sample also experienced a decrease in strength; however, its strength remained above 26 MPa, equivalent to 63% of the power achieved at room temperature.

Geopolymer was produced using environmentally friendly alkaline activators (based on Rice Husk Ash and potassium hydroxide). The aluminosilicate particles, carbon fibers, and ceramics were used as reinforcing materials. The effect of reinforcement materials on bending strength, linear shrinkage, thermo-physical properties and micro-geopolymer structures in the room and high temperature (1200 ° C) were studied. The result showed that the composite toughness increased by 110.4% for geopolymer which reinforced with ceramic fiber (G-AF) at room temperature. The presence of particles improved by 265% on bending behavior for geopolymer which reinforced by carbon fibers after exposure to 1200 ºC. Linear-shrinkage occurred in geopolymer which reinforced by ceramic. Its particle grains and G-AF geopolymer compared with reference samples (without fibers and particles) increased by 27.88% and 7.88% respectively at 900 ° C. The geopolymer material which developed in this work was a porous material with low thermal conductivity and excellent mechanical properties with potential thermal insulation applications for building applications.

Based on an observation done by [34] the usage of hot-stage TEM could be used to observe the geopolymer microstructure. The researcher explained that four hour geopolymer heat exposure at 1000°C could stabilize the Nano sized microstructure. This researcher tried to compare three different factor of curing, pressure less, warm, and highly-pressured sample in an autoclave with the purpose to find out the effects on the diminution of entrapped air, gain of reactivity, improvement of compound dissolution, and the degree of workability.
A review of [35] suggested that the standard curing temperature was at the range of 60°C to 90°C for 24 hours. [36] agreed to that statement and conducted the research at temperature of 65°C for 24 hours inside oven. A slightly different method of curing performed by [31]. The researcher placed the samples in an electric oven at 40°C for 48 hours. This curing treatment will increase the strength up to 70% within three to four hours.

4 Conclusion

Local material waste could be potentially used to substitute fly ash in geopolymer binders. It could improve geopolymer concrete setting time and compressive strength. Another factor that determines the compressive strength was the fineness of the material since it enhances the activation during the polymerization process. Yet, some other parameters such as alkaline molarity, activator ratio, alkaline composers, superplasticizer addition and any other way play important role for geopolymer concrete's performances. The addition of silica fume will also increase the compressive strength rapidly. In addition, the addition of silica fume can also create a denser and more homogeneous sample, which supports the strong performance. The usage of the seashell in OPC substitution could disrupt the composition of calcium which could potentially change the microstructure and distract the polymerization process. Yet, seashell substitution in OPC could increase the flexural and tensile strength. Bottom ash cannot be used to partly or wholly substitute fly ash due to its bad homogeneity and density. Therefore, the incorporation of other material such as FGDG or silica-rich material is highly recommended. Geopolymer concrete has a better ability than conventional concrete with high-temperature exposure.

IOP Conf. Series: Materials Science and Engineering **267** (2017) 012001 doi:10.1088/1757-899X/267/1/012001

5 Reference

1. Van Chanh N, Trung BD, Van Tuan D, editors. Recent research geopolymer concrete. The 3rd ACF International Conference-ACF/VCA Vietnam; 2008.
2. Kamseu E, à Moungam LB, Cannio M, Billong N, Chaysuwan D, Melo UC, et al. Substitution of sodium silicate with rice husk ash-NaOH solution in metakaolin based geopolymer cement concerning reduction in global warming. Journal of Cleaner Production. 2017;142:3050-60.
3. Nazari A, Bagheri A, Riahi S. Properties of geopolymer with seeded fly ash and rice husk bark ash. Materials Science and Engineering: A. 2011;528(24):7395-401.
4. Bayuaji R, Sigit Darmawan M, Wibowo B, Ahmad Husin N, Subekti S, Ekaputri JJ, editors. The Influence of Chloride Environment on Compressive Strength of Geopolymer Concrete with Fly Ash Using Taguchi Approach. Applied Mechanics and Materials; 2015: Trans Tech Publ.
5. Dwijayanti LY, Ekaputri JJ, Bayuaji R, editors. Lightweight geopolymer binder with abaca fiber in different curing. Materials Science Forum; 2016: Trans Tech Publ.
6. Fansuri H, Fatmawati A, Utomo WP, Supriadi W, Bayuaji R, editors. Cd 2+ and Cr 3+ Cation Immobilization by Using Geopolymer Based on PT. IPMOMI Fly Ash. Materials Science Forum; 2016.
7. Supriadi W, Bayuaji R, Burhan R, Fansuri H, editors. The Effect of Pb 2+ and Cd 2+ Addition to Mechanical Properties of Fly Ash Geopolymer Paste. Materials Science Forum; 2016.
8. Rangan BV. Fly ash-based geopolymer concrete. 2008.
9. Ilmiah R. Pengaruh Penambahan Abu Sekam Padi sebagai Pozzolan pada Binder Geopolimer Menggunakan Alkali Aktivator Sodium Silikat (Na2SiO3) serta Sodium Hidroksida (NaOH). Surabaya.: Institut Teknologi Sepuluh November; 2017.
10. Kurniasari PT. Pemanfaatan Penggunaan Fly Ash dan Bottom Ash sebagai Pozzolan pada Binder Geopolymer. Surabaya: Institut Teknologi Sepuluh Nopember; 2017.
11. M A. Studi Penggunaan Abu Ampas Tebu dan Fly Ash pada Pasta Geopolimer. Surabaya: Institut Teknologi Sepuluh November; 2017.
12. Restu FS. Pengaruh Penggunaan Limbah Cangkang Kerang dan Fly Ash pada Binder Geopolimer. Surabaya.: Institut Teknologi Sepuluh Nopember; 2017.
13. Rizky TA. Studi Pemanfaatan Fly Ash dan Limbah Sandblasting (Silica Fume) pada Binder Geopolimer. Surabaya: Insitut Teknologi Sepuluh Nopember; 2017.
14. Samlistiya. PN. Studi Pemanfaatan Limbah Karbit pada Binder Geopolimer. Surabaya Institut Teknologi Sepuluh Nopember: Surabaya.; 2017.
15. Duan P, Yan C, Zhou W. Compressive strength and microstructure of fly ash based geopolymer blended with silica fume under thermal cycle. Cement and Concrete Composites. 2017;78:108-19.
16. Olivia M, Mifshella AA, Darmayanti L. Mechanical properties of seashell concrete. Procedia Engineering. 2015;125:760-4.
17. Lertwattanaruk P, Makul N, Siripattarapravat C. Utilization of ground waste seashells in cement mortars for masonry and plastering. Journal of environmental management. 2012;111:133-41.
18. Castaldelli VN, Tashima MM, Melges JLP, Akasaki JL, Monzó J, Borrachero M, et al., editors. Preliminary studies on the use of sugar cane bagasse ash (SCBA) in the manufacture of alkali activated binders. Proceedings of the 14th International Conference on Non-Conventional Materials and Technologies, João Pessoa, Brazil; 2013.
19. Pereira A, Akasaki JL, Melges JL, Tashima MM, Soriano L, Borrachero MV, et al. Mechanical and durability properties of alkali-activated mortar based on sugarcane bagasse ash and blast furnace slag. Ceramics International. 2015;41(10):13012-24.
20. Bahurudeen A, Santhanam M. Influence of different processing methods on the pozzolanic performance of sugarcane bagasse ash. Cement and Concrete Composites. 2015;56:32-45.
21. Shinohara Y, Kohyama N. Quantitative analysis of tridymite and cristobalite crystallized in rice husk ash by heating. Industrial Health. 2004;42(2):277-85.

IOP Conf. Series: Materials Science and Engineering **267** (2017) 012001 doi:10.1088/1757-899X/267/1/012001

22. Castaldelli VN, Akasaki JL, Melges JL, Tashima MM, Soriano L, Borrachero MV, et al. Use of slag/sugar cane bagasse ash (SCBA) blends in the production of alkali-activated materials. Materials. 2013;6(8):3108-27.
23. Hwang C-L, Huynh T-P. Effect of alkali-activator and rice husk ash content on strength development of fly ash and residual rice husk ash-based geopolymers. Construction and Building Materials. 2015;101:1-9.
24. Sturm P, Gluth G, Brouwers H, Kühne H-C. Synthesizing one-part geopolymers from rice husk ash. Construction and Building Materials. 2016;124:961-6.
25. Phetchuay C, Horpibulsuk S, Suksiripattanapong C, Chinkulkijniwat A, Arulrajah A, Disfani MM. Calcium carbide residue: Alkaline activator for clay–fly ash geopolymer. Construction and Building Materials. 2014;69:285-94.
26. Palomo A, Grutzeck M, Blanco M. Alkali-activated fly ashes: a cement for the future. Cement and concrete research. 1999;29(8):1323-9.
27. Du F-P, Xie S-S, Zhang F, Tang C-Y, Chen L, Law W-C, et al. Microstructure and compressive properties of silicon carbide reinforced geopolymer. Composites Part B: Engineering. 2016;105:93-100.
28. Hanjitsuwan S, Phoo-ngernkham T, Damrongwiriyanupap N. Comparative study using Portland cement and calcium carbide residue as a promoter in bottom ash geopolymer mortar. Construction and Building Materials. 2017;133:128-34.
29. Sathonsaowaphak A, Chindaprasirt P, Pimraksa K. Workability and strength of lignite bottom ash geopolymer mortar. Journal of Hazardous Materials. 2009;168(1):44-50.
30. Lee W, Van Deventer J. The interface between natural siliceous aggregates and geopolymers. Cement and Concrete Research. 2004;34(2):195-206.
31. Boonserm K, Sata V, Pimraksa K, Chindaprasirt P. Improved geopolymerization of bottom ash by incorporating fly ash and using waste gypsum as additive. Cement and Concrete Composites. 2012;34(7):819-24.
32. Kong DL, Sanjayan JG. Effect of elevated temperatures on geopolymer paste, mortar and concrete. Cement and concrete research. 2010;40(2):334-9.
33. Dao D, Forth J, editors. Investigation of the behaviour of geopolymer mortar after heating to elevated temperatures. Proceedings of the 3rd International Conference on Sustainable Construction Materials and Technologies, Kyoto, Japan; 2013.
34. Kriven WM, Bell JL, Gordon M. Microstructure and microchemistry of fully-reacted geopolymers and geopolymer matrix composites. Ceramic Transactions. 2003;153(1994).
35. Zerfu K, Ekaputri JJ, editors. Review on Alkali-Activated Fly Ash Based Geopolymer Concrete. Materials Science Forum; 2016.
36. Islam A, Alengaram UJ, Jumaat MZ, Bashar II. The development of compressive strength of ground granulated blast furnace slag-palm oil fuel ash-fly ash based geopolymer mortar. Materials & Design. 2014;56:833-41.

IOP Conf. Series: Materials Science and Engineering **267** (2017) 012002 doi:10.1088/1757-899X/267/1/012002

The Effects of Bottom Ash on Setting Time and Compressive Strength of Fly Ash Geopolymer Paste

B A Affandhie[a], P T Kurniasari[a], M S Darmawan[a], S Subekti[a], B Wibowo[a], N A Husin[a], R Bayuaji[1,a], S Irawan[b]

[a] Departement of Civil Infrastructure Engineering, Institut Teknologi Sepuluh Nopember, Surabaya Indonesia.
[b] Universiti Teknologi Petronas, Bandar Sri Iskandar, Malaysia

E-mail: bayuaji@ce.its.ac.id

Abstract. This research is to find out the contribution of waste energy utilization of fly ash and bottom ash coal as binding agent of geopolymer concrete. This research methodology uses experimental approach in laboratory by making cylinder paste test object with dimension diameter of 2.5 cm x height 5 cm with some combination of fly ash and bottom ash mix with time setting test (ASTM C 191-04a) and compressive strength (ASTM C 39-04a). The research concludes that the effect of bottom ash on fly ash-based geopolymer paste shows good results in setting time and compressive strength.

1. Introduction

Coal has a major role as a primary energy source for industrial processes and has a significant effect on Indonesia. As noted in 2016, Indonesia's coal production of 434 million tons of total reserves is estimated at 54 million tons per year. The abundant presence of coal is directly proportional to the number of industries that use coal as a source of energy, which will continue to develop [1]. This encourages national and international researchers to utilize waste coal, fly ash (FA) on building materials and construction materials.

Previous studies [2-7] had demonstrated significant FA for the use of building materials and construction materials. However, the problems in the field may arise, since the coal quality in terms of geography, shipping process and coal combustion process in the industry make the quality of FA is very varied. This makes difficulties in previous studies to assurance the quality of binder and geopolymer concrete. In addition to FA quality issues, the main problem of binder and geopolymer concrete is the setting time. The hardening time is quite short especially FA with quite high CaO content. Therefore, it is necessary to conduct research to substitute FA with other local waste material in order to avoid dominance in quantity. This study is not only expected to give the effect of quality assurance of binder and concrete geopolymer based on FA but also increase the setting time binder and geopolymer concrete.

Bottom ash (BA) is another waste of coal combustion process. The ratio of coal waste composition between FA: BA is 1: 4. BA has potential as an alternative substance of FA-based geopolymer binder substitution; Dark gray particles, granules and porous particles, lower reactivity than FA, 10-60% grain size passes on 0.6 mm (No. 40), SiO_2, Al_2O_3 and Fe_2O_3 compounds are similar to FA. Therefore, this study further explores the potential of BA as an FA substitute material on geopolymer binders to improve setting time and compressive strength.

---

[1] bayuaji@ce.its.ac.id

2. Methodology

The methodology used in this study is laboratory experiments. The binder test object used the sample size of 2.5cm x 5cm. The main ingredients of FA and BA were used waste from PT Ipmomi Paiton (Table 1). From result of XRF test, it was known that FA used in this research was Si + Al + Fe> 70%. CaO <10%. Thus, the fly ash used is the F type. CaO content of fly ash used in this study was 5.83%, according to ASTM [8] which has less than 10% CaO content classified into F fly ash type. Figures 1 and 2 are the description of FA and BA particles respectively. BA particle used in this research was not grinded and sieved as it was in the initial condition in field.

Table 1. Fly ash and Bottom Ash Oxide Compounds

Material	SiO_2	Al_2O_3	Fe_2O_3	CaO	Na_2O	K_2O	others
Fly Ash (C)	47,1	24,25	16,07	5,83	0,645	1,64	4,465
Bottom Ash	89,91	2,93	4,71	0,73	0,0802	0,679	0,9608

Figure 1. Scanning electron microscope *Fly Ash*

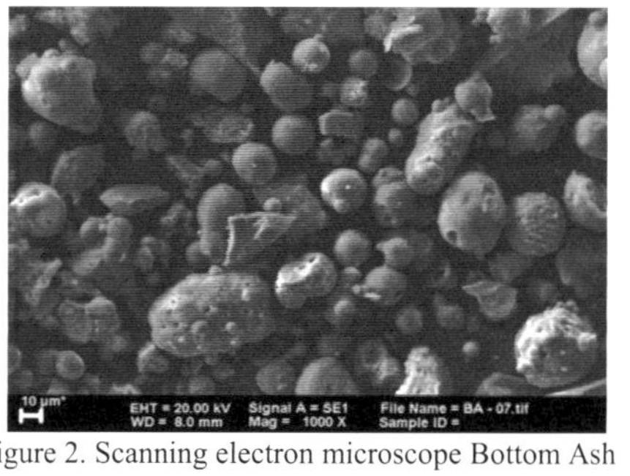

Figure 2. Scanning electron microscope Bottom Ash

The X-Ray Diffraction (XRD) technique was used to analyze the crystalline phases of FA and BA in the study. The representation of FA and BA patterns are shown in Figures 3 and 4.

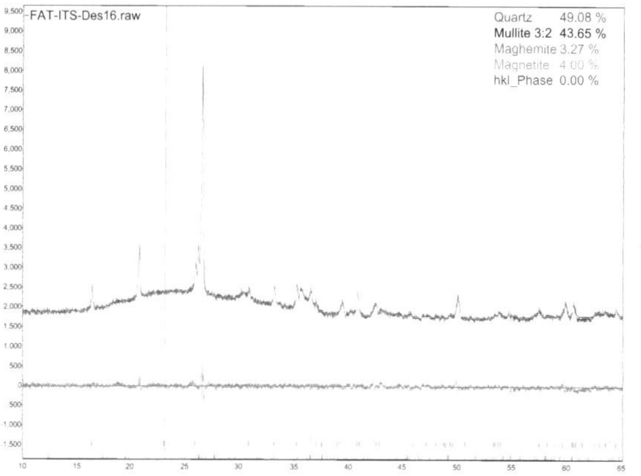

Figure 3. *Fly Ash* pattern from XRD test

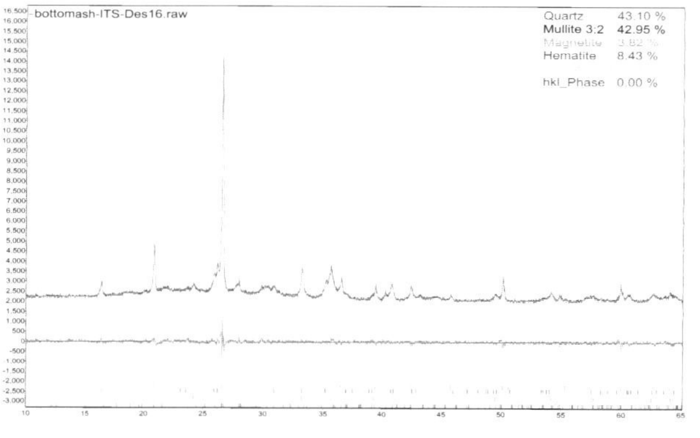

Figure 4. Bottom Ash pattern from XRD test

IOP Conf. Series: Materials Science and Engineering **267** (2017) 012002 doi:10.1088/1757-899X/267/1/012002

The types of activator material used were Sodium Silicate (Na2SiO3) and Sodium Hydroxide (NaOH) with molarity 12 M. The composition of geopolymer binder included 6 combinations which include: variation of silica alumina material composition and activator ratio variation. Details of the main composition of silica alumina material included: 100% BA, 100% FA and 50% BA + 50% FA. For the composition of 100% FA and 50% FA + 50% BA using a 74% ratio of alumina silicate material + 26% activator, 100% BA composition using 55% BA + 45% activator ratio. 2 ratios of activator material between NaOH to Na2SiO3 are 0.5 and 1.5. Details of the geopolymer paste composition are described in Table 2.

Table 2 Composition of Pasta Geopolymer

No	Binder code	Information
1	BA12 - 0,5	BA 100%, NaoH 12M, Na_2SiO_3/NaOH = 0,5
2	BA12 - 1,5	BA 100%, NaoH 12M, Na_2SiO_3/NaOH = 1,5
3	FA12 - 0,5	FA 100%, NaoH 12M, Na_2SiO_3/NaOH = 0,5
4	FA12 - 1,5	FA 100%, NaoH 12M, Na_2SiO_3/NaOH = 1,5
5	BF12 - 0,5	BA 50% + FA 50%, NaoH 12M, Na_2SiO_3/NaOH = 0,5
6	BF12 - 1,5	BA 50% + FA 50%, NaoH 12M, Na_2SiO_3/NaOH = 1,5

After the preparation of the tools and finished material, the next step was the manufacture of test specimens that included: mixing, printing and maintenance until the age testing reached 3, 28 and 56 days. Along with the mixing and printing process, "setting time" was tested with the standard [9] to determine the effect of BA on the lifting paste matrix to the strength compound. Implementation of the manufacture of specimens, time setting test and compressive strength test [10] were carried out in Material Laboratory and Building Structure

3. Result and Discussion

3.1. Setting time

Table 1 described the result of geopolymer binder setting time with 6 combinations and illustrated the recapitulation of the whole setting time. The longest setting time result occurred on the bottom ash 12M of the activator with 1.5 ratios where the initial binding occurred within 30 minute with a 25mm drop and the final binding took place at 85 minutes with a 3mm drop. At the fly ash composition, the longest setting time occurred on the comparison of activator 0.5 with the initial binding at 22.5 minutes with a 25mm reduction and the final binding occurred at 55 minutes with a 2mm drop. On the compositions of bottom ash and fly ash, the longest setting time result in comparison of activator 0.5 in which the initial binding occurs at 45 minutes with a 25mm drop and the final declined at 70 minutes with a 2mm drop. The result of setting time test showed that the comparison of activator 0.5 had longer binding time than 1.5.

The higher the ratio of mass ratio between NaOH and Na_2SiO_3 solution, the faster the final binding time took place. This was because the amount of Na_2SiO_3 presented in the binder mixture was greater than the amount of NaOH thus accelerated the polymerization reaction (Djwantoro 2005). The result of this setting time test also supported by the previous researcher's statement that Na_2SiO_3 were used in geopolymer paste served to speed up the polymerization reaction, therefore, if the levels were high, the crystallization process would also take place more quickly. Particle granules and BA reactivity could increase the timing of the geopolymer paste setting, compared to BA12-1,5 which had the initial setting time longer than the composition of FA12-1,5.

Table 3. Time for Binding in the beginning and end of the geopolymer paste

No	Binder code	Early binding time (min)	Last binding time (min)
1	BA12 - 0,5	26.67	75
2	BA12 - 1,5	30	85
3	FA12 - 0,5	22.5	55
4	FA12 - 1,5	12.5	40
5	BF12 - 0,5	45	70
6	BF12 - 1,5	12.5	55

3.2. Compressive strength

Figure 1 to 3 described the compressive strength of geopolymer paste using BA 100%, FA 100% and BA50% +FA50%. BA-based geopolymer paste showed the lowest compressive strength compared to the FA and FA + BA geopolymer paste. The BA12-0,5 geopolymer paste composition showed better compressive strength than BA12-1.5 at all ages. At 56 days of age the compressive strength of BA12-0.5 indicated a 300% strength of the lye press of BA12-1.5. In contrast to the compressive strength of the FA-based geopolymer paste showed FA12-1.5 was greater than FA12-0.5.

The effect of BA on the FA-based geopolymer paste could be shown in the highest compressive strength in the mixture composition of BA 50% and FA 50% with the comparison of activator 1.5 reaching 41.02 MPa at age 28 days and 49,71 at age 56 days. This result was better than the 100% FA-based geopolymer paste with a compressive strength of 34.9 MPa aged 28 days and 43.32 MPa aged 56 days.

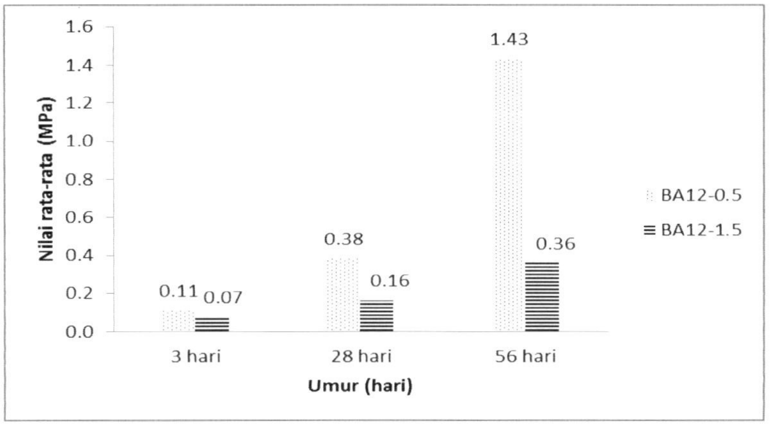

Figure 5. Compressive strength of BA geopolymer paste

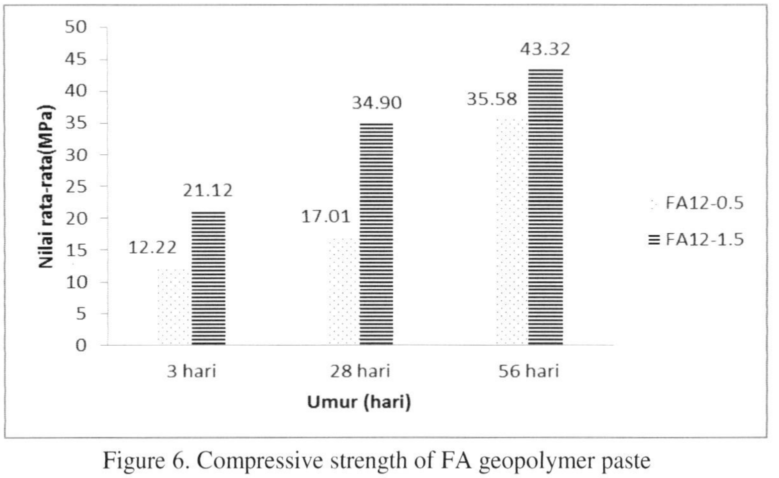

Figure 6. Compressive strength of FA geopolymer paste

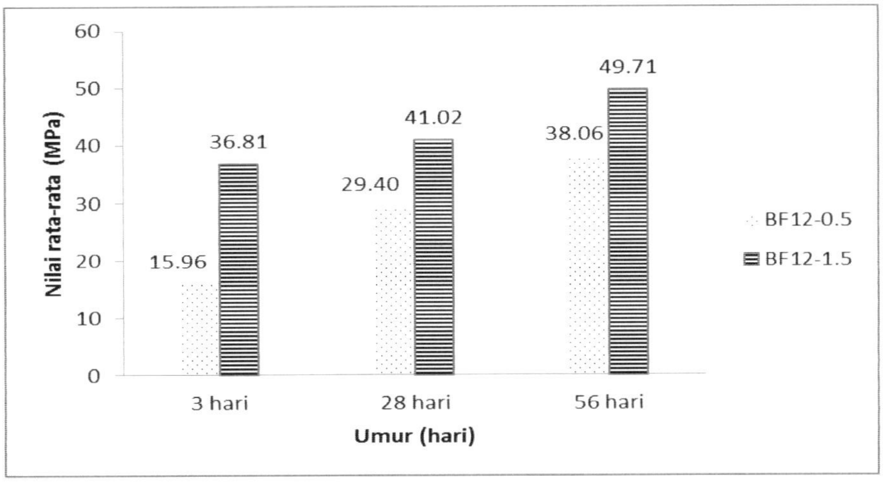

Figure 7. Compressive strenght of BF geopolymer paste

In every concrete study, BA is widely used as a filler material because of its physical and chemical properties. However, in this study, the combination of physical and chemical BA characters can show better collaboration of performance on the FA-based geopolymer paste, which slow the binding time thus it becomes one solution to solve the problem of geopolymer concrete with fast binding time due to one of the FA characters contain with quite high level of CaO and the high level ratio of Na_2SiO_3 and NaOH. The last factor is to achieve a high compressive strength target on geopolymer concrete. Strong factor that is considered from the physical character of BA, its shape is almost round and its dimension is bigger than FA. This factor proves that the increase of binding time of BA composition 100% and FA 50% + BA50% is longer than FA 100%. Of course, further research is needed, beside BA physical factors, there must be a chemical mechanism that needs to be proven in further research to strengthen BA's influence on the performance of the FA-based geopolymer paste.

To maximize the performance more of BA in the polymerization mechanism with FA, it is proposed in subsequent studies to perform a more gradual composition from the smallest percentage to the optimum composition for all aspects of geopolymer paste performance as a geopolymer concrete binder. The economical treatment of BA also needs to be proposed in subsequent research, grinding, sieving or other treatment processes to maximize its application function in geopolymer paste. In addition to setting time, BA is also able to increase the value of compressive strength of FA-based geopolymer paste by reducing 50% FA content and substituted with BA, this composition results strength similar to the FA100% geopolymer paste. This fact illustrates that BA contributes to strength despite the previous literature [11] BA reactions on hydration mechanisms are less reactive than FA. The effect of BA on fly ash geopolymer paste was tabulated in Table 2.

Table 4. The effect of bottom ash on fly ash geopolymer paste

Binder code	Performance effect			
	Setting Time		Compressive Strength	
	Poor	Good	Poor	Good
BA12 - 0,5		√	√	
BA12 - 1,5		√	√	
FA12 - 0,5	√			√
FA12 - 1,5	√			√
BF12 - 0,5		√		√
BF12 - 1,5	√			√

4. Conclusion

The best of composition ratio is found in mixture BA 50% + FA 50% with the ratio of activator 1.5 which has compressive strength reached at 49.71 MPa. The higher the mass ratio of the sodium silicate and sodium hydroxide solutions, they do not always result in a high compressive strength which is particularly occured in bottom ash. The use of bottom ash in a binder in a particular composition can increase the value of compressive strength and tends to slow the setting time on the binder so as to provide opportunities fly ash in order not to harden quickly.

5. Acknowledgment

Researchers would like to thank the Directorate of Research and Community Service of the Ministry of Research, Technology and Higher Education for research assistance and research results that can be successfully published.

6. References

1. Kemenperin. *Target Produksi Batu Bara Adaro Naik*. 2017; Available from: http://www.kemenperin.go.id/artikel/14262/Target-Produksi-Batu-Bara-Adaro-Naik.
2. Kurniasari, P.T., *Pemanfaatan Penggunaan Fly Ash dan Bottom Ash sebagai Pozzolan pada Binder Geopolymer". Surabaya : Tugas Akhir Diploma III* in *Civil Infrastructure Engineering* 2017, Institut Teknologi Sepuluh Nopember: Surabaya.
3. Ridho Bayuaji, M.S.D., Boedi Wibowo, Nur Ahmad Husin, Srie Subekti, Januarti Jaya Ekaputri, *The Influence of Chloride Environment on Compressive Strength of Geopolymer Concrete with Fly Ash Using Taguchi Approach.* Applied Mechanics and Materials, 2015. **754-755**: p. 400-405
4. Bayuaji, R., et al. *Mechanical Properties of MIRHA-Fly Ash Geopolymer Concrete.* in *Materials Science Forum*. 2015. Trans Tech Publ.
5. MS Darmawan, et al., *The Effect of Chloride Environment on Mechanical Properties Geopolymer Binder with Fly Ash.* Key Engineering Materials, 2014. **594**: p. 648-655.
6. Bayuaji, R., et al. *The Effect of High Temperature on Compression Strength of Geopolymer Paste.* in *MATEC Web of Conferences*. 2017. EDP Sciences.
7. Bayuaji, R., et al. *The Effect of Corrosive Environment on Geopolymer Concrete Tensile Strength.* in *MATEC Web of Conferences*. 2017. EDP Sciences.
8. ASTMC618-03, *Standard Specification for Coal Fly Ash and Raw or Calcined Natural Pozzolan for Use in Concrete*. 2003: Philadelphia, USA.
9. ASTMC191-13, *Standard Test Methods for Time of Setting of Hydraulic Cement by Vicat Needle*. 2013, ASTM International.
10. ASTMC39/C39M-17, *Standard Test Method for Compressive Strength of Cylindrical Concrete Specimens,*. 2017, ASTM International.
11. ul Haq, E., S.K. Padmanabhan, and A. Licciulli, *Synthesis and characteristics of fly ash and bottom ash based geopolymers–a comparative study.* Ceramics International, 2014. **40**(2): p. 2965-2971.

IOP Conf. Series: Materials Science and Engineering **267** (2017) 012003 doi:10.1088/1757-899X/267/1/012003

Physical and Chemical Character of Fly Ash of Coal Fired Power Plant in Java

Triwulan[11], K A Priadana[1], J J Ekaputri[1] and R Bayuaji[2]

[1]Civil Engineering Department, Institut Teknologi Sepuluh Nopember Surabaya, Indonesia
[2]Civil Infrastructure Engineering Department, Institut Teknologi Sepuluh Nopember Surabaya, Indonesia

E-mail: triwulan_marwan@yahoo.com[1]

Abstract. Quality of fly ash is varying widely in the field, it depends on the combustion process and the quality of the basic ingredients, namely coal. It will affect the physical and mechanical properties of the concrete mixtures used. This study used 12 samples of fly ash. The physical and chemical properties and finesse modulus were analyzed. The fly ash was mixed with OPC (Ordinary Portland Cement) with the proportion of 20% fly ash and 80% OPC. The specimens were form with mortar dimension of 5cm x 5 cm. The test was affected by the correlation of fly ash fineness modulus to compressive strength, correlation density of fly ash to compressive strength, and correlation of carbon content to the compressive strength.

1. Introduction

The potential of fly ash (FA) utilization in order to respond environmentally friendly alternative cement solution describes the positive development. This fact was supported by the data from the Ministry of Energy and Mineral Resources. About 40% of Indonesian power plants still rely on coal as a source of energy up until 2019. During this process, there were 15,027,100 tons of A was produced [1]. Power plant industry utilizes about 780 billion tons of coal and produced FA of 7.8 billion tons each year worldwide [2].

In reality, the quality of fly ash is very varied, the variation greatly affects the standard quality in the application of geopolymer concrete. Therefore, until this paper is being written, the standard for application of fly ash in geopolymer concrete technology industry was still in development phase. The main problem to map fly ash characters is the complexity of the process [3] starting from coal as its base material. Coal is a combination of biological, chemical, and physical processes that occur due to the loading effect of covering sediments, temperature, pressure, and time on the organic component.
Therefore, the quality of coal can not be equated to adjust the surrounding environment that shapes it. Each location has its own character traits based on the process. Second, the process of energy utilization of coal was in accordance with the needs of industries that use it. The combination of transportation and coal usage management are sometimes considered for cost optimization. The final problem in the management of hoarding and fly ash transportation from fly ash sources is utilizing the concrete industry. One character that needs to be noted is that the quality of coal determines the chemical composition of fly ash as summarized in table 1.

Table 1 The chemical composition of fly ash by the used coal [4]

[1] triwulan_marwan@yahoo.com.

Compounds	Bituminous Coal (%)	Sub-Bituminous Coal (%)	Lignite (%)
SiO_2	20-60	40-60	15-45
Al_2O_3	5-35	20-30	10-25
Fe_2O_3	10-40	4-10	4-15
CaO	1-12	5-30	15-40
MgO	0-5	1-6	3-10
SO_3	0-4	0-2	0-10
Na_2O	0-4	0-2	0-6
K_2O	0-3	0-4	0-4
LOI	0-15	0-3	0-5

FA character mapping has been developed by the Indonesian Geopolymer Research Consortium (KORIGI) since 2014 with the outcome of geopolymer paste [5, 6], geopolymer mortar and geopolymer concrete [7]. Hardjito [8] had utilized F class fly ash waste into geopolymer concrete. The compressive strength of the concrete can reach 80 MPa. Ekaputri [9] made geopolymer grouting material with Na_2SiO_3 activator mass ratio to 8M NaOH of 1.5, resulting in achieving a compressive strength of 48.59 MPa within 28-day.

From the description above, it shows that the use of fly ash as a mixture of concrete material, base material of geopolymer concrete, and raw material of hybrid alkaline has double benefits. It also reduces CO_2 emissions, utilize waste, and reduce energy consumption for hydration cement kilns. On the other hand, the quality of fly ash is so varied that it gives impact to the nature of the concrete and the mixture that uses it. For that we need to mapping the physical, chemical and mechanical characterization of fly ash in order to facilitate the use of fly ash based on its quality. In addition, the results of this characterization also provide information on the use of fly ash especially as a geopolymer binder.

2. Methodology

The research method used to answer fly ash characterization mapping: the literature studies about materials and coal processing and stage interviews. The first phase was the coal-burning process interviews, the regulations used in the production process, quality control, quantity data produced by power plant, the spread of consumers, the management of waste handling and transport procedures to the consumer. Then the second phase was interviews to consumers include fly ash sources and quality control. The field data collection was conducted to obtain data related to sample tests that had been done by the PLTU (steam power plant) in processing fly ash and supporting data as stated in the above questionnaire. The data group was divided into 2 sources i.e. fly ash SB power plant (77 data) and fly ash PLTU-P power plant (12 data).

Each data group will be calculated based on the amount of data that pass ASTM C-618 standard, then divided by the amount of data contained in each data group and multiplied by100%. To obtain the fineness impact graph with concrete compressive strength (cubed mortar size 5x5x5 cm3), the field data obtained were sorted by fineness (ASTM C430) from the smallest value to the largest one, while density test use the ASTM C188 standard. There were two parts of discussion on following chemical properties in this study, namely data on the oxide levels in fly ash obtained through XRF test and data on minerals in fly ash obtained through the XRD test. The data on fly ash physical properties were obtained from fly ash sieve analysis, fly ash weight analysis, mortar making, and pasta making.

3. Result and Discussion

3.1 Fineness

The Data collection of 89 data from SB power plant and PLTU-P power plant can be summarized in Table 2. The standard used for categorization is ASTM C618. Figure 1 illustrates the fineness effects of fly ash on strength

Table 2 Confirmation of field data with ASTM C618

No	Origin	Auto Clave	Fineness + 325 mesh	Strength Activity Index	Water Requirement max 105%(Class F or C)	Water Requirement max 115% (Class N)	Total Data
1	SB	100%	84.40%	98.70%	100%	100%	77
2	PLTU-P	100%	100%	100%	100%	100%	12

From the data above, it can be concluded that the sample variance obtained from PLTU-P is better than the SB power plant in relation to ASTM C-618 comparison.

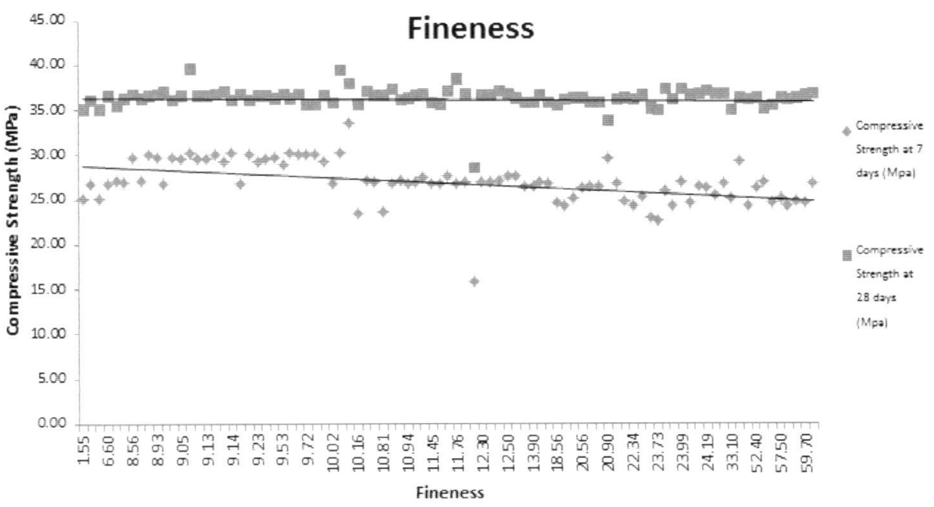

Figure 1. Fineness

The graph above shows the decrease of mortar compressive strength at the age of 7 days which followed by the increasing of fineness. Thus, it can be concluded that the rougher fly ash the less ability it has as a filler, causing a decrease in compressive strength and vice versa. However, the graph does not show a drastic decline rate at 28 days. This was attributed to the fly ash fineness value which still within the range allowed by ASTM that was 34% maximum stuck in filter no.325 in most data.

3.2 Density

Figure 2 shows the increased compressive strength at 7 days as density increases. In accordance with the function of fly ash as filler or cavity filler in concrete, it can be concluded that the denser fly ash, the more solid concrete. Thus, it can increase the strength of concrete.

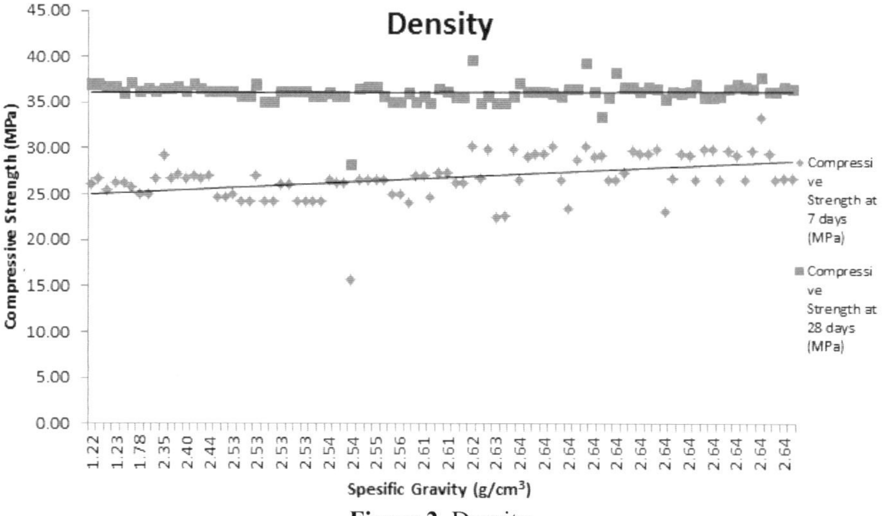

Figure 2. Density

3.3 Data of Oxide Compounds

The data on the chemical properties of fly ash are obtained from PLTU-P as the producer and SB as consumers.

Table 3 Confirmation of field chemical data with ASTM and CSA

No	COMPANY	SiO_2	Al_2O_3	Fe_2O_3	SO_3	$SiO_2 + Al_2O_3 + Fe_2O_3$	CaO	ASTM		CSA	
								Class	Index	Class	Index
1	PLTU-P	34.97	14.23	20.05	7.49	69.25	13.92	F	0.7	CI	0.5
2	PLTU-P	68.05	10.98	6.39	6.1	85.42	5	F	0.9	F	1
3	PLTU-P	47.55	20.28	10.34	5.23	78.17	9.37	F	0.8	CI	0.5
4	PLTU-P	48.01	19.19	9.38	6.7	76.58	9.96	F	0.8	CI	0.5
5	PLTU-P	61.12	15.63	15.32	2.36	92.07	2.11	F	1	F	1
6	PLTU-P	60.47	16.51	15.19	2.26	92.17	2.12	F	1	F	1
7	SB	50.18	28.11	9.06	0.46	87.81	3.66	F	0.9	F	1
8	SB	50.54	31.2	8.21	1.41	91.36	5.34	F	1	F	1

Information :

Not linear with ASTM C-618 (SO_3> 5%)

From the table above, it can be concluded that fly ash derived from PLTU-P in general can be classified as F class fly ash according to ASTM C-618 and class F / CI according to CSA. However, 4 of 8 data indicate that fly ash at PLTU-P Power Plant contains a relatively high SO_3 (> 5%) Based on Table 3.3 on the chemical composition of fly ash, it can generally be said that fly ash which produced by PLTU-P power plant has a tendency to derive from coal of Bituminous Coal type because it has high Al2O_3 content (> 10%) and has qualified range of allowed oxide.

IOP Conf. Series: Materials Science and Engineering **267** (2017) 012003 doi:10.1088/1757-899X/267/1/012003

3.4 Mineral in Fly Ash

From the result of graph comparison between amorphous global and mortar / pasta mortar strength index showed similar pattern. Therefore, it can be concluded that the level of amorphous fly ash greatly affects the strength of mortar / pasta. Here are the results of pure fly ash XRD according to the code: Perbandingan global amorphous terhadap index fly ash

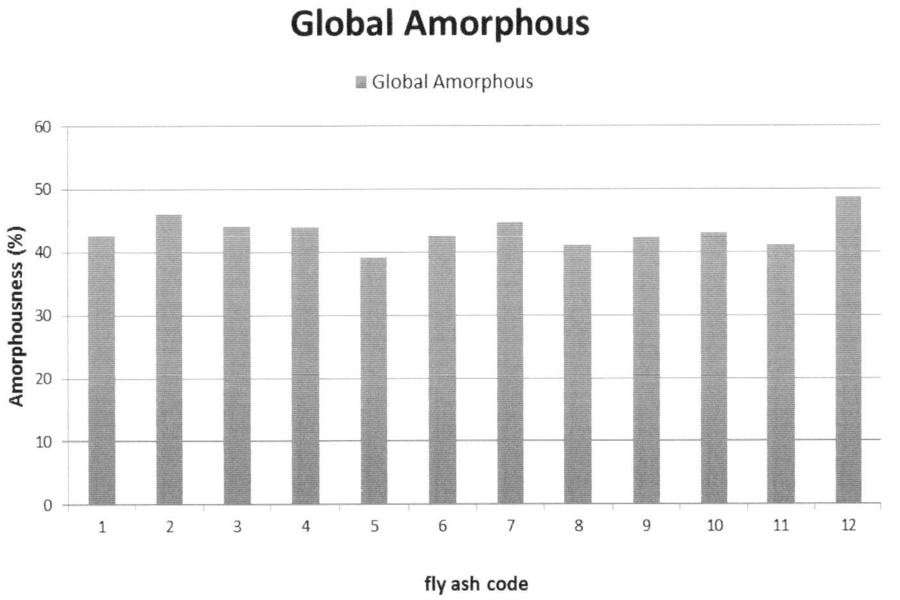

Figure 3. Global Amorphous comparison of fly ash.

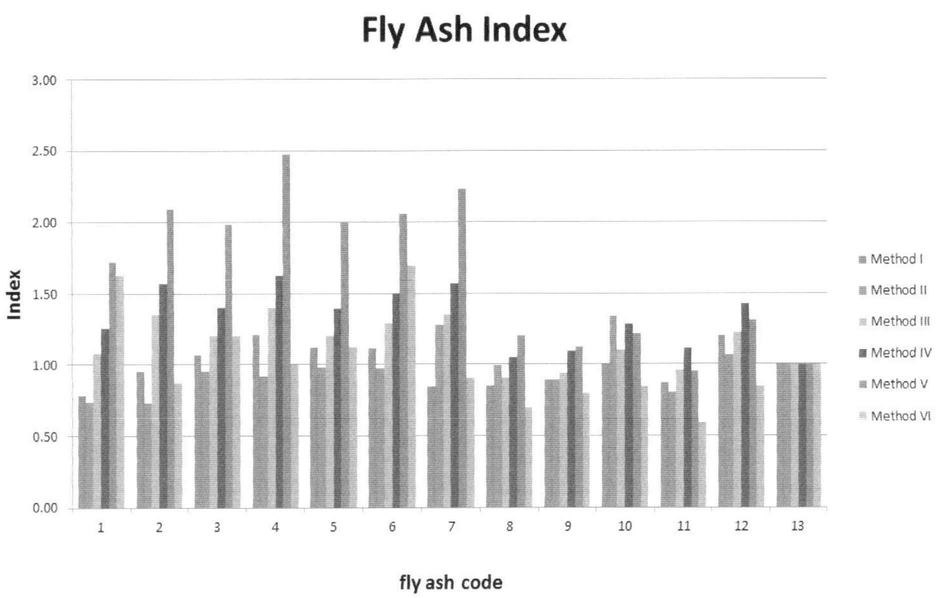

Figure 4. Fly ash index comparison of fly ash.

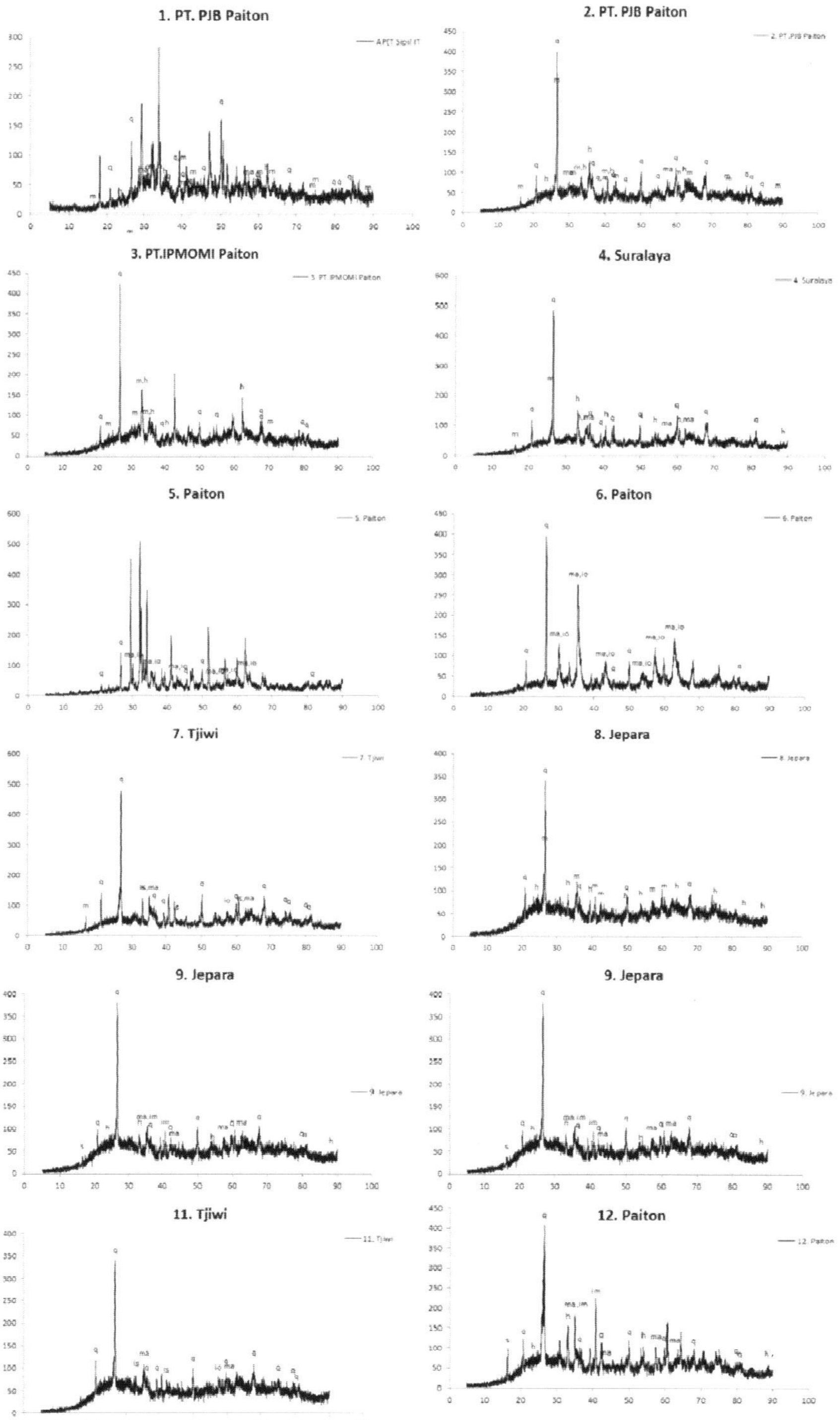

Figure 5. Results of XRD fly ash test

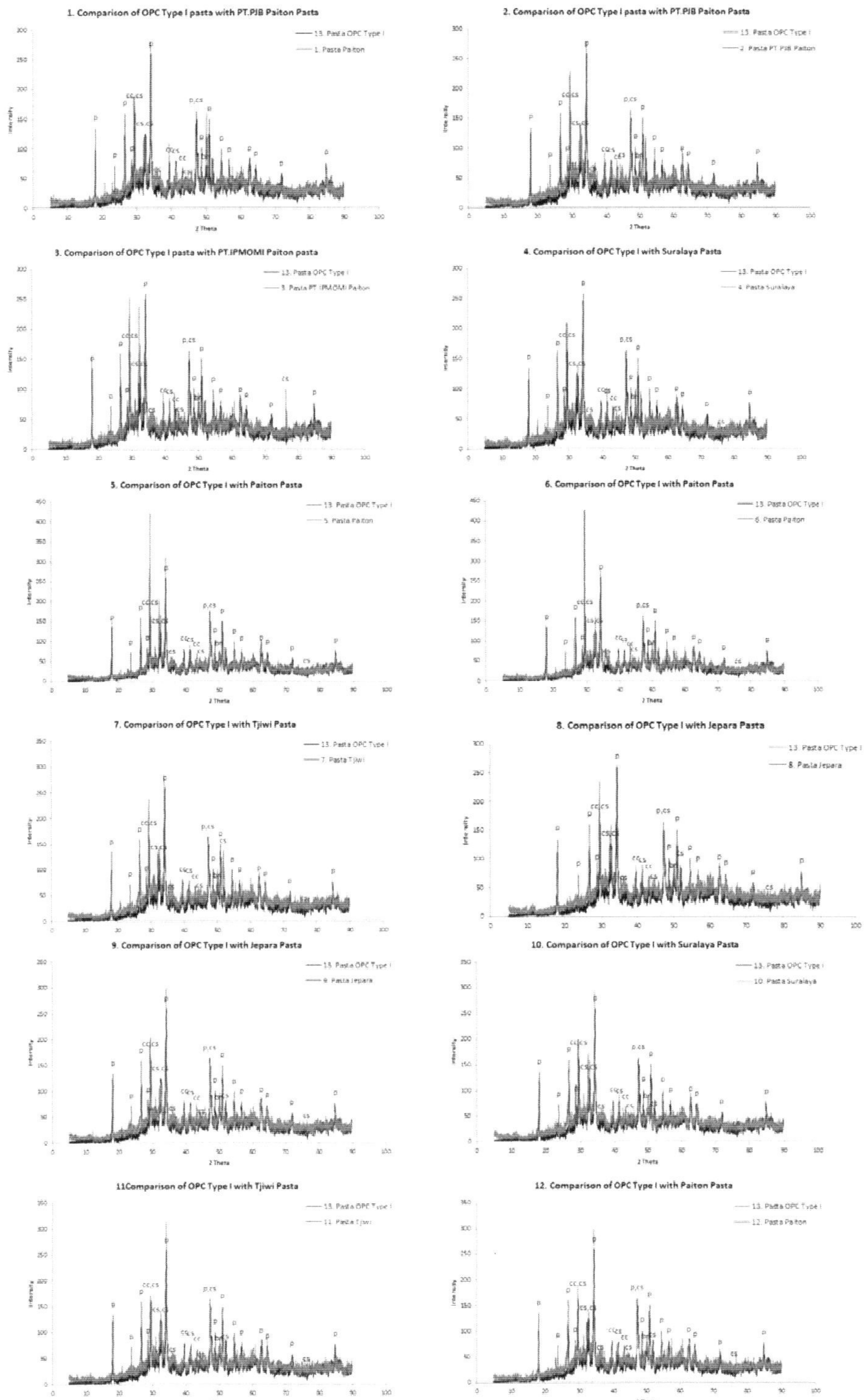

Figure 6. Comparison of OPC paste type I with other pasta

There are 3 main mineral compositions in fly ash which are minerals formed from SiO_2, $Al2O_3$, and Fe_2O_3. Formation of the percentage of mineral composition is generally proportional to the percentage of oxide composition that is formed. These minerals have different shapes. Generally, the main constituent minerals in fly ash are Quartz (SiO_2), Mullite ($Al_6Si1_2O_3$), and Hematite (Fe_2O_3). In the pattern of flyogy minerals usually Quartz has the highest peak and followed by other minerals. The strength of mortar and pasta is not only influenced by the peaks alone, but also influenced by the number of peaks that arise (Global Amorphous).

In general, the minerals in OPC type I paste are portlandite (Ca (OH) 2), calcium carbonate, and calcium silicate. The above graph shows that with the addition of fly ash to the paste mixture, it will reduce the intensity of the minerlandite minerals in certain peaks and will increase the mineral intensity of calcium silicate. This is related with the reaction of fly ash in a paste utilizing Ca $(OH)_2$ formed from the cement hydration process to form calcium silicate. Fly ash can be said to have high reactivity rates if fly ash can react with many portlandite marked by significant Portland peak decline.

4. Conclusion

The secondary field data of fly ash shows the variation of physical properties of the best samples are derived from PLTU-P power plant in relation to comparison with ASTM C-618 and chemically classified under class F / CI according to CSA. In general, fly ash from PLTU-P Power Plant contains high $Al2O_3$ (> 10%) and has fulfilled the required oxide range requirement. In laboratory tests (primary data), the physical properties of sieved fly ash proven to obtain better compressive strength result because most of the carbon will be retained on filter number 200 and reduce the cement water factor. The higher amorphous level of a fly ash will give a good impact on the strength of the concrete chemically, and vice versa. The strength of mortar and paste under review is not only influenced by the highest peak, but also influenced by the number of peaks that arise (Global Amorphous).

5. Acknowledgments

Researchers would like to thank the Directorate of Research and Community Service of the Ministry of Research, Technology and Higher Education for research assistance and research results can be successfully published.

6. References

1. ESDM, *Indonesia Electricity Development Plan And Indonesia Coal-Ash Management Implementation*, in *International Coal Based Power Conferences*. 2016: New Delhi.
2. Heidrich, C., H.-J. Feuerborn, and A. Weir. *Coal combustion products: a global perspective*. in *Proceedings of the World of Coal Ash Conference*. 2013.
3. Bayuaji, R., et al., *Utilization of High Calcium Content Fly Ash: Flexural Strength of Geopolymer Concrete Beams in Sea Water Environment.* The Open Civil Engineering Journal, 2016. **10**(1).
4. ASTMD388-15, *Standard Classification of Coals by Rank*. 2015.
5. Dwijayanti, L.Y., J.J. Ekaputri, and R. Bayuaji. *Lightweight Geopolymer Binder with Abaca Fiber in Different Curing*. in *Materials Science Forum*. 2016. Trans Tech Publ.
6. MS Darmawan, et al., *The Effect of Chloride Environment on Mechanical Properties Geopolymer Binder with Fly Ash.* Key Engineering Materials, 2014. **594**: p. 648-655.
7. Ridho Bayuaji, M.S.D., Boedi Wibowo, Nur Ahmad Husin, Srie Subekti, Januarti Jaya Ekaputri, *The Influence of Chloride Environment on Compressive Strength of Geopolymer Concrete with Fly Ash Using Taguchi Approach.* Applied Mechanics and Materials, 2015. **754-755**: p. 400-405
8. Hardjito, D., Wallah, S. E., Sumajouw D. M.J., and Rangan, B.V., *Factors Influencing The Compressive Strength OfFly Ash-Based Geopolymer Concrete.* Dimensi Teknik Sipil 2004. **Vol 6, No. 2**(September 2004): p. 88 – 93.
9. Ekaputri, J.J. *Geopolymer Grout Material*. in *Materials Science Forum*. 2016.

A review in high early strength concrete and local materials potential

A K Yasin[1], R Bayuaji[1], T E Susanto [2]
[1] Department of Civil Infrastructure Engineering, Institut Teknologi Sepuluh Nopember, JL. Menur 127, Surabaya, Indonesia
[2] Department of Research and Product Application, PT Semen Indonesia (persero) Tbk, JL. Veteran, Gresik, Indonesia

E-mail: abdul.karim13@mhs.ce.its.ac.id

Abstract. High early strength concrete is one of the type in high performance concrete. A high early strength concrete means that the compressive strength of the concrete at the first 24 hours after site-pouring could achieve structural concrete quality (compressive strength > 21 MPa). There are 4 (four) important factors that must be considered in the making process, those factors including: portland cement type, cement content, water to cement ratio, and admixture. In accordance with its high performance, the production cost is estimated to be 25 to 30% higher than conventional concrete. One effort to cut the production cost is to utilize local materials. This paper will also explain about the local materials which were abundantly available, cheap, and located in strategic coast area of East Java Province, that is: Gresik, Tuban and Bojonegoro city. In addition, the application of this study is not limited only to a large building project, but also for a small scale building which has one to three-story. The performance of this concrete was apparently able to achieve the quality of compressive strength of 27 MPa at the age of 24 hours, which qualified enough to support building structurally.

1. Introduction

High early strength concrete was created to solve the setting time problems that occur in the world of buildings and infrastructures construction. Generally, it takes 7 to 14 days to achieve the full concrete hardening and reach the minimum compressive strength of 0.7 to 0.85 fc'. This setting time generated on concrete was generally not fast enough to achieve the minimum quality required. This circumstances might concern public on cost production, time spent and efficiency [8]. To solve this problem, some research on concrete technology were conducted to produce high strength concrete in early age with several approach, such as: use several types of cement (type I cement, type III cement, rapid set cement); high cement content (410 to 740 kg/m^3); use several types of admixture (calcium chloride and polycarboxylate ether-PCE); low water to cement ratio (0.2 to 0.4). Thus, the result of compressive strength may reach 39 to 66 MPa at 24 hours [1,2,3,8]. In accordance with its high performance, the production cost is estimated to be 25 to 30% higher than conventional concrete. One effort to cut the production cost is to utilize local materials. With these innovations, the cost of making high early strength concrete can be cheaper, so people do not have to worry about spending a lot of

[1] Corresponding author. Tel.: +62 8 5815377227, +62 8 970592080
Email: abdul.karim13@mhs.ce.its.ac.id (A K Yasin), a3ridho@gmail.com (R Bayuaji),
[2] Corresponding author. Tel.: +62 8 155205317
Email: tri.susanto@semenindonesia.com (T E Susanto)

IOP Conf. Series: Materials Science and Engineering **267** (2017) 012004 doi:10.1088/1757-899X/267/1/012004

money to buy it. Then, another benefit is to improve the community economy around the area of materials resources [8].

2. Definition

High early strength concrete is one of the type in high performance concrete. A high early strength concrete means that the compressive strength of the concrete at the first 24 hours after site-pouring could achieve structural concrete quality (compressive strength > 21 MPa) [4]. Furthermore, high early strength concrete is a concrete that is capable of reaching 50% *fc'* at 24 hours of age, with water to cement ratio ranging from 0.3 to 0.4 [1].

3. Previous research on high early strength concrete

The following are brief data from some previous research:

3.1 High Early Strength Concrete by Strategic Highway Research Program (1993)

In 1993, Strategic Highway Research Program provided some data on material specifications, mix designs, and mechanical properties from 4 (four) different regions origin of concrete aggregates in America: Crushed Granite (CG), Marina Marl (MM), Rounded Gravel (RG), and Dense Crushed Limestone (DL) [1]. Its composition and strength properties are presented in the (Tab.1) below:

Table 1. High Early Strength Concrete Mix Designs by Strategic Highway Research Program [1].

Materials	CG Lilington	MM Lilington	RG Memphis	DL Van Buren
Type III Cement (Kg/m^3)	522	522	522	522
Coarse Aggregate (Kg/m^3)	1032	942	990	1008
Fine Aggregate (Kg/m^3)	576	588	540	618
W/C Ratio	0.32	0.32	0.34	0.34
HRWR (Naphtalene Based) (mL/100 Kg) cementitious materials	26	26	26	16
Calcium Nitrite (gr/m^3)	2.4	2.4	2.4	2.4
Air Entraining Agent (mL/100 Kg) cementitious materials	9	1	1	4
Slump (cm)	2.54	17.2	17.8	7.7
Compressive strength-1day (MPa)	37	38.7	39	36.5

3.2 High Early Strength Concrete by Federal Highway Administration (2001)

In 2001, Federal highway administration has conducted various experiments in making high early strength concrete with several types of mix designs [2]. Its composition and other information are presented in the (Tab.2) below:

Table 2. High Early Strength Concrete Mix Designs by Federal Highway Administration [2].

Materials	Type I (GADOT)	Type III (Fast Track I)	Type III (Fast Track II)	RSPC	Rapid Set Cement
Cement (Kg/m^3)	450	400	450	350	400
Fly ash (Kg/m^3)	-	45	50	-	-
Coarse Aggregate (Kg/m^3)	1070	830	780	1010	1070
Fine Aggregate (Kg/m^3)	610	810	780	830	600
W/C Ratio	0.4	0.4-0.48	0.4-0.48	0.41	0.45
Water Reducer	-	Use	Use	-	-
Air Entraining Agent	6 ±2%				

Materials	Type I (GADOT)	Type III (Fast Track I)	Type III (Fast Track II)	RSPC	Rapid Set Cement
$CaCl_2$ % from Cement Weight	1.0	-	-	-	-
Open Traffic Time	4 hours	24 – 72 hours	12 – 24 hours	4 hours	4 – 6 hours

3.3 Research of Products Application Department PT. Semen Indonesia (2015)

In 2015, Research of Product Application PT. Semen Indonesia, developed a high early strength concrete that called *Rapid Strength Concrete*. In this study, the variables studied is the effect of adding doses of admixture type F which is a HRWR (High Range Water Reducer)-based on polycarboxylate ether [3]. Its composition and strength properties are presented in the (Tab.3) below:

Table 3. High Early Strength Concrete Mix design by R&D Department of Product Application PT. Semen Indonesia [3].

Materials	Mix 1	Mix 2	Mix 3	Mix 4
Cement (Kg/m^3)	740	740	740	740
Coarse Aggregate (Kg/m^3)	1007.50	1007.50	1007.50	1007.50
Fine Aggregate (Kg/m^3)	542.50	542.50	542.50	542.50
W/C Ratio	0.2	0.25	0.2	0.25
Polycarboxylate Ether (%) from Cement Weight	0.6	0.6	0.8	0.8
Compressive Strength (Kg/cm^2):				
- 8 Hours	229.28	223.49	255.63	183.45
- 24 Hours	600.59	593.712	663.93	654.57
Slump flow (cm)	50	56	56	60

4. Factors to be considered in making high early strength concrete

From some of the studies described in the previous chapter, there are several factors that influence the high initial strength mechanisms in concrete:

4.1 Types of Portland Cement

The following are some types of portland cement commonly used in making high early strength concrete:

4.1.1 Portland cement type III

This type of portland cement quickly hardened, commonly used in the application of concrete structures that require high initial strength and also concretion at low temperatures. The mechanism in the cement which can accelerate the hydration process, followed by acceleration of hardening and strength development are: C3S and C3A content of a higher portland cement type I, and also the finess value with the blaine tool which is higher than portland cement type I [1,2]. The following are some of compressive strength results on high early strength concrete in previous research that used portland cement type III:

Table 4. Compressive strength and the age test of high early strength concrete using portland cement type III [1,2].

Researcher	Compressive strength	Age
SHRP 1993	36.5 MPa – 39 MPa	24 hours
FHWA 2001	Min. 13.5 MPa	24hours

4.1.2 Portland cement type I (Ordinary Portland Cement)
This type of cement is a hydraulic cement. It widely used for general construction, such as building construction that do not require special requirements for: sulphate resistance, hydration heat, and high initial strength [2,3]. The following are the compressive strength results on high early strength concrete in previous research that used portland cement type I:

Table 5. Compressive strength and the age test of high early strength concrete using portland cement type I [2,3].

Researcher	Compressive strength	Age
FHWA 2001	Min. 13.5 MPa	4 hours
Research of Product Application Department PT. Semen Indonesia	18.35 MPa – 25.56	8 hours
	59.37 MPa – 66.39 MPa	24 hours

4.1.3 Rapid Set Cement (RSC)
This type is a type of hydraulic cement or hydraulic blend cement that has a rapid strength advantage over the first 24 hours during the hydration process. The cement has been simultaneously milled specifically with non-chloride accelerator materials, so it is safe for reinforced concrete structures. This type has a very fast final setting time of 15 to 35 minute, and can achieve structural strength within 1 hour. In its application, rapid set cement is used for: improvement and creation of structural elements. The concrete price is more expensive than other type of cement [2]. The following are some of compressive strength results on high early strength concrete in previous research that use rapid set cement:

Table 6. Compressive strength and the age test of high early strength concrete using rapid set cement [2].

Reseacher	Compressive strength	Age
FHWA 2001	Min. 13,5 MPa	4 to 6 hours

4.2 Cement Content
Each concrete grade has different cement content levels. The following are the cluster of cement content based on it levels and applications:

Table 7. Several type of content cement based on levels and applications [5].

Content cement (Kg/m^3)	Levels	Application
200 – 400	Medium cement content	Normal concrete
> 400 – 600	High cement content	High strength concrete
> 600	Very high Cement Content	Very high strength concrete

According to (Tab.7) above, it can be seen that in the manufacture of high strength concrete, the cement content used ranges from 400 Kg/m^3 to 600 Kg/m^3. For very high strength concrete, it used more than 600 Kg/m^3. But, in application of high strength concrete in field, high and very high strength concrete do not use more than 600 Kg/m^3 of cement content. The following table is previous research data of cement content that used in high strength concrete, very high strength concrete and high early strength concrete:

Table 8. Cement content that used in several *"high strength"* concrete manufacturing projects [6].

Projects	**Compressive strength (MPa)**	**Cement content**
Ens, Cachan Prancis	80	449
LCPC University of Sher Brooke Canada	120	505
Bridge PLN, ASG	75	424
Arche De La Depense	60	425
Pretviset Bridge	86	430
Hassan Ll's Mosque in Cassablanka	95	465
Kwang Tong By Pass, Bridge	80	500
Two Union Square, Seattle	119	555
ITS Concrete Lab. 1	100	578.25
ITS Concrete Lab. 2	90	589.24
Typical Concrete in French	89.55	584.73
PPC Concrete of PT. Semen Gresik	41.5	487
Average cpmpressive strength (MPa)	**86.34**	**499.35**
Range cement content (Kg/m^3)	**424 – 589.24**	

According to (Tab.8) above, it is explained that the range of cement content for high strength concrete and very high strength concrete in applications in the field ranges from 424 Kg/m^3 to 589.25 Kg/m^3. In the (Tab.9) below is the level of cement content used in previous research on high early strength concrete:

Table 9. Cement content levels of high early strength concrete in several past research [1,2,3].

Researcher	**Compressive strength at 24 hours (MPa)**	**Cement content (Kg/m^3)**	**Level of cement content**
SHRP 1993	39	522	High
FHWA 2001	Min. 13.5	410	High
Research of Product Application Department PT. Semen Indonesia 2015	66.39	740	Very High

According to (Tab.9) above, the average cement level of content used in high early strength concrete is included in high level of cement content.

4.3 Water to Cement Ratio

By reducing the water to cement ratio, it reduces the volume of the air cavity and increases the density of the concrete .This transformation contributes significantly to the strength and durability of the concrete [5]. The following table is classifications of water to cement ratio levels and several results from the concrete:

Table 10. Classifications of water to cement ratio [5].

Water to cement ratio	**Classification**
> 0.45	Conventional Concrete
0.45 – 0.30	High Strength Concrete
< 0.3	Very High Strength Concrete

Table 11. Classifications of water to cement ratio of high early strength concrete in several past research [1,2,3].

Researcher	Water to cement ratio	Classification
SHRP 1993	0.32	High Strength
FHWA 2001	0.40	High Strength
Research of Product Application Department PT. Semen Indonesia 2015	0.20	Very High Strength

According to (Tab.11) above, the water to cement ratio used in ranges 0.2 to 0.4 *(high to very high strength classification).*

4.4 Types of Admixtures

In high early strength concrete manufacturing, there are several types of admixture used for different purposes: accelerating and water reducing admixture and high range water reducing admixture. The following table is several types of chemical admixture that used in making high early strength concrete:

Table 12. Typical of chemical admixture in making high early strength concrete [4].

Admixture types	Function	Based on	Type
Accelerating admixture and Water Reducing	1. Accelerate cement hydration process. 2. Reducing water content by 5 to 10%.	Calcium Chloride	E (1st Generation)
		Aluminum Chloride	
		Sodium Sulphate	
		Aluminum Sulphate	
		Calcium Nitrite	
High Range Water Reducing	1. Accelerate cement hydration process. 2. Reducing water content by 20% (generation 2) 3. Reducing water content by 40% (generation3).	Ligno Sulphonate	F (2nd Generation)
		Gluconate	
		Naphtalene Sulphonate	
		Melamine	
		Vinyl Copolymers	F (3rd Generation)
		Polycarboxylate Ether	
		Modifed Polycarboxylate	

Table 13. Admixtures are used depend on water to cement ratio [5].

Water to cement ratio	Admixture
$0.3 < w/c \leq 0.45$	Water Reducing Admixture
$w/c < 0.3$	High Range Water Reducing Admixture

Table 14. Type of admixtures that used in several past research [1,2,3].

Researcher	W/C	Strength at 24 hours (MPa)	Chemical compounds base	Admixtures type
SHRP 1993	0.32	39	Naphtalene	High Range Water Reducing Admixture (2nd Generation)
FHWA 2001	0.40	Min 13.5	Calcium Chloride, $CaCl_2$	Water Reducing Admixture (1st Generation)
Research of Product Application Department PT. Semen Indonesia 2015	0.20	66.39	Polycarboxylate Ether	High Range Water Reducing Admixture (3rd Generation)

5. High early strength concrete using local materials

In some areas in northern coast (*Pantura*) of east java province, namely: Tuban, Bojonegoro, and Gresik proved to have a potential source of high early strength concrete materials. These materials are: *kethak* (calcite) stone in Merakurak-Tuban and Ujung Pangkah-Gresik, then Bengawan Solo river sand in Bojonegoro. With the formation of the strategic material source location, the concrete can be implemented. In 2016, Institut Teknologi Sepuluh Nopember (ITS) and PT. Semen Indonesia has conducted research on high early strength concrete with utilizing the local materials [8]. The following table is brief several specifications of the material and the performance of the concrete:

5.1 Specification of "kethak" (calcite) stone and Bengawan Solo river sand

Table 15. Material testing results of *kethak* (calcite) stone from Tuban and Gresik [8].

Testing Parameters	Unit	Results	
		***Kethak* (calcite) Stone**	**Bengawan Solo River Sand**
Physical Properties			
Specific gravity	ton/m^3	2.72	2.84
Moisture content	%	2.63	2.08
Absorption	%	4.24	0.27
Bulk density	kg/m3	1330.00	1626.80
Materials finer than 75-μm (No.200)	%	1.13	4
Organic impurities index	Color Index	-	2
Abrasion	%	32.00	-
Grading Zone	-	Max. 20 mm	3
Fines modulus	-	7.5	3
XRD Analysis			
Andesite	%	Up to 50	
Calcite	%	Up to 90.28	

5.2 High Range Water Reducer

Table 16. Technical specification of High range water reducer based on Polycaroboxylate ether [8].

Chemical compound	color	Dosage

IOP Conf. Series: Materials Science and Engineering **267** (2017) 012004 doi:10.1088/1757-899X/267/1/012004

Modifed polycarboxcylate Copolymer	*Turbid white liquid*	0.3-0.8% *(soft plastic)* *0,8 – 2% (SCC)*

5.3 Performance of high early strength concrete with utilizing local materials

Table 17. High Early Strength Concrete using local materials test results [8].

No.	Testing Parameters	Units	Results	
1.	Initial Slump ($T_{0\ minutes}$)	*cm*	30.00	
2.	Slump ($T_{60\ minutes}$)	*cm*	8.00	
			Ages	
			1 day	**28 day**
3.	Average of compressive strength	*MPa*	27.73	64.03
4.	Ultrasonic Pulse Velocity (UPV)	*Km/s*	3.80	4.30

According to the (Tab.17) above, the concrete produces a compressive strength value that qualifies for high early strength concrete, which has a structural compressive strength (above 210 Kg/cm^2) at 1 day [4]. In addition, from the UPV test parameters, it produces a *good* index of V values between 3.5 – 4.5 Km/s according to IS 13311-1-1992 [7].

6. Conclusions

From some of the explanations in this paper, we can get a conclusion:

a) High early strength concrete was created to solve the setting time problems that occur in the world of buildings and infrastructures construction. Generally, it takes 7 to 14 days to achieve the full concrete hardening and reach the minimum compressive strength of 0.7 to 0.85 *fc'*. High early strength concrete means that the compressive strength of the concrete at the first 24 hours after site-pouring could achieve structural concrete quality (compressive strength > 21 MPa)
b) There are 4 (four) important factors that must be considered, those factors including: portland cement type, cement content, water to cement ratio, and admixture type.
c) The performance of high early strength concrete with local materials (*kethak* or calcite stone from Merakurak-Tuban and Ujung Pangkah-Gresik, then Bengawan Solo river sand from Bojonegoro) was apparently able to achieve the quality of compressive strength of 27MPa at the age of 24 hours, which qualified enough for the structural compressive strength > 210 MPa.

7. References:

[1] P. Zia, S. H.Ahmad, J. J. Schemmel and R. P. Ellion, "*Mechanical Behavior of High Performance Concrete, Volume 4, High Early Strength Concrete,*" Strategic Highway and Research Program, National Academy of Sciences, Washington DC, 1993.

[2] Federal Highway Administarion, "*PCC Pavement Evaluation and Rehabilitation,*" Federal Highway Administration, Washington DC, 2001.

[3] Research and Development Department of Products Application PT Semen Indonesia (persero) Tbk, "*Rapid Strength Concrete,*" Research and Development Department of Products Application PT Semen Indonesia (persero) Tbk, Gresik, 2015.

[4] Sika Services AG, *Sika Concrete Handbook*, Zurich: Sika Services AG, 2013.

[5] Portland Cement Association, "*Concrete Technology Today : High-Strength Concrete,*" *High Strength Concrete,* vol. **15**, no. 1, pp. 1-8, 1994.

IOP Conf. Series: Materials Science and Engineering **267** (2017) 012004 doi:10.1088/1757-899X/267/1/012004

[6] A. Subakti, B. Picesa and M. Irawan, *Teknologi Beton dalam Praktik*, Surabaya: ITS Press, 2012.

[7] Indian Standarts, "*Non-Destructive Testng of Concrete-Methods of Test, Part 1: Ultrasonic Pulse Velocity*," Indian Standarts, New Delhi, 1992.

[8] A. K. Yasin, A. B. K. Suharso, R. Bayuaji and T. E. Susanto, "*Engineering on High Early Strength Concrete with Utilizing Kethak (Calcite) Stone Tuban-Gresik and Bengawan Solo River Sand*," Institut Teknologi Sepuluh Nopemner, Surabaya, 2016.

IOP Conf. Series: Materials Science and Engineering **267** (2017) 012005 doi:10.1088/1757-899X/267/1/012005

Synthesis of Nano-Ilmenite (FeTiO3) doped TiO2/Ti Electrode for Photoelectrocatalytic System

[1]Hikmawati, [1]A H Watoni, [1]D Wibowo, [1]Maulidiyah, [1]M Nurdin.
[1]Department of Chemistry, Faculty of Mathematics and Natural Sciences, Universitas Halu Oleo, Jl. H.E.A. Mokodompit Kampus Baru Anduonohu, Kendari 93232 – Southeast Sulawesi, Indonesia

E-mail: mnurdin06@yahoo.com

Abstract. Ilmenite ($FeTiO_3$) doped on Ti and TiO_2/Ti electrodes were successfully prepared by using the sol-gel method. The structure, morphology, and optical properties of $FeTiO_3$ are characterized by XRD, UV-Vis DRS, and SEM. The $FeTiO_3$ and TiO_2 greatly affect the photoelectrocatalysis performance characterized by Linear Sweep Voltammetry (LSV) and Cyclic Voltammetry (CV). The characterization result shows a band gap of $FeTiO_3$ is 2.94 eV. XRD data showed that $FeTiO_3$ formed at 2θ were 35.1° (110), 49.9° (024), and 61.2° (214). The morphology of $FeTiO_3$/Ti and $FeTiO_3$.TiO_2/Ti using SEM shows that the formation of $FeTiO_3$ thin layer signifies the Liquid Phase Deposition method effectively in the coating process. Photoelectrochemical (PEC) test showed that $FeTiO_3$.TiO_2/Ti electrode was highly oxidation responsive under visible light compared to the $FeTiO_3$/Ti electrodes i.e. 7.87×10^{-4} A and 9.87×10^{-5} A. Degradation test of $FeTiO_3$/Ti and $FeTiO_3$.TiO_2/Ti electrodes on titan yellow showed that the percentages of degradation with photoelectrocatalysis at 0.5 mg/L were 41% and 43%, respectively.

1. Introduction

The management and disposal of wastewater needed attention for increasing the awareness and the main concern for public health in a modern era [1,2]. The wastewater treatment has categorized in physical, chemical, and biological processes to improve an excellent method of eliminating aspects in wastewater pollutants [3,4]. The handling efficiency and low-cost factors are one of ways to solve the wastewater pollutants [5,6]. Chemical processes are a way of dealing with environmental pollution using semiconductor materials such as TiO_2, ZnO, Fe_2O_3, etc [7,8].

The pollution control requires the development of new infrastructure technology as a degradation instrument in overcoming the level of wastewater pollution in the environment. Therefore, it is needed a material to support the realization of the instrument design. In recent decades, environmentally friendly materials have been developed such as TiO_2. It has various benefits such as environmental protection, water purification, sensors, and photovoltaic [9,10]. The utilization of TiO_2 based on the photocatalyst process as categorized of a very strong oxidizing agent when exposed to ultra violet (UV) light with a wavelength of 365-385 nm [11,12]. UV light irradiation will experience separation of electrons in the conduction band and the valence band to form a hole (h+) [13,14]. In the valence band will produce a hydroxyl radical (•OH) an oxidizing agent to overcome the wastewater [15,16].

In the last decade, TiO_2 photocatalyst study still has various constraints namely the recombination of electrons and holes which decrease the performance of TiO_2 semiconductors [17,18]. Recent development performed by Nurdin et al. using the TiO_2 semiconductor by merging photocatalysis and electrochemical methods into photoelectrocatalysts that have the advantages for degrading of organic pollutants into harmless compounds such as H_2O and CO_2. It is more efficiently in the use of chemicals and energy consumption [19,20].

The current TiO_2 photoelectrocatalyst method becomes an advantage in the development of TiO_2 as a degradation agent of organic matter and Chemical Oxygen Demand (COD) sensors [15,21]. However, the disadvantage of TiO_2 is only active by using UV light irradiation because of the relatively large value of TiO2 energy gap (Eg) of 3.2 eV for the anatase phase, whereas UV rays entering the Earth only about 3-5% and low yield efficiency (<10%). One of attempts to activate TiO_2 in visible light which have been performed in some researches is by adding dopants (both metals and nonmetals) [9]. Researchers have done by immobilizing metal elements in TiO_2, including gold (Au), iron (Fe), and silver (Ag) show the success of activating TiO_2 in visible light to enhance the photoelectrocatalytic performance [9,21]. Research for non-metallic immobilization, which have been carried out among others are Nitrogen (N), Carbon (C), Sulphur (S) have been shown to effectively shift TiO_2 optical absorption to lower energy levels [13]. The function of dopant addition is to modify of TiO_2 crystals so it can change the electronic structure, optical properties, and high stability [10].

A recent study by Yu et al. has performed Fe-TiO_2 nanotube doping and its application in photoelectrocatalysis that the Fe doping is best used because the radius of Fe^{3+} (0.64Å) ion is almost the same as the radius of Ti^{4+} (0.68Å) ion. Fe^{3+} ions will more easily substitute Ti^{4+} ions in TiO_2 crystals [22]. Raghavender et al. reported that Fe^{3+} doping on TiO_2 with $Fe(NO_3)_3$ composition tends to form ilmenite, which can overcome environmental pollution by using visible light with band gap value of 2.58-2.9 eV [23]. The iron element (Fe) in ilmenite with structure of TiO_2 perovskite for upgrading the titanium lattice, then yield rutile-type TiO_2 that can be active in visible light with Eg <3.0 eV [22].

Based on the previous literature, in this research has fabricated Fe-doped TiO_2 working electrode with of $Fe(NO_3)_3$ as a Fe dopant to form ilmenite ($FeTiO_3$/Ti) with a sol-gel method to degrade Titan Yellow (TY) test compound. The focus of this research to assess the performance of $FeTiO_3$/Ti electrodes which would be compared with the $FeTiO_3.TiO_2$/Ti electrode to study the efficiency level in degrading TY compound presents in water. The TY organic compound has chosen because it is one of the harmful triazine dyes that irritate human skin when contained in water because of contains one or some relatively stable benzene rings [24].

2. Experimental Method

2.1. Fabrication of TiO_2/Ti by anodizing

The Ti (titanium) plate was performed by cutting Ti plate with 99% purity and 0.5 mm thickness with a size of 4 cm × 0.5 cm then sanded using fine sandpaper size of 1200CC until the surface was clean and shiny then washed by using a detergent solution, water, and distilled water (D.I. H_2O). After being air-dried, the Ti plate was etched using a mixture solution of HF, HNO_3, and D.I. H_2O with a ratio of 1: 3: 6 for 2 minutes.

The Ti plate was inserted into the probe which has contained electrolyte solution of 0.27 M NH_4F and D.I. H_2O in 98% glycerol. The anodizing process was performed by placing Ti plate as anode and Cu plate as a cathode and providing bias potential of 25 Volts connected to the power supply. Anodizing process performed for 4 hours and calcined for 1.5 hours at a temperature of 500°C to evaporate the remaining electrolyte solution on Ti plate and to obtain TiO_2 anatase crystals which have high-photodegradation activity.

2.2. Synthesis of $FeTiO_3$ by sol-gel method

The $FeTiO_3$ sol-gel was made with mixing two solutions in a reflux method. Solution 1 was prepared in colloidal solution of TiO_2 using controlled hydrolysis of 4 mL TTIP in 0.5 mL of acetyl acetonate and 15 mL of 99% ethanol. Solution 2 was prepared by 15 mL of 99% ethanol and 2 mL of D.I. H_2O by adding 1 mL of 0.1 M acetic acid. Then two solutions was mixed and stirred using a magnetic stirrer for 3 h at 50°C and added $Fe(NO_3)_3$ to produce $FeTiO_3$ sol. The resulting solvent was evaporated at room temperature for 48 hours to form a gel and heated to 80°C in the oven for 30 minutes. Ti and TiO_2/Ti anodizing plates were immobilized with sol-gel by coating the $FeTiO_3$ sol-gel on the Ti and TiO_2 plate using Liqiud Phase Deposition (LPD) method then sintered at 150°C for 15 minutes to obtain $FeTiO_3$/Ti and $FeTiO_3.TiO_2$/Ti electrodes.

2.3. Photoelectrocatalytic degradation test
Determination of TY standard curve with a concentration of 0.5; 2.0; 4.0; and 5.0 ppm dissolved with 0.1 M $NaNO_3$ as an electrolyte solution. The wavelength was determined and absorbance value using UV-Vis spectrophotometer. TY dye degradation tests of 0.5 ppm (+ 0.1 M $NaNO_3$) were performed using a Multi Pulse Amperometry (MPA) method with a duration of 10 min and a potential difference of 0.5 Volt under variations of UV and visible lights. For 10 minutes the absorbance was determined to obtain decrease of concentration. Measurements were variated against $FeTiO_3$/Ti and $FeTiO_3.TiO_2$/Ti electrodes.

3. Results and Discussion
3.1. Fabrication of TiO_2/Ti by anodizing
Preparation of TiO_2/Ti nanotubes by electrochemical oxidation was performed by using copper (Cu) metal at cathode and Ti metal at the anode for 4 hours with a bias potential of 25.0 Volt. The function of high voltage in the fabricating TiO_2/Ti thin film is to form nodules or particulates on Ti metal. The anodizing process will form nanomaterials with sizes between 1-100 nm [3]. The reaction process in anodizing methods containing F^- ions can be seen in Figure 1.

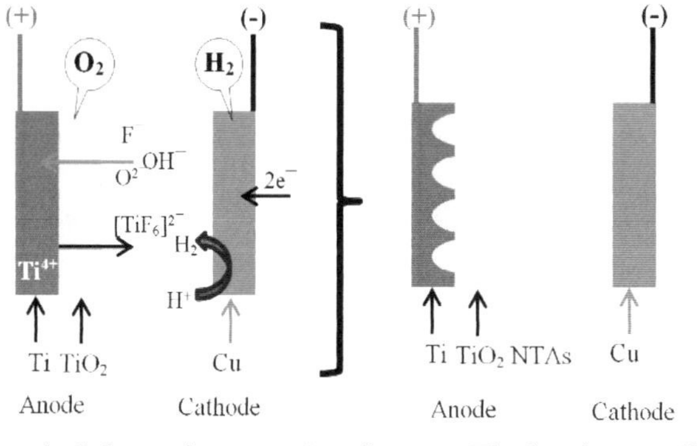

Figure 1. Schematic nanotubes form on Ti plate by anodizing

3.2. $FeTiO_3$ doped TiO_2/Ti electrode
The sol-gel method is a method for synthesizing non-metallic or metal-doped TiO_2 nanoparticles. This method has many advantages in terms of purity, homogeneity, flexibility, easily controlled stoichiometry, easily processing and compositions which are controlled quickly, economically, simply and accurately [10].

The basic composition in the preparation of $FeTiO_3$ sol-gel used was (TTIP) compound in sol-gel method for dopant ion distributor medium or matrix and forming a nanoparticle, acetyl acetone acting as a chelating ligand which will produce an exothermic reaction and make the yellow solution [10]. The creation of $FeTiO_3$ sol by adding $Fe(NO_3)_3$ compounds may increase the activity of photocatalysis which acts as trap of electron photoregeneration and hole so it can suppress the occurrence of electron-hole recombination. The modification of TiO_2 using $FeTiO_3$ doping can produce high photocatalytic activity under visible light. Visually the results obtained in $FeTiO_3$/Ti and $FeTiO_3.TiO_2$/Ti doping techniques are shown in Figure 2.

IOP Conf. Series: Materials Science and Engineering **267** (2017) 012005 doi:10.1088/1757-899X/267/1/012005

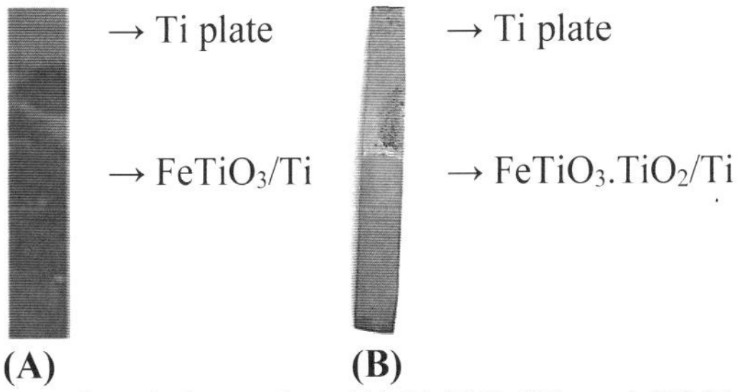

Figure 2. The results of electrodes; (A) $FeTiO_3/Ti$, and (B) $FeTiO_3.TiO_2/Ti$

3.3. Characterizations

3.3.1. Bandgap $FeTiO_3$

The UV-Vis Diffuse Reflectance Spectrophotometer (UV-Vis DRS) is used to determine the energy gap (Eg) produced by semiconductor from the synthesized $FeTiO_3$ solid. Eg values of the resulting semiconductor can affect the performance of the semiconductor in exciting electrons from the valence band to conduction band region [2]. This condition is depending on the bandwidth gap width generated by the semiconductor. Small Eg values will make it easier for electrons to move from the valence band to the conduction band so that the flow of electrons will get easier.

$$Eg(eV) = \frac{1239.8}{\lambda}$$

The Eg value was plotted in the equation above to obtain the wavelength area of $FeTiO_3$ performance and it has been obtained the wavelength at 423.13 nm [15].

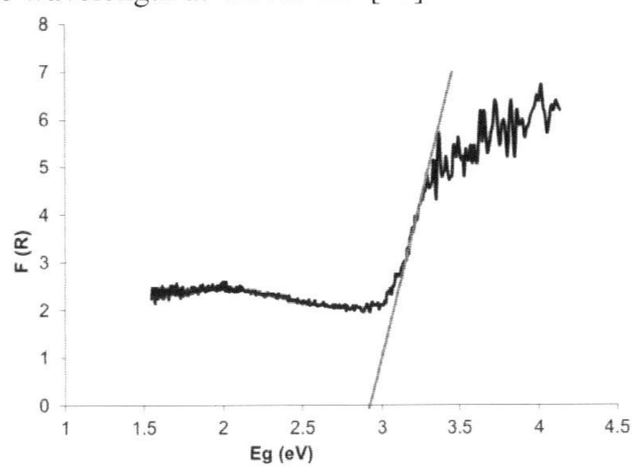

Figure 3. The bandgap of $FeTiO_3$

3.3.2. X-ray Diffraction (XRD)

Measurements using XRD aims to obtain crystal structure information of $FeTiO_3/Ti$ and $FeTiO_3.TiO_2/Ti$ working electrodes. The appearance of XRD diffractogram peaks of 35.1° (110), 49.9° (024) and 61.2° (214) indicates the $FeTiO_3$ has coated on the substrate. It has been compared with the JCPDS standard No. 29-0733, and 76-1821. The peak of 42.5° (400) shows the Fe_3O_4 magnetic corresponding to JCPDS No. 19-0629.

The anatase crystal in TiO_2 at 2 theta 25° (101) does not absence to indicate that the possibility of rutile crystals is formed. The presence of Fe dopant in TiO_2 crystal that possibility of rutile crystals formed. the Ti element as conducting substrate has appeared at 2 theta 37.5°, 39°, dan 52°. The diffractogram peaks between $FeTiO_3/Ti$ and $FeTiO_3.TiO_2/Ti$ plate crystal are shown in Figure 4.

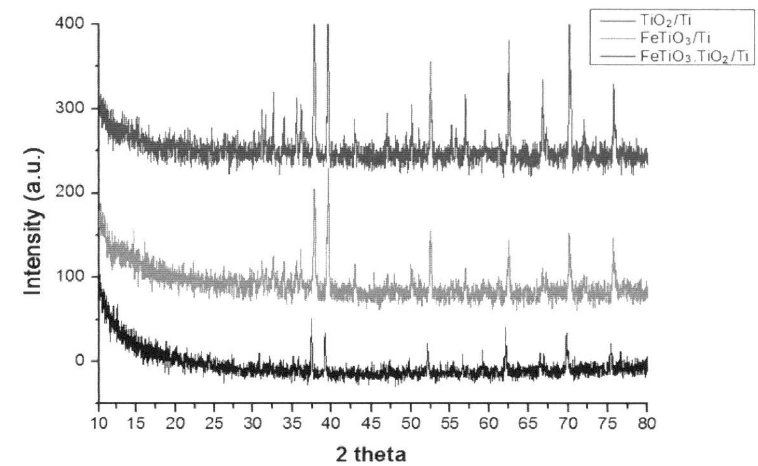

Figure 4. The diffractogram of $FeTiO_3/Ti$ and $FeTiO_3.TiO_2/Ti$ electrodes

3.3.3. Scanning Electron Microscopy (SEM)

The result of characterization using SEM aims to look at the morphology on the surface of the electrode that has been coated with $FeTiO_3$ sol-gel. The result of SEM characterization can be seen in Figure 5.

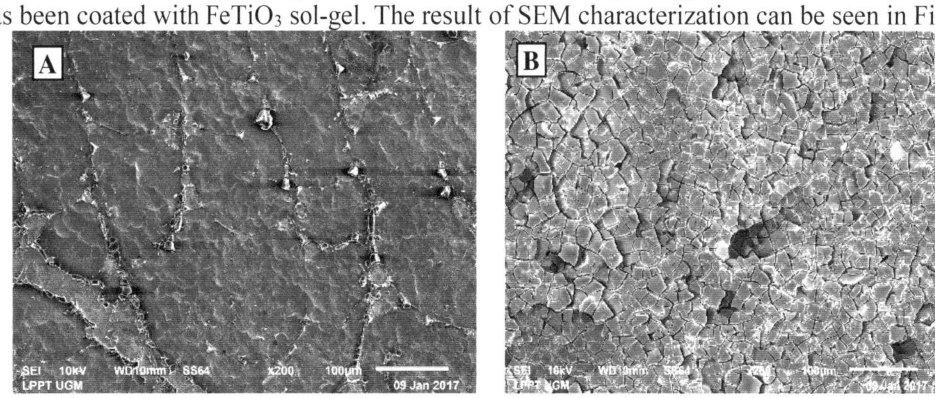

Figure 5. Morphology on electrodes, (A) $FeTiO_3/Ti$, (B) $FeTiO_3.TiO_2/Ti$

Figure 5A the morphology of $FeTiO_2/Ti$ electrode indicated that the sol-gel method has the thin layer formed and evenly distributed. Based on literature that allows the formation of tubes in Ti metals during etching process using HF, HNO_3, and D.I. H_2O solvents. The formed tube improves the adhesion of sol-gel onto Ti metal surface. Figure 5B shows a high regular array of crystals in random type. The coated by the sol-gel method on the $FeTiO_3.TiO_2/Ti$ electrode exhibits uniform results with an excellent crystalline order. The role of TiO_2/Ti as a template is excellent for the sol-gel coating process and provides a small probability of covering the surface of TiO_2/Ti nanotubes [25]. According to Liu et al. the design of TiO_2 electrode by anodizing method acts as a sol-gel adhesive agent to improve the electronic properties of the material [26]. These results show that the LPD technique is very effective in the $FeTiO_3$ coating process on the substrate surface.

3.4. Photoelectrochemical test

Measurements using the LSV method to measure the proportional of photocurrent to the potential difference function. The photocurrent is current that can be observed when the electrode is irradiated by UV and visible light. This current is also a measure of the charge transfer rate between semiconductors/electrolytes which can be used as a measure of •OH formation rate on the surface of catalyst. Figure 6A and 6B the LSV data shows that the activity of good photocurrent response when it is irradiated visible light. The electrons move from the valence band (VB) to the conduction band (CB), so the electrons are trapped by the Fe metal, thus minimizing the occurrence of electron and hole recombination [27]. Visible light irradiated required the lower energy, that the electrons can move from VB to the CB.

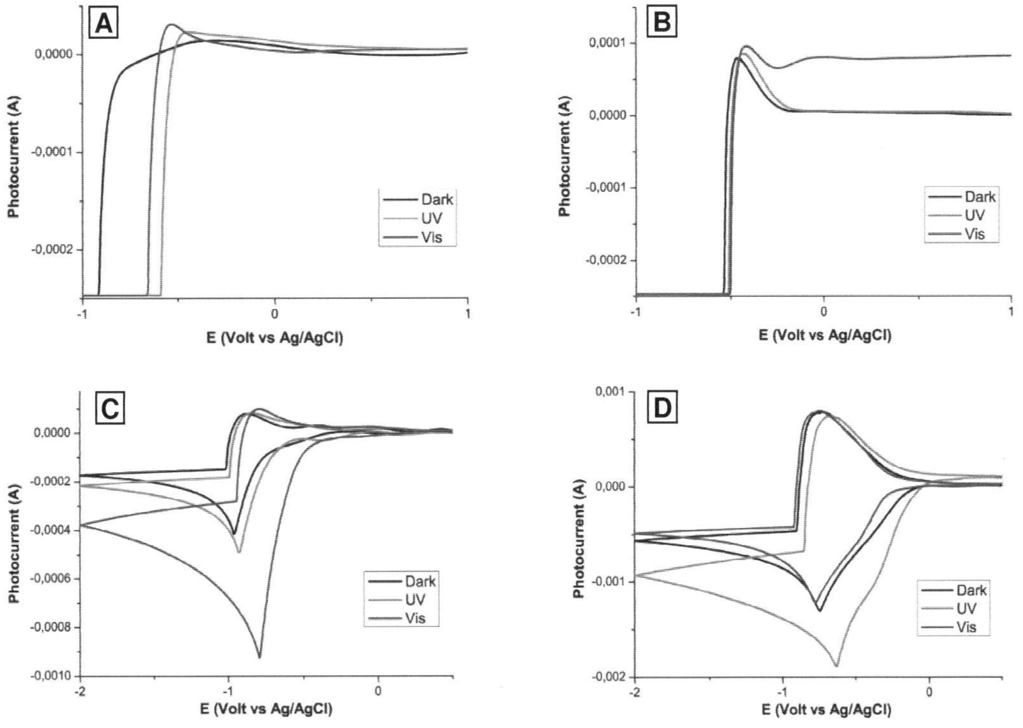

Figure 6. Photoelectrochemical graph, (A) LSV of $FeTiO_3$/Ti electrode, (B) LSV of $FeTiO_3.TiO_2$/Ti electrode, (C) CV of $FeTiO_3$/Ti electrode, and (D) CV $FeTiO_3.TiO_2$/Ti electrode

The CV test is an electrochemical analysis method based on the measurement of the electric current value as a function of potential flow which is given back and forth on the electrochemical cells. Test using CV connected to a potentiostat and connected to a computer with bias potential setting of 0.5 volt and scanrate of 0.05, the movement of potential bias was from -2 volts to 0.5 volts and sensitivity of 1.0 $\times 10^{-3}$ A/V. Performed by test variation to see the peak current differences during the redox reaction. Determination of photocurrent for each electrode by CV technique (Figure 6C and 6D) indicates that the high oxidation process based photocurrent response be greater using $FeTiO_3.TiO_2$/Ti electrode.

3.5. Photoelectrocatalytic Degradation

Determining the maximum wavelength of TY dye solution at 0.5 ppm was characterized using a UV-Vis spectrophotometer. The results of the absorbance spectrum of the test dye solution are shown in Figure 7.

IOP Conf. Series: Materials Science and Engineering **267** (2017) 012005 doi:10.1088/1757-899X/267/1/012005

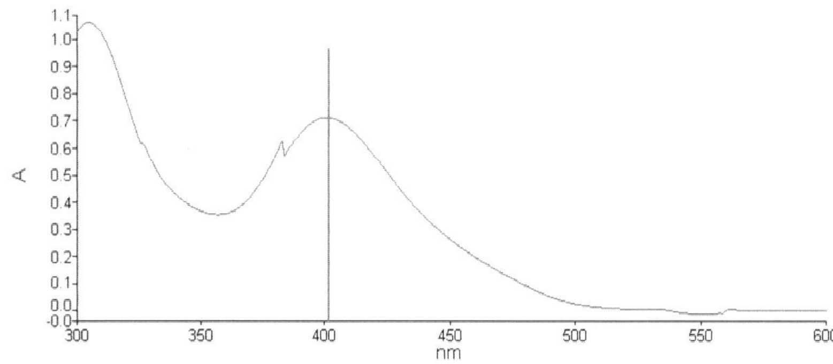

Figure 7. The wavelength of Titan yellow organic compound

The calibration curve was obtained by plotting concentration with absorbance so that the equation of straight line is $y = 0.2003x + 0.0092$. Photocatalysis is a catalyst that can work with the aid of light (photons). The occurence of TY concentration decreases in photocatalysis method is shown in Figure 8A and 8B.

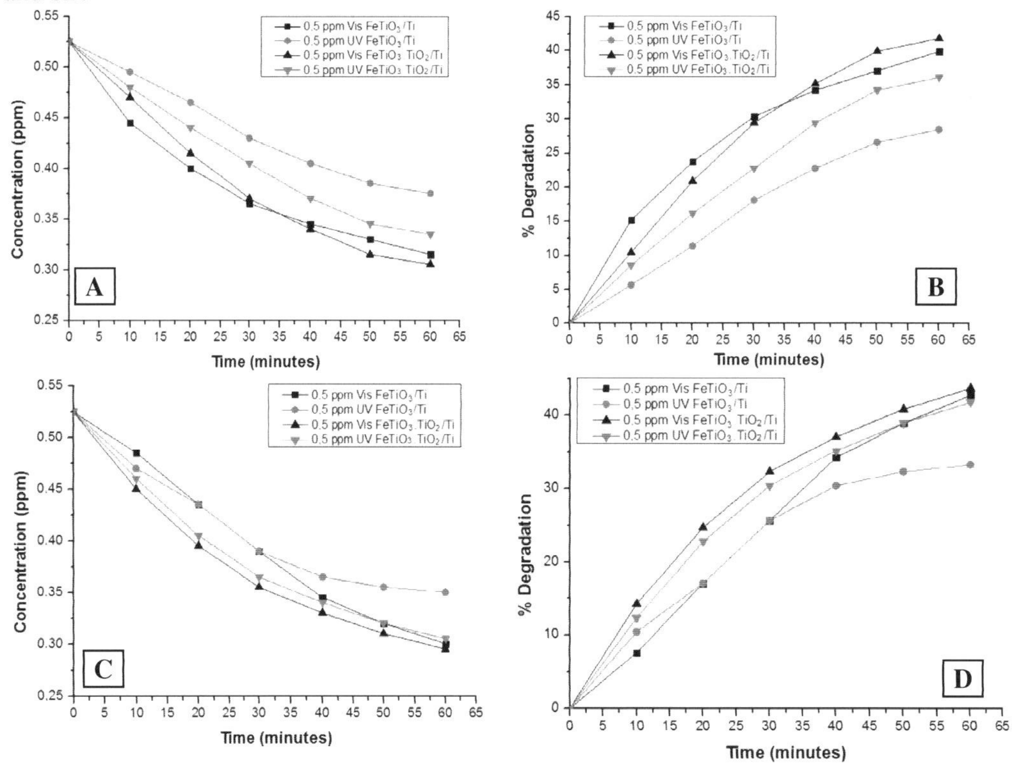

Figure 8. Comparison graph of photocatalytic and photoelectrocatalytic performance (A) Photocatalytic degradation, (B) % degradation of photocatalytic, (C) Photoelectrocatalytic degradation, (D) % degradation of photoelectrocatalytic

The process of photoelectrocatalysis (Figure 8C and 8D) will form many active spaces on the catalyst, theoretically the degradation process will run well. This method using $FeTiO_3/Ti$ and $FeTiO_3.TiO_2/Ti$ working electrodes illuminated UV and visible lights produce pairs of electrons and holes. The hole will initiate the oxidation reaction on the electrode surface, while the electrons are flown through the back contact to counter electrode and transferred to the existing electron capturer in the solution [5].

4. Conclusion

The photoelectrochemical response test of $FeTiO_3$/Ti and $FeTiO_3.TiO_2$/Ti electrodes showed that the photocurrent values of visible light irradiance on each electrode were 9.87×10^{-5} A and 7.87×10^{-4} A, respectively. The photoelectrocatalytic activity test to degrade TY organic compound showed that the percent of degradation under visible light at 0.5 ppm were 41% and 43%, respectively.

5. Acknowledgement

We acknowledge for financial support of the DRPM-Ministry of Research, Technology and Higher Education, the Republic of Indonesia

6. References

[1] Nurdin M, Wibowo W, Supriono, Febrian M B, Surahman H, Krisnandi Y K, Gunlazuardi J 2009 *Makara, Sains,* **13** 1.
[2] Nurdin M 2014 *Int. J. Pharm. Bio. Sci.* **5** 360.
[3] Nurdin M, Maulidiyah 2014 *Int. J. Sci. Technol. Res.* **3** 122.
[4] Sari A A, Kurniawan H H, Nurdin M, Abimanyu H 2015 *Energy Procedia* **68** 254.
[5] Maulidiyah, Ritonga H, Salamba R, Wibowo D, Nurdin M 2015 *Int. J. ChemTech Res.* **8** 645.
[6] Dahnum D, Tasum O S, Triwahyuni E, Nurdin M, Abimanyu H 2015 *Energy Procedia* **68** 107.
[7] Maulidiyah, Nurdin M, Erasmus, Wibowo D, Natsir M, Ritonga H, Watoni A H 2015 *Int. J. ChemTech Res.* **8** 416.
[8] Maulidiyah, Nurdin M, Widianingsih E, Azis T, Wibowo D 2015 *ARPN J. Eng. & Appl. Sci.* **10** 6250.
[9] Maulidiyah, Nurdin M, Wibowo D, Sani A 2015 *Int. J. Pharma. Pharmaceu. Sci.* **7** 141.
[10] Nurdin M, Muzakkar M Z, Maulidiyah M, Nurjannah M, Wibowo D 2016. *J. Mater Environ. Sci.* **7** 3334.
[11] Nurdin M, Maulidiyah, Watoni A H, Abdillah N, Wibowo D 2016 *Int. J. ChemTech Res.* **9** 483.
[12] Nurdin M, Zaeni A, Maulidiyah, Natsir M, Bampe A, Wibowo D 2016 *Orient. J. Chem.* **32** 2713.
[13] Maulidiyah, Wibowo D, Hikmawati, Salamba R, Nurdin M 2015 *Orient. J. Chem.* **31** 2337.
[14] Nurdin M, Ramadhan L O A N, Darmawati, Maulidiyah, Wibowo D 2017. *J. Coatings Technol. & Res.* DOI: 10.1007/s11998-017-9976-8.
[15] Ruslan, Mirzan M, Nurdin M, Wahab A W 2016 *Int. J. Appl. Chem.* **12** 399.
[16] Maulidiyah, Tribawono D S, Wibowo D, Nurdin M 2016 *Anal. Bioanal. Electrochem.* **8** 761.
[17] Maulidiyah, Ritonga H, Faiqoh C E, Wibowo D, Nurdin M 2015 *Biosci. Biotechnol. Res. Asia* **12** 1985.
[18] Arham Z, Nurdin M, Buchari B 2016 *Int. J. ChemTech Res.,* **9** 113.
[19] Mursalim L O A M, Ruslan, Safitri R A, Azis T, Maulidiyah, Wibowo D, Nurdin M 2017 *IOP Conference Series: Materials Science and Engineering* xx xx.
[20] Nurhidayani, Muzakar M Z, Maulidiyah, Wibowo D, Nurdin M 2017 *IOP Conference Series: Materials Science and Engineering* xx xx.
[21] Wibowo D, Ruslan, Maulidiyah, Nurdin M 2017 *IOP Conference Series: Materials Science and Engineering* xx xx.
[22] Yu J, Wu Z, Gong C, Xiao W, Sun L, Lin C 2016 *Nanomaterials* **6** 107.
[23] Raghavender A T, Hong N H, Lee K J, Jung M H, Skoko Z, Vasilevskiy M, Cerquira M F, Samantilleke A P 2013 *J. Magnesium Materials* **331** 129.
[24] Maulidiyah, Imran, Muntu W, Nurdin M 2016 *Int. J. Appl. Chem.* **12** 347.
[25] Roy P, Berger S, Schmuki P 2011 *J. Chem.* **50** 2904.
[26] Liu Y, Wang L, Wang H, Xiong M, Yang T, Zakharova G S 2016 *Sensors and Actuators B: Chem.* **236** 529.
[27] Weber U, Weinmann M, Natter H, Bahre D 2015 *J. Appl. Electrochem.* **45** 591.

IOP Conf. Series: Materials Science and Engineering **267** (2017) 012006 doi:10.1088/1757-899X/267/1/012006

Synthesis and Photoelectrocatalytic Performance of Mn-N-TiO_2/Ti Electrode for Electrochemical Sensor

L O Mursalim A M[1], Ruslan[2], R A Safitri[1], T Azis[1], Maulidiyah[1], D Wibowo[1], M Nurdin*[1]

[1]Department of Chemistry, Faculty of Mathematics and Natural Sciences, Universitas Halu Oleo, Jl. H.E.A. Mokodompit, Kendari 93232 – Southeast Sulawesi, Indonesia.
[2]Department of Chemistry, Faculty of Mathematics and Natural Sciences, Univesitas Tadulako, Jl. Soekarno-Hatta, Palu 94118 – Central Sulawesi, Indonesia

E-mail: *[1] mnurdin06@yahoo.com

Abstract. Preparation of manganese (Mn) and nitrogen (N)-doped TiO_2/Ti electrode as a COD sensor was successfully performed to enhance photoelectrocatalytic activity under visible light. The preparation of Mn and N using the sol-gel method then coated on TiO_2 thin layer hidrothermal by dip-coating technique. Characterization of Mn-N-TiO_2/Ti electrode using FTIR showed that indicated the Ti-O, Ti-N, N-O, and Mn-O bonds. SEM-EDX showed a thin layer formed of Mn and N on TiO_2/Ti surface with the Mn, N, Ti, O spectrums at 0.277 KeV, 0.677 KeV, 4.508 KeV, and 0.579 KeV, respectively. The photoelectrochemical test using a potentiostat was indicated that Mn-N-TiO_2/Ti electrode active under visible light which evidenced by photocurrent response.

1. Introduction

Titanium dioxide (TiO_2) is a semiconductor material widely used for photocatalysis, microelectronics, optical cells, inactivation of microorganisms, anti-fogging and self-cleaning [1,2]. TiO_2 phase in anatase crystal was the most effective form and widely used as a photocatalyst because it has high chemical and physical stability, non-toxic, low-cost, and higher photoactivity compared to other semiconductor types [3,4]. The disadvantage of TiO_2 was it could only activated by UV light irradiation with a band gap energy (Eg) of 3.2 eV [5,6]. The limitations of TiO_2 photoactivity can be overcome to modified TiO_2 material by using inserted element (dopants) to activate it under visible light [7,8].

The doping method is one of TiO_2 modification techniques performed by adding another dopant into TiO_2 crystal lattice. Modification of TiO_2 has developed to increase TiO_2 activity in visible light and separate it from electrons and holes [9]. The common non-metallic dopants which generally added to TiO_2 are Nitrogen (N) [10], Sulfur (S) [11], Phosphorus (P) [12], Carbon [13] and Fluor (F) [14]. According to Yin et al. a variety of non-metallic elements as dopants which showed N dopant is effective to lowers TiO_2 band gap energy [15]. This phenomenon was increased TiO_2 thin film activity in visible light. The efficiency of photocatalysts with single N-TiO_2 dopant under visible light irradiation is still very limited [16].

IOP Conf. Series: Materials Science and Engineering **267** (2017) 012006 doi:10.1088/1757-899X/267/1/012006

In the last few decades, transition metal dopants were studied to increase the TiO_2 photocatalysis activity, such as Fe [17,18], Co [19], Ni [20], Cu [4], and Mn [21]. Manganese (Mn) is one of transition metals which has electropositive charge and has excellent electrical conductivity properties [22], it was potentially use as metal ion dopant [23]. Choudhury et al. have tested TiO_2 photocatalyst activity by Mn metal ions doped under visible light shows better results compared to TiO_2 without doping of Mn metal ions [24]. Hu et al. and Zhao et al. succeeded in synthesizing Mn and N-doped TiO_2 (Mn-N-TiO_2) using a sol-gel method which can decrease the energy gap and show strong wavelength absorption at 400 ~ 650 nm in photocatalysts performance [25,26]. Mn and N dopants also improve thermal stability in TiO_2 structure and high-photocatalytic activity to degrade organic dyes.

High-photocatalytic activity of Mn-N-TiO_2 catalysts could not apply to degrade organic dyes. Therefore, it is necessary for high ordered Mn-N-TiO_2 application to determine the Chemical Oxygen Demand (COD) sensor by photoelectrocatalysis [27,28]. The approach to the photoelectrochemical transfer by Faraday Law $(Q = nVFC)$ as photocurrent response was proportional to increase the concentration organic dyes. Nurdin et al. has developed a photoelectrocatalysis based on a COD sensor system with TiO_2/Ti working as electrode on Indium Tin Oxide (ITO) conductive glass giving it a high-performance in determining COD values [29]. Based on literatures, TiO_2 modification by metal and non-metallic dopants to develop COD sensors is still rarely explored. In this study, we have synthesized Mn-N-TiO_2 to develop the photoelectrocatalyst electrochemical sensor.

2. Experimental

2.1. Preparation of Titanium (Ti) foil

Preparation of Ti plate was performed by cutting Ti plate with purity of 99.7% and 0.5 mm thickness to size of 4 cm × 0.5 cm. It was sanded with 1200CC fine sandpaper until the surface was clean, shiny, and washed with the detergent solution, water and distilled water (D.I. H_2O). The foil was etched in HF, HNO_3, and D.I. H_2O in ratio 1:3:6 for two minutes. All chemicals were used to remove fat and oxide layer on Ti foil. The final stage was to rinse Ti foil with D.I. H_2O in order to remove residual etching solution on the surface [30].

2.2. Fabrication of TiO_2 hydrothermal electrode

A highly ordered TiO_2 electrode was synthesized by hydrothermal method. The oxidation process Ti foil was carried out in saucer porcelain and added by D.I. H_2O until it is well immersed. Then it is annealing at the temperature of 500°C to oxidation process by D.I. H_2O-assisted to form the TiO_2 in anatase structure.

2.3. Preparation of Mn-N-TiO_2/Ti by Sol-gel method

Aqueous TiO_2 colloid was prepared by the hydrolysis of titanium tetra-isopropoxide (TTIP) according to the method described by Maulidiyah et al [8]. Sol-gel was prepared by adding the 30 mL ethanol, 2 mL D.I. H_2O, 4 mL TTIP, 0.5 mL acetylacetonate (AcAc), and 1 mL acetic acid (0.5M). The solution was mixed and refluxed for 3 hours at temperature 50°C and stirred using a magnetic stirrer. After 1 hour followed by adding of 2 mL $CO(NH_2)_2$ 0.5 M and 1 mL $MnCl_2.4H_2O$ 2.5% solution to obtain the sol-gel N and Mn. The resulting sol was evaporated at room temperature for 48 hours to form a gel. Furthermore, the gel was heated in the oven at temperature 80°C for 30 minutes. The coating process was done by dip-coating technique on the TiO_2/Ti areas than calcined for 15 minutes at temperature of 150°C.

2.4. Photocurrent response in Electrochemical

Photocurrent response test was done by using Mn-N-TiO_2 electrode as working electrode, Pt as a counter electrode, and Ag/AgCl as a reference electrode were carried out in 0.1M $NaNO_3$ as electrolyte solution by Multi-Pulse Amperometry (MPA) I-t with bias potential 0.5 V, run-time 60 s under visible light irradiation.

3. Results and Discussion

3.1. Synthesis of TiO_2 thin film by hydrothermal method

Hydrothermal oxidation is a synthesis of TiO_2 thin film, which usually used by using D.I. H_2O at high temperature. Ti foil was prepared then annealed in 500°C to vapour the H_2O in order to remove residual solution. The annealing aimed to grow the TiO_2 by oxidation reaction from H_2O and roasting (O_2) process, where the mechanism can be seen in Figure 1:

$$2H_2O + Ti\ (surface) + O_2 \rightarrow TiO_2 + 2H_2\uparrow + O_2$$
$$Ti\ (surface) + 2H_2\uparrow + O_2 \rightarrow TiO_2$$

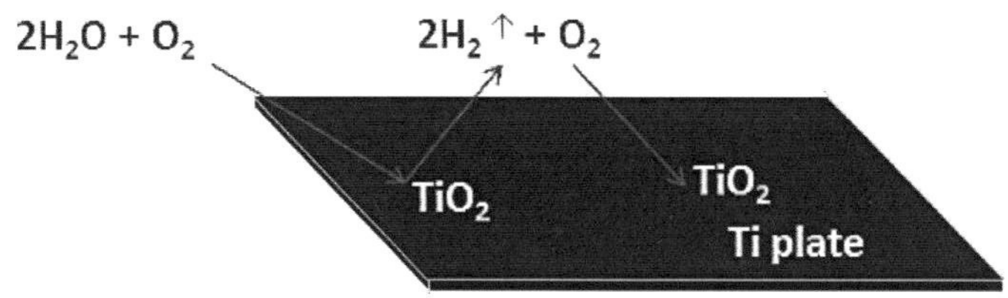

Figure 1. Schematic mechanism reaction on Ti foil by hydrothermal method

The remaining O_2 in the hydrothermal system serves to increase the amount of oxygen to accelerate Ti metal oxidation process [22]. Nurdin et al. have reported that the observed anatase crystalline form occurring on heating TiO_2 at temperature of 120°C could reach its perfect temperature at 500°C [31]. The long process of sintering need to be considered as it could affect the formed crystal phase. At the same temperature with different sintering times, it will produce different crystal phases. This research conducted sintering process at 500°C temperatures for 90 minutes to produce an anatase crystalline structure which has high-photocatalyst activity compared to other crystal structures [32,33]. The profile of the obtained TiO_2/Ti hydrothermal electrode can be seen in Figure 2.

Figure 2. TiO_2 thin film hydrothermal on Ti foil

The Ti plate surface has been grown by TiO_2 underwent a color change to purplish blue. According to Xiong et al. the color change of titanium metal shows the formation of TiO_2 oxide crystals on the surface and the probability of Ti^{3+} ion concentration in TiO_2 crystals to provide characteristics in the visible light spectrum [33].

3.2. TiO_2/Ti hydrothermal doped Manganese (Mn) and Nitrogen (N)

The coating of TiO_2/Ti thin film with Mn and N was performed by sol-gel method. The sol-gel method has advantages compared to other techniques for thin film deposition, i.e. it is excellent in composition control and to produce high-surface area on TiO_2 thin film substrate [10]. The precursors in sol-gel which used in this study was TTIP as a medium for distributing dopant ions to form Mn-N-TiO_2/Ti,

AcAc as a stabilizer and inhibit hydrolyzed TTIP when reacted with D.I. H_2O and ethanol. Precursors of TTIP, AcAc, ethanol, acetic acid and D.I. H_2O were refluxed for three hours at 50°C to increase the interaction of the compounds [34]. The addition of acetic acid was carried out to control the hydrolysis rate and assist the pore formation process to the TiO_2 mesoporous of anatase phase [10].

The addition of $MnCl_2.4H_2O$ as the Mn source and $CO(NH_2)_2$ as the N source have been done previously by Zhao et al. TiO_2/Ti hydrothermal electrode was coated with Mn and N sol-gel by dip-coating technique, then the electrode was sintered at 150°C [26,35]. This condition aimed to attach Mn-N-TiO_2 thin layer will be stronger bonds, to remove the water solvent and can form TiO_2 crystals of anatase phase [36]. TiO_2/Ti hydrothermal electrode which was doped by Mn and N can be seen in Figure 3.

Figure 3. The Mn-N-TiO_2/Ti electrode

3.3. Characterization of Mn-N-TiO2/Ti electrode

3.3.1. X-Ray Diffraction (XRD)

Characterization by XRD has carried out to investigate the composition, phase structure, and crystalline of the as-synthesized sample. The material type has determined by comparing XRD characterization data with the synthesized Mn-N-TiO_2/Ti pattern. Figure 4 shows diffraction pattern of Mn-N-TiO_2/Ti hydrothermal electrode.

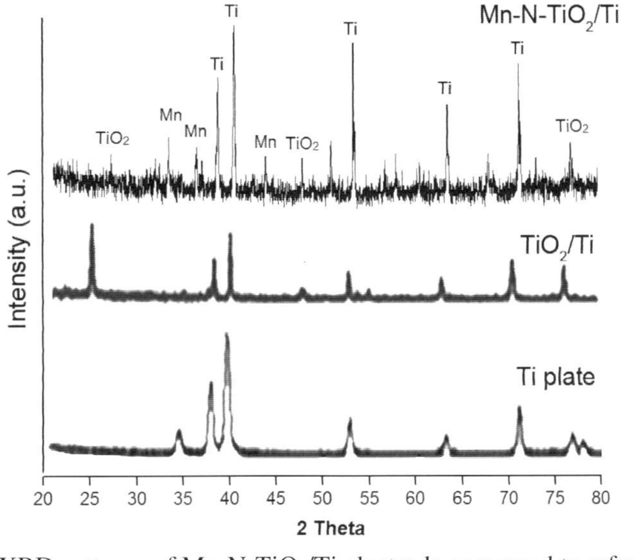

Figure 4. XRD patterns of Mn-N-TiO_2/Ti electrode compared to reference [31]

The XRD diffractogram of Mn-N-TiO_2/Ti electrode shows the specific peaks has been identified MnO_2, TiO_2, and Ti metal crystals. The Mn-N-TiO_2/Ti hydrothermal electrode from synthesis results is a rutile crystal type at 2 theta of 27°. The formation of rutile crystals is caused by inserting Mn atoms in TiO_2 matrix, thus changing the TiO_2 anatase phase in crystal structure. MnO_2 crystals formed on TiO_2 by sol-gel method at 2 theta of 32°, 37°, and 44° [37].

Mn and TiO_2 peaks appearing on XRD data indicate Mn and N sol-gel are well attached to TiO_2/Ti hydrothermal surface.

3.3.2. Energy Dispersive X-ray (EDX)

Characterization performed by using EDX on electrode surface to determine the presence of Mn and N elements. The results of EDX characterization on Mn-N-TiO_2/Ti electrode surface can be seen in Figure 5.

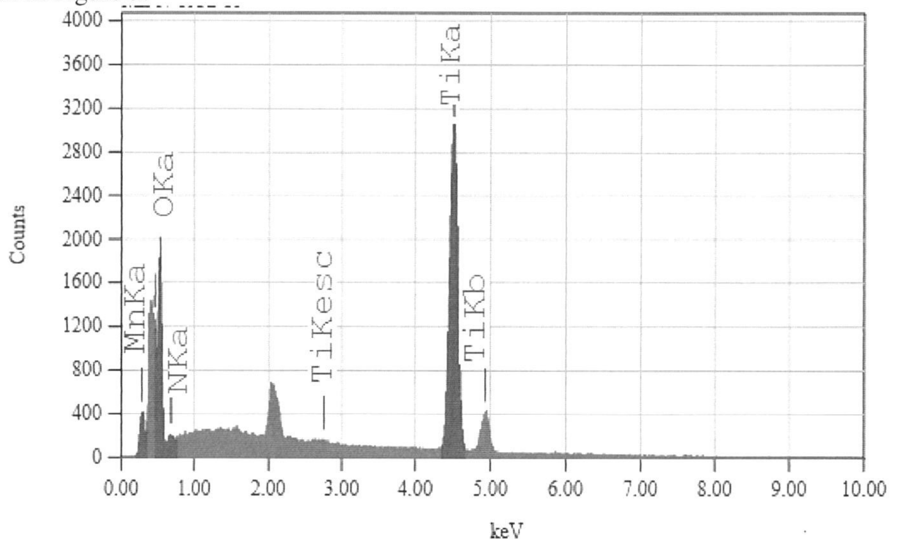

Figure 5. EDX spectra analyst of Mn-N-TiO_2/Ti electrode

EDX spectra showed the Mn peaks at 0.277 KeV, N at 0.677 KeV, Ti at 4.508 KeV and O at 0.579 KeV with the composition amount of 2.05 w%; 2.55 w%; 58.38 w%, and 37.02 w%, respectively. The calculated empirical formula from EDX data is the formation $Mn_1N_{0.3}Ti_{24.8}O_5$. The high Ti value existed from Ti plate substrate giving the highest intensity. The peaks of Mn and N indicate the presence of Mn and N which shows the success of Mn metal and N non-metal dopings by dip-coating technique.

3.3.3. Fourier Fransform Infra-Red (FTIR)

Characterization with FTIR was performed to obtain the structure of a molecule especially the functional groups seen in specific areas. The characterization result of Mn-N-TiO_2/Ti electrode with FTIR can be seen in Figure 6.

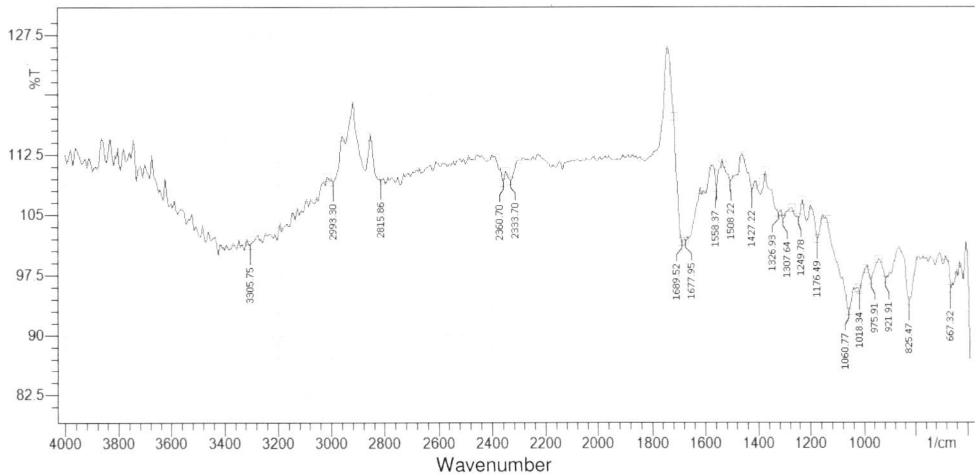

Figure 6. FTIR Spectra on Mn-N-TiO_2/Ti electrode

The FTIR spectrum above indicated the specific regions which presence of molecular vibrations in Mn-N-TiO_2/Ti electrode. Figure 6 the absorption in 1558.37 cm^{-1} shows the vibrations of O-H and N-H. The uptake appears in 1427.22 cm^{-1} the presence of N-O bonding vibrations. Wavenumber of 3305.75 cm^{-1} shows the O-H stretching from H_2O molecule vibration trapped in Mn-N-TiO_2/Ti electrode [38]. According to Yang et al. absorptions that appear on wavenumber of 1249.78 cm^{-1} and 1176.49 cm^{-1} indicate the presence of Ti-N bonding vibrations [39]. While, absorptions at the wavenumber of 2330.70 cm^{-1} and 2360.70 cm^{-1} indicate the presence of Ti-O bond vibration [26]. Absorption below 1000 cm^{-1} indicates Mn-O bond [40]. The absorption at wavenumber of 1060.77 cm^{-1} the existence of manganese metal ion attached to electrode surface [26].

3.4. Photocurrent Response by Amperometry (I-t)

Photocurrent response measurements were conducted by using a potentiostat DY2100B with a Multi-Pulse Amperometry (MPA) technique at a bias potential of 500 mV. This condition to obtain the high-potential bias and photocurrent response. The range detection was arranged 1 s to obtain the photocurrent value every 1 s for 60 s of measurement time. It was expected the stability of data and minimized the possibility of electrode heat due to exposure of visible light. Photocurrent response was obtained when the potentiostat has turned on during 10 s in dark condition. This condition to obtain the data with uniformity of illumination time between photocurrent. After 10 s the visible lamp turned on exhibit the excited initial photocurrent due to separate of electron-hole and initiating a reduction-oxidation reaction.

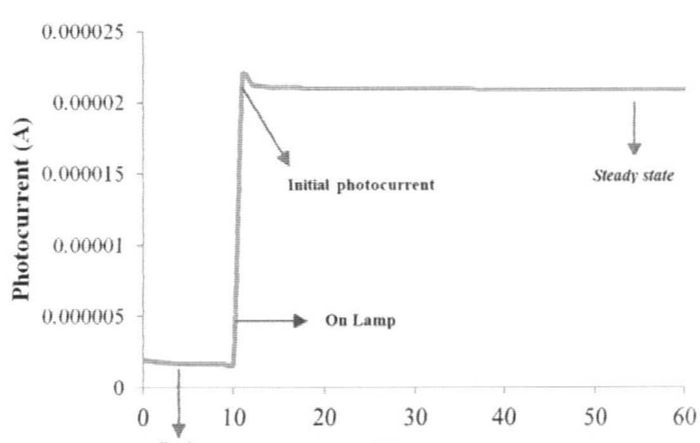

Figure 7. Photoelectrocatalysis profile using Mn-N-TiO_2/Ti in $NaNO_3$ 0.1M solution

Figure 7 shows the profile of photocurrent during measurement time ranges. Measurements with visible light off will result in background current which is the initial state of the solution. When the lights were switched on, there was a large current surge that indicated the occurrence of electron movement due to the reaction on surface of Mn-N-TiO_2/Ti electrode. The resulting current will decrease as the time of measurement increases, resulting in steady state. Steady state currents are formed because there is an equilibrium between the analytical fraction in surface of anode and diffusion of analyte from body of the solution [29,41].

4. Conclusion

We have studied the possibility of using the double doped i.e. Mn as metal and N as non-metal on the TiO_2/Ti hydrothermal electrode to be applied to COD sensor. The existence of N on TiO_2/Ti characterized by using FTIR showed the presented of Ti-O, Ti-N, N-O, and Mn-O bonds. Supported by SEM-EDX showed a thin layer formed of Mn and N on TiO_2/Ti surface with the spectrum of Mn, N, Ti, O at 0.277 KeV, 0.677 KeV, 4.508 KeV, and 0.579 KeV, respectively. The photoelectrochemical test using a potentiostat was indicated that Mn-N-TiO_2/Ti electrode active under visible light.

5. Acknowledgement

We acknowledge the financial support of the DRPM – Ministry of Research, Technology and Higher Education, the Republic of Indonesia

6. References

[1] Dzinun H, Othman M H D, Ismail A F, Puteh M H, Rahman M A, Jafar J 2015 *Chem. Engineer. J.* **269** 255.
[2] Maulidiyah, Ritonga H, Faiqoh C E, Wibowo D, Nurdin M 2015 *Biosci. Biotechnol. Res. Asia* **12** 1985.
[3] Zhang Q, Wang H, Fan X, Lv F, Chen S, Quan X 2016 *Surface & Coatings Technol.* **298** 45.
[4] Nurdin M 2014 *Int. J. Pharm. Bio Sci.* **5** 360.
[5] Maulidiyah, Ritonga H, Salamba R, Wibowo D, Nurdin M 2015 *Int. J. ChemTech Res.* **8** 645.
[6] Maulidiyah, Nurdin M, Erasmus, Wibowo D, Natsir M, Ritonga H, Watoni A H 2015 *Int. J. ChemTech Res.* **8** 416.
[7] Maulidiyah, Nurdin M, Widianingsih E, Azis T, Wibowo D 2015 *ARPN J. Eng. & Appl. Sci.* **10** 6250.
[8] Maulidiyah, Nurdin M, Wibowo D, Sani A 2015 *Int. J. Pharma. Pharmaceu. Sci.* **7** 141.
[9] Shirsath S R, Pinjari D V, Gogate P R, Sonawane S H, Pandit A B 2013. *Ultrasonics Sonochem.* **20** 277.
[10] Nurdin M, Muzakkar M Z, Maulidiyah M, Nurjannah M, Wibowo D 2016. *J. Mater Environ. Sci.* **7** 3334.
[11] Anitha B, Ravidhas C, Venkatesh R, Raj A M E, Ravichandran K, Subramanian B, Sanjeeviraja C 2017 *Phys. E:Low-dimensional Systems & Nanostruc.* **91** 148.
[12] Sacco O, Vaiano V, Han C, Sannino D, Dionysiou D D 2015 *Appl. Catal. B: Environ.* **164** 462.
[13] Maulidiyah, Wibowo D, Hikmawati, Salamba R, Nurdin M 2015 *Orient. J. Chem.* **31** 2337.
[14] Karamian E, Abdellahi M, Khandan A, Abdellah S 2016 *J. Alloys & Compounds* **679** 375.
[15] Yin W J, Chen S, Yang J H, Gong X G, Yan Y, Wei S H 2010 *Appl. Phys. Let.* **96**.
[16] Quan F, Hu Y, Zhang X, Wei C 2014 *Appl. Surface Sci.* **320** 120.
[17] Nurdin M, Maulidiyah, Watoni A H, Abdillah N, Wibowo D 2016 *Int. J. ChemTech Res.* **9** 483.
[18] Nurdin M, Zaeni A, Maulidiyah, Natsir M, Bampe A, Wibowo D 2016 *Orient. J. Chem.* **32** 2713.
[19] Gaikwad P N, Hankare P P, Wandre T M, Garadkar K M, Sasikala R 2016 *Materials Sci. & Eng.:B* **205** 40.

[20] Nurdin M, Ramadhan L O A N, Darmawati, Maulidiyah, Wibowo D 2017. *J. Coatings Technol. & Res.* DOI: 10.1007/s11998-017-9976-8
[21] Xin W, Zhu D, Liu G, Hua Y, Zhou W 2012 *Int. J. Photoenergy*, **1**.
[22] Ruslan, Mirzan M,Nurdin M, Wahab A W 2016 *Int. J. Appl. Chem.* **12** 399.
[23] Mahadik M A, Shinde S S, Hunge Y M, Mohite V S, Kumbhar S S, Moholkar A V, Rajpure K Y, Bhosale C H 2014 *J. Alloys & Compounds* **611** 446.
[24] Choudhury B, Paul S, Ahmed G A, Choudhury A 2015 *IOPScience* **2** 096.
[25] Hu Y, Zhang X, Wei C 2009 *Materials Sci. Forum* **620** 683.
[26] Zhao Y F, Li C, Lu S, Liu R X, Hu J Y, Gong Y Y, Niu L Y 2016 *J. Solid State Chem.* **235** 160.
[27] Yao N, Wang J, Zhou Y 2014 *Sensors.* **14** 9949.
[28] Wibowo D, Ruslan, Maulidiyah, Nurdin M 2017 *IOP Conference Series: Materials Science and Engineering* xx xx.
[29] Nurdin M, Wibowo D, Supriyono, Febrian M B, Surahman H, Krisnandi Y K, Gunlazuardi J 2009 *MAKARA SAINS*, **13(1)**: 1.
[30] Maulidiyah, Tribawono D S, Wibowo D, Nurdin M 2016 *Anal. Bioanal. Electrochem.* **8** 761.
[31] Nurdin M, Maulidiyah 2014 *Int. J. Sci. Technol. Res.* **3** 122.
[32] Sharma S, Chaudhary S, Kashyap S C, Sharma S K 2011 *Journal of Applied Physics* **109** 1.
[33] Xiong L-B, Li J-L, Yu Y 2012 *J. Nanomaterials* **2012** 1.
[34] Purnama A, Priatmoko S, Wahyuni S 2013 *Indonesian J. Chem. Sci.* **2** 200.
[35] Hikmawati, Watoni A H, Wibowo D, Maulidiyah, Nurdin M 2017 *IOP Conference Series: Materials Science and Engineering* xx xx.
[36] Zhao K, Wu Z, Tang R, Jiang Y 2013 *J. the Korean Chemical Soc.* **57** 176.
[37] Larios A P, Lartundo L, Mantilla A, Gordillo A H 2016 *Catal. Today* **266** 9.
[38] Qian J, Cui G, Jing M, Wang Y, Zhang M, Yang J 2012 *Int. J. Photoenergy* **1**.
[39] Yang G, Jiang Z, Shi H, Xiao T, Yan Z 2010 *J. Materials Chem.* **20** 5301.
[40] Xin W, Zhu D, Liu G, Hua Y, Zhou W 2012 *Int. J. Photoenergy* **1**.
[41] Arham Z, Nurdin M, Buchari B 2016 *Int. J. ChemTech Res.,* **9** 113.

Determination of COD based on Photoelectrocatalysis of FeTiO3.TiO2/Ti Electrode

D Wibowo[1], Ruslan[1], Maulidiyah[1], M Nurdin[1]
Department of Chemistry, Faculty of Mathematics and Natural Sciences, Universitas Halu Oleo, Jl. H.E.A. Mokodompit Kampus Baru, Kendari 93232 – Southeast Sulawesi, Indonesia

E-mail: mnurdin06@yahoo.com

Abstract. Iron infrastructure technology of (Fe)-doped TiO_2 nanotubes arrays (NTAs) was prepared for COD photoelectrocatalysis sensor. Fe-TiO_2 NTAs was prepared using sol-gel method and coated with TiO_2/Ti electrode by immersion technique. The optimization of COD photoelectrocatalytic sensor against Rhodamine B, Methyl Orange, and Methylene Blue organic dyes using photoelectrochemical system in a batch reactor. The high ordered $FeTiO_3.TiO_2$/Ti NTAs to determine COD value showed the high photocurrent response linearity and sensitivity to MO organic dye from the concentration of 5 ppm to 75 ppm with an average RSD value of 3.35. The development in this research is to utilize ilmenite mineral as model applied to COD sensor.

1. Introduction

The aquatic environment has become a phenomenal issue in recent years because wastewater generated from human activities caused environmental pollution [1,2]. The high content of harmful chemical compounds such as dissolved organic substances in water bodies causes various diseases [3,4]. The needed techniques to quickly detect the water pollution level in aquatic environment is a determinant factor of wastewater quality [5,6]. According to Zhang et al. one of the indicators of water quality as a polluted water control index is by determining Chemical Oxygen Demand (COD) [7]. The determination of COD values is a national standard requirements for determining the pollution content of organic waste in water bodies as applied in developed countries such as Japan and China [8].

The technique of determining COD values is still experiencing problems such as potassium dichromate ($K_2Cr_2O_7$) usage as a strong oxidizing agent. It has an effect on dangerous experimental conditions such as high temperature and H_2SO_4, high-toxicity exposure (Hg(II) and Cr(VI)), expensive reagent (Ag_2SO_4), and automation difficulty [9]. In last years the determination of COD values has been developed and studied intensively by Zhang et al. using the electrocatalytic oxidation technique electrochemically by utilizing an environmental-friendly titanium dioxide (TiO_2) semiconductor material [10]. Determination of COD value based on photoelectrocatalytic is an infrastructure

[1] To whom any correspondence should be addressed.

technology that will be developed, which is more likely to be considered than conventional methods which can be harmful for humans.

The determination of COD photoelectrocatalytic is considered a strong oxidizing agent characterized by the number of electrons transferred when induced in photon energy [11,12]. This phenomenon excite the electron-hole on TiO_2 semiconductor which initiated a reduction-oxidation (redox) reaction [13,14]. This is evidenced by Nurdin et al. in utilization of TiO_2 semiconductor as degradation of organic compounds in water sample [15].

A COD photoelectrocatalytic system using TiO_2 semiconductors was measured by correlating the measured analytical signal PEC as electro catalytic and photocatalytic approach to overcome the oxygen demand problems in organic wastewater [16,17]. Based on Gracia-Munoz et al. the development of TiO_2 semiconductor from ilmenite ($FeTiO_3$) compound has a strong oxidation agent [18]. It can be developed as an electrode for determining COD value by PEC system.

$FeTiO_3$ is a semiconductor belonging to iron-titanium oxide mineral that can be obtained naturally with extraction method. It was obtained through two distinct ways: iron sand extraction [13,14] and sol-gel synthesis [19]. The extraction method has constraint, i.e. the high solubility properties in acid solvent makes it hard to obtain high-purity yields, hence it is necessary to develop the synthesis method as a model of natural $FeTiO_3$ immobilized on the working electrode. According to Wilson et al. (2005) the substitution of Fe^{3+} ions in TiO_2 crystal can occur because the radius of Fe^{3+} ion is smaller than the radius of Ti^{4+} (Fe 0.64 Å; Ti 0.68 Å) ions which allows $FeTiO_3$ to form [20,21].

In this research, $FeTiO_3$ has been synthesized using the sol-gel method then immobilized on TiO_2-NTAs to determine COD value based on PEC of RhB, MO, and MB organic dyes. The organic dyes are often used in the coloring process within textile industry and easy to accumulate aquaticly. The pollution resultant of organic substances can cause unpleasant odors because it contains several elements of Nitrogen (N) and high-stability elements due to some delocalization of electron in benzene rings which are difficult to degrade so that good oxidation agent is needed by using $FeTiO_3$ electrode.

2. Experimental Method

2.1. Fabrication TiO_2/Ti by anodizing

The preparation of Ti (titanium) plate was performed by cutting Ti plate with 99.7% purity with thickness of 1 mm with a size of 4 cm × 0.5 cm then sanded using fine sandpaper of 1200CC until the surface was clean and shiny. Then, it was washed using a detergent solution, water and distilled water (D.I. H_2O). Ti plate was etched using a mixture of HF, HNO_3 and D.I. H_2O with ratio of 1: 3: 6 for 30 seconds. The final step, Ti plate was rinsed with D.I. H_2O to remove the residual etching solution [22].

The Ti plate was put into probe which had been filled with electrolyte solution of NH_4F 0.27 M and D.I. H_2O in 98% glycerol. The anodizing process was carried out by placing Ti plate as anode and Cu plate as cathode using magnetic stirrer, and provided a potential bias of 25 Volts connected to the power supply for 4 hours. The final stage, Ti plate was calcined for 1.5 hours at a temperature of 500°C to evaporate the residual electrolyte solution and to obtain TiO_2 anatase crystal that has high-photodegradation activity [23].

2.2. Synthesis of $FeTiO_3$ using sol-gel

The synthesis of $FeTiO_3$ by modifying Raghavender et al. by using sol-gel method [19]. Solution 1: colloidal solution of TiO_2 was prepared by hydrolyzing in controlled using 4 mL of TTIP compound in 0.5 mL of acetyl acetonate and 15 mL of 99% ethanol. Solution 2: 15 mL of 99% ethanol and 2 mL of D.I. H_2O with addition of 1 mL of 0.1 M acetic acid. The solution mixture was refluxed for 3 hours at 50°C using a magnetic stirrer followed by adding 1 mL of 0.5 M $Fe(NO_3)_3$ to produce $FeTiO_3$ sol. The resulting solvent was evaporated at temperature room for 48 hours to form gel, then heated at a temperature of 80°C in oven for 30 minutes. The coating process of $FeTiO_3$ sol-gel used immersion technique on TiO_2-NTAs surface. The immersion technique was performed in 1 day which aim to evaporate the solvent agent and form $FeTiO_3$ crystal on TiO_2-NTAs electrode. Finally, it was heated for 15 minutes at a temperature of 80°C to form $FeTiO_3.TiO_2$/Ti.

2.3. Determination of COD value using photoelectrocatalytic system

The concentration of RhB, MO, and MB organic dyes were 5, 10, 20, 30, 40, 50, 60, 75 and 100 ppm (+0.1M $NaNO_3$) then it was taken 15 mL of the volume into glass probe in the PEC batch system. Subsequently, the working electrode was assembled using a three-electrode system (working electrode, Ag/AgCl electrode, and Pt electrode). The measured process used a portable potentiostat DY2100B in Multi-Pulse Amperometry (MPA) with a time duration of 60s, scan rate 1×10^{-4} V/s and bias potential of 0.5 Volt. The first 10s in dark condition obtained the stability of current in dark condition and the same condition, then turned on a UV light to obtain the initial photocurrent and stadystate condition. The data were interpreted by determining the linearity graph of net charge (I_{net}) to each concentration.

$$I_{net} = I_{sample} - I_{blank} \tag{1}$$

The linear graph was made in the concentration to obtain slope, $y = ax + b$, where y is the value of I_{net}, a and b are constants and x is the COD value of photoelectrocatalytic.

3. Results and Discussion

3.1. Determination of COD value by photocatalytic system

The organic dyes used RhB, MO, and MB organic dyes because it has used and easily found on dyeing process. The impact of organic dyes had potential as an aquatic pollutant in the environment due to easy accumulation in water and difficult to degrade naturally. Maulidiyah et al have reported that the organic dyes could degrade using un-doped and doped of (M (Metal) – N (Nitrogen)) into TiO_2/Ti electrode [16]. In this research, the determination of COD values based on photoelectrocatalytic was done by using $FeTiO_3.TiO_2$/Ti electrode as material for photooxidation of compound test.

The COD determination test was performed in a batch system with varied concentration in 5, 10, 20, 30, 40, 50, 60, 75 and 100 ppm (+0.1M $NaNO_3$). This system was chosen to obtain the ability of oxidation level and linearity limit of COD measurement using $FeTiO_3.TiO_2$/Ti electrode. The solution test took 15 mL to be put into glass probe in three electrode systems connected to the potentiostat. The dark condition in 10 s obtained the uniformity initial photocurrent and reduced measurement errors due to non-uniformity of the start time in amperomogram profile.

The oxidation current of the organic compound could be determined by reducing I_{sample} to I_{blank}. The resulting current (Figure 1A) was very low (<1.0 × 10^{-5} A) indicating the difficulty to be degraded by photoelectrocatalytic because RhB compound belongs to cationic compounds which could affect hole formation and increase saturation on working electrode due to a similar charge in the system and it was affected by dyes that could absorb photon so the incoming intensity was not too large [9].

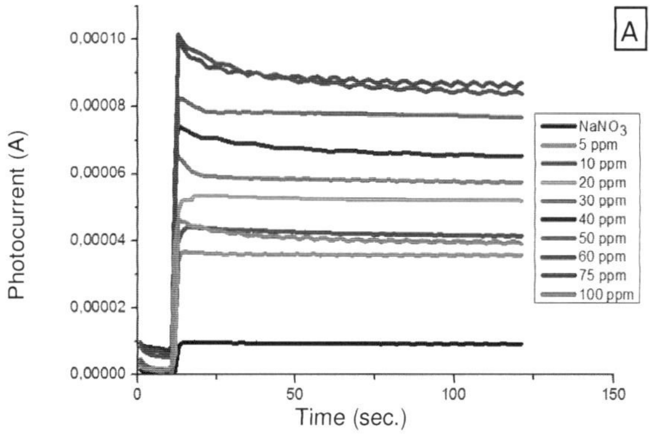

IOP Conf. Series: Materials Science and Engineering **267** (2017) 012007 doi:10.1088/1757-899X/267/1/012007

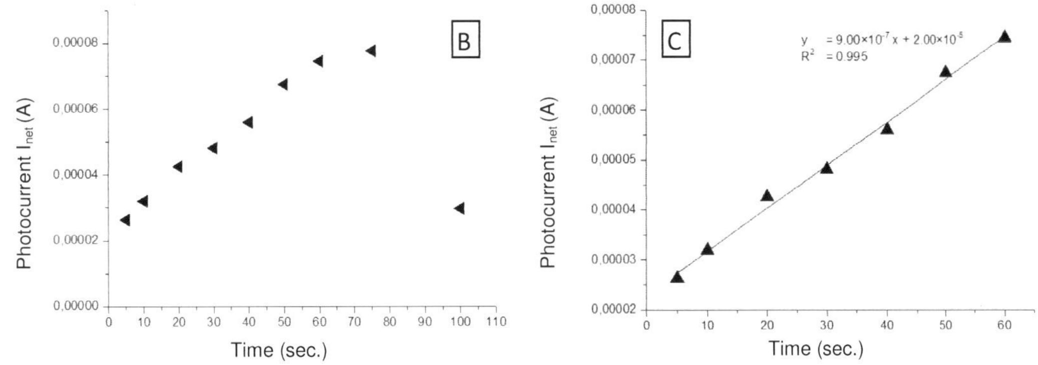

Figure 1. Photoelectrocatalytic profile of RhB; (A) Amperomogram of RhB, (B) Inet value of RhB (C) the linearity performance of RhB.

Figure 1B shows that at concentrations above 60 ppm there is a decrease in photocurrent response due to the difficulty of capturing photo-hole by RhB as characteristic of the cationic compounds and reducing the active side of a catalyst. Figure 1 shows that the optimal performance area of the working electrode to determine RhB COD value is <60 ppm, whereas for > 75 ppm the decrease in electrode activity occured (Figure 1C).

The MO compound was organic dye anionic derived from a hydrophobic group of benzene ring and hydrophilic group derived from S atom binding to an O atom [24]. This group was rich in electrons (negatively charged groups) so it could form hydrogen bonds when dissolved in water. The photocurrent response of analyte (Figure 2A) at a concentration of 75 ppm has a low initial photocurrent compared to the others because the electron excitation ability was affected by the analyte concentration in system.

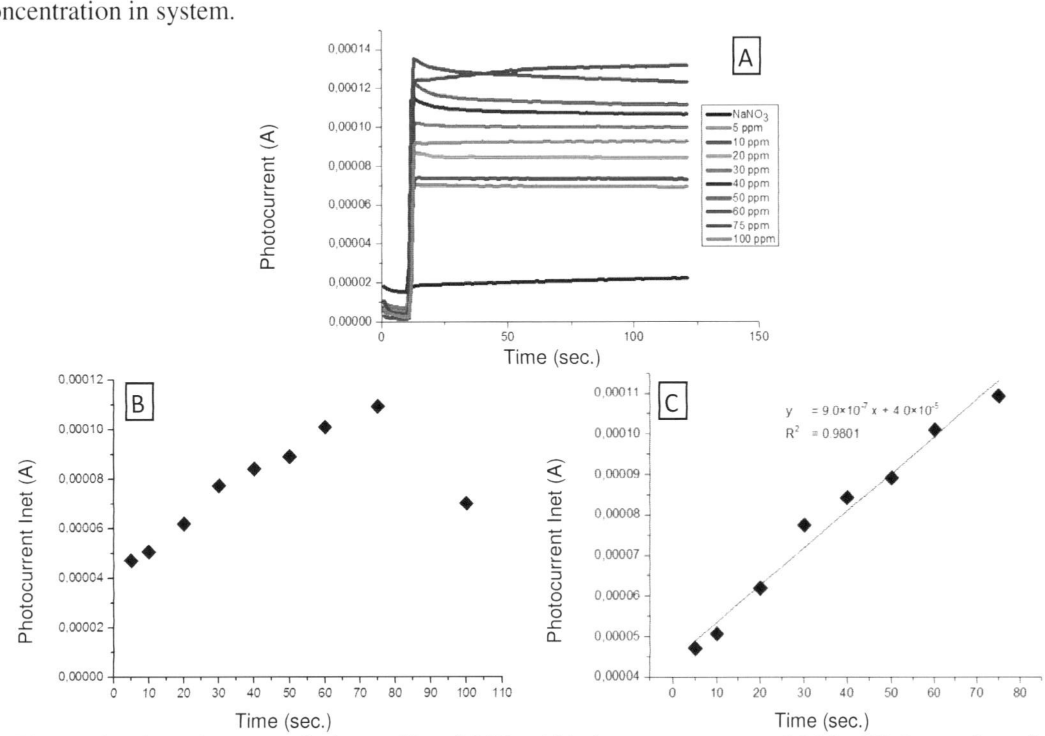

Figure 2. Photoelectrocatalytic profile of MO; (A) Amperomogram of MO, (B) Inet value of MO (C) the linearity performance of MO.

Figure 2 shows the reduced I_{net} values between I_{sample} and I_{blank}. The result showed that the increase of I_{net} is very stable up to a concentration of 75 ppm (Figure 2C), but at concentration of 100 ppm there has been a decrease of I_{net} caused by higher concentration of analyte which can affect the performance of active side of electrode surface with I_{net} value of 7.02×10^{-5} A.

The photocurrent response test in MB solution had a relatively low photocurrent response value compared to MO test compound. Amperomogram data shows that the concentration is increased and it is proportional to photocurrent response which increased as well. Amperomogram in Figure 3A has been lower because the photoelectrocatalytic oxidation process of MB solution resulting in ammonium ions (NH_4^+) caused to affect the performance of reference electrodes. The difficulty of being adsorbed on the electrode surface reduced the oxidation process of MB compounds by holes while on the other side the reduction process was at Pt electrode.

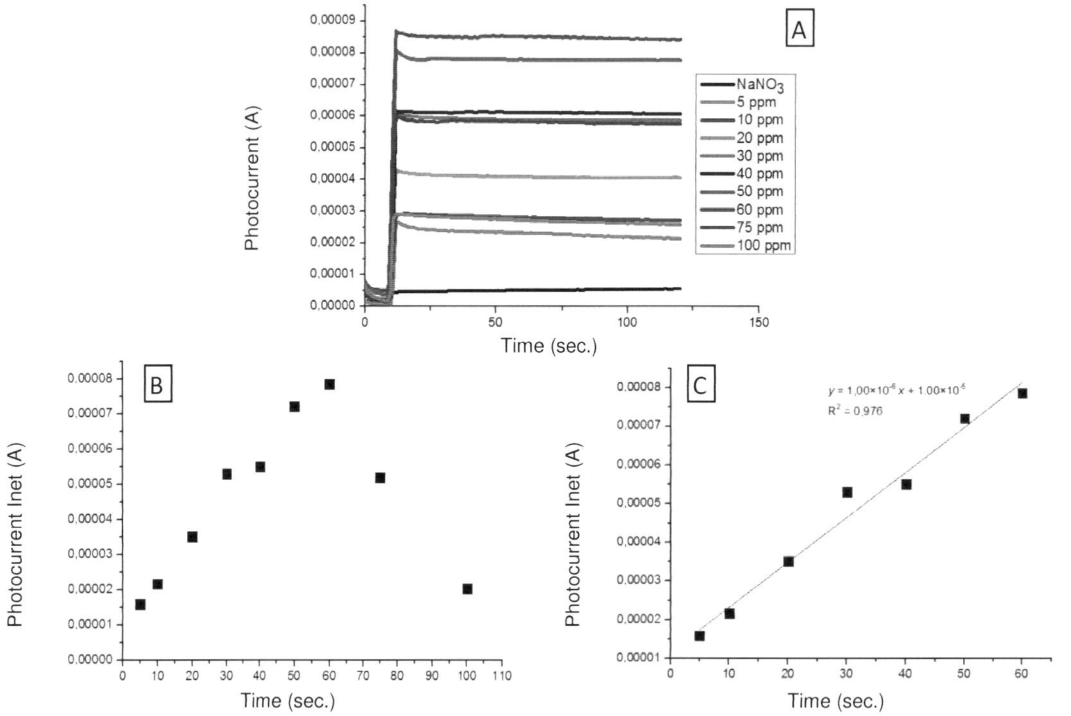

Figure 3. Photoelectrocatalytic profile of MB; (A) Amperomogram of MB, (B) Inet value of MB (C) the linearity performance of MB.

Figure 3 shows I_{net} values are the reduced results of I_{sample} and I_{blank} values. The results shows that the concentrations of 75 ppm and 100 ppm, there was a decrease of photocurrent responses. This is because the saturation on the electrode and intensity of UV light is difficult to penetrate to working electrode. These data show that electrode performance area of MB test compound is limited to 60 ppm.

3.2. Parameters of working electrode

3.2.1. Reproducibility.
Figure 4 is the results of 10 repetitions of a 5 ppm RhB analyte solution for 120 s. The results show that repeated photocurrent responses obtained a highly stable photoelectrochemical profile and met the reproducibility standard with an average photocurrent responses of 3.43×10^{-5} A with a STD value of 6.2×10^{-7}. Figure 4 shows that $FeTiO_3.TiO_2/Ti$ electrode has good precision capability against photocurrent response to compound test. The repeatability of measurement of photoelectrocatalytic system has decreased analytical concentration since this system was applied in the degradation process

of organic compounds electrochemically [25]. In short, representation of proportional electrochemical measurement results is comparable to the analyte concentration in the system.

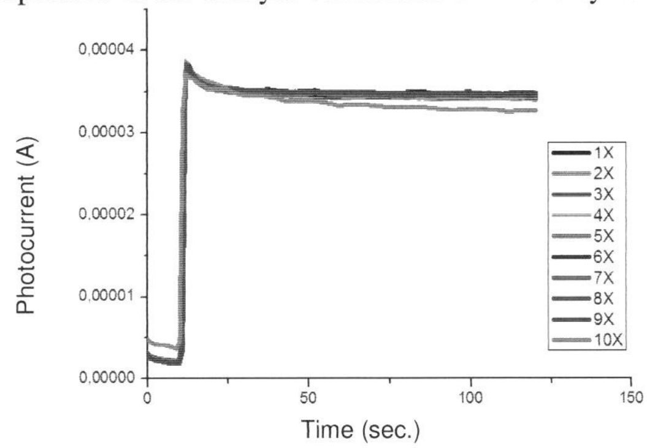

Figure 4. the reproducibility amperomogram of $FeTiO_3.TiO_2/Ti$ electrode.

3.2.2. Sensitivity

Figure 5 shows the comparison of photocurrent responses to various compound tests. The working electrode has a sensitivity to MO compound characterized by the increased photocurrent response which is greater than test compounds of RhB and MB. This factor is due to the highly degradable photoelectrocatalytic presented to the emerging photocurrent charge [26]. MO test compound has good COD proximity from 5 ppm to 60 ppm compared to RhB and MB with average STD values of 3.35, 45.78, and 11.60, respectively.

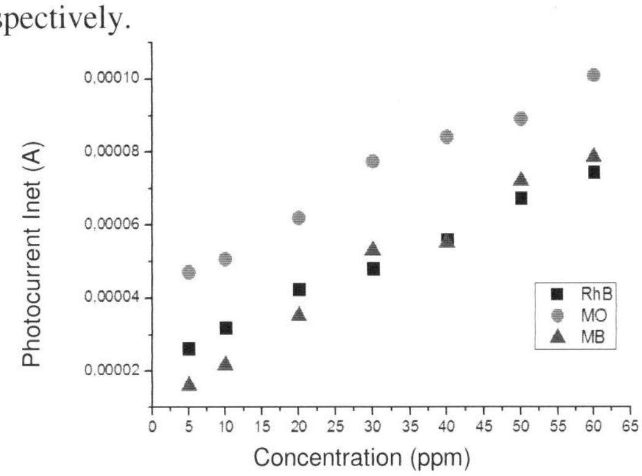

Figure 5. The sensitivity graph of test compounds.

3.2.3. Linearity

Figure 6 represents the linearity of data of COD values measurement on each of the test compounds of RhB, MO, and MB using $FeTiO_3.TiO_2/Ti$ electrode to see the theoretical COD value approach with photoelectrocatalytic. Figure 6A compares theoretical COD values and COD PEC values as the RhB concentration increases from 5 ppm to 100 ppm. The data shows that the increase in the concentration of RhB compound makes COD PEC values are away from linearity to the theoretical COD values, so probability of linearity was observed at RhB concentration from 5 ppm to 20 ppm (Figure 6B). Figure 6C shows $FeTiO_3.TiO_2/Ti$ electrode can work well on MO test compound with concentration ranges from 5 ppm to 75 ppm. The linearity data shows the proximity of theoretical COD value of MO compound with photoelectrocatalytic system is not much different, so it can be assumed that $FeTiO_3.TiO_2/Ti$ electrode worked very well on MO compound (Figure 6D).

Figure 6E shows photoelectrocatalytic COD values of MB test compound also experienced a change in linearity from theoretical COD measurement. This case is similar to RhB compound with linearity of MB compound from concentration 5 ppm to 30 ppm. Compared to test compounds of RhB and MB. The effectiveness of linearity is still better in MB compound. This phenomenon suggests that the possible roles of the test compound properties are anionic and cationic, as well as the magnitude of the test compound so as to rapidly cover the active side of catalyst on electrode surface.

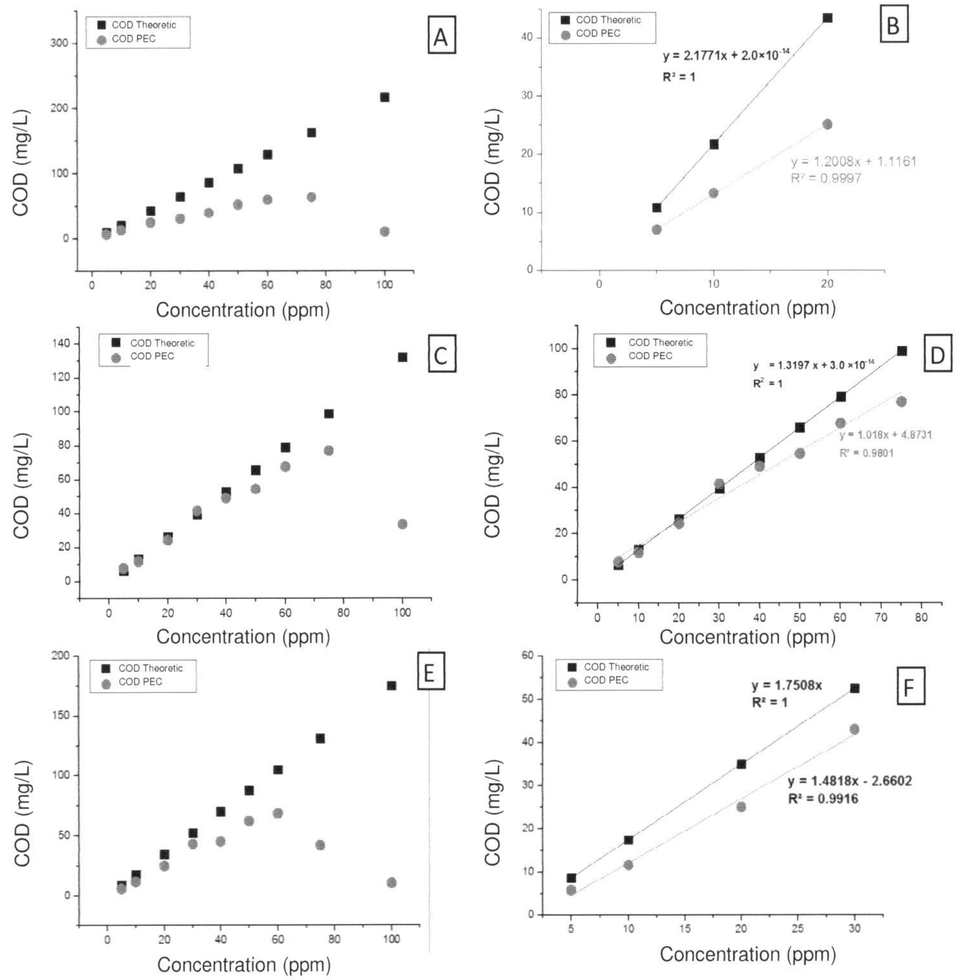

Figure 6. Determination of linearity values (A) COD values of RhB (B) proximity of RhB linearity (C) COD values of MO, (D) proximity of MO linearity (E) COD values of MB, (F) proximity of MB linearity.

4. Conclusion

The determination of COD values used $FeTiO_3.TiO_2/Ti$ electrode on various compound tests of RhB, MO, and MB showed the performance linearity of each compound test i.e. RhB at concentration of 5 ppm, 10 ppm and 20 ppm with COD values of 7.00 mg/L, 13.29 mg/L, and 25.07 mg/L. MO compounds at concentrations of 5 ppm, 10 ppm, 20 ppm, 30 ppm, 40 ppm, 50 ppm, 60 ppm, and 75 ppm with COD values of 7.86 mg/L, 11.862 mg/L, 24.37 mg/L, 41.63 mg/L, 54.57 mg/L, 67.733 mg/L, and 77.00 mg/L, respectively. While MB compounds at concentrations of 5 ppm, 10 ppm, 20 ppm, 30 ppm with COD values of 5.91 mg/L, 11.66 mg/L, 25.04 mg/L, and 43.04 mg/L with average value of STD sensitivity of 45,78; 3.35; and 11.60, respectively.

5. References

[1] Nurdin M, Wibowo W, Supriono, Febrian M B, Surahman H, Krisnandi Y K, Gunlazuardi J 2009 *Makara, Sains,* **13** 1.
[2] Arham Z, Nurdin M, Buchari B 2016 *Int. J. ChemTech Res.,* **9** 113.
[3] Maulidiyah, Ritonga H, Faiqoh C E, Wibowo D, Nurdin M 2015 *Biosci. Biotechnol. Res. Asia* **12** 1985.
[4] Maulidiyah, Tribawono D S, Wibowo D, Nurdin M 2016 *Anal. Bioanal. Electrochem.* **8** 761.
[5] Ruslan, Mirzan M, Nurdin M, Wahab A W 2016 *Int. J. Appl. Chem.* **12** 399.
[6] Nurdin M, Ramadhan L O A N, Darmawati, Maulidiyah, Wibowo D 2017. *J. Coatings Technol. & Res.* DOI: 10.1007/s11998-017-9976-8
[7] Zhang S, Li L, Zhao H 2009 *Environment Science and Technology* **43** 7810.
[8] Wang J, Yao N, Li M, Hu J, Chen J, Hao Q, Wu K, Zhou Y 2015 *Microchimia Acta* **182** 515.
[9] Maulidiyah, Nurdin M, Erasmus, Wibowo D, Natsir M, Ritonga H, Watoni A H 2015 *Int. J. ChemTech Res.* **8** 416.
[10] Zhang Z, Chang X, Chen A 2016 *Sensor and Actuators B* **223** 664.
[11] Maulidiyah, Wibowo D, Hikmawati, Salamba R, Nurdin M 2015 *Orient. J. Chem.* **31** 2337.
[12] Nurdin M, Muzakkar M Z, Maulidiyah M, Nurjannah M, Wibowo D 2016 *J. Mater Environ. Sci.* **7** 3334.
[13] Nurdin M, Maulidiyah, Watoni A H, Abdillah N, Wibowo D 2016 *Int. J. ChemTech Res.* **9** 483.
[14] Nurdin M, Zaeni A, Maulidiyah, Natsir M, Bampe A, Wibowo D 2016 *Orient. J. Chem.* **32** 2713.
[15] Nurdin M 2014 *Int. J. Pharm. Bio. Sci.* **5** 360.
[16] Maulidiyah, Nurdin M, Wibowo D, Sani A 2015 *Int. J. Pharma. Pharmaceu. Sci.* **7** 141.
[17] Maulidiyah, Nurdin M, Widianingsih E, Azis T, Wibowo D 2015 *ARPN J. Eng. & Appl. Sci.* **10** 6250.
[18] Gracia-Munoz P, Pliego G, Zazo J A, Bahamonde A, Casas J A 2016 *J. Environ. Chemical Engineering,* 4:542-548.
[19] Raghavender A T, Hong N H, Lee K J, Jung M-H, Skoko Z, Vasilevskiy M, Cerqueira M F, Samantilleke A P 2013 *J. Magnetism and Magnetic Materials* **331** 129.
[20] Yu J, Wu Z, Gong C, Xiao W, Sun L, Lin C 2016 *Nanomaterials* **6** 107.
[21] Hikmawati, Watoni A H, Wibowo D, Maulidiyah, Nurdin M 2017 *IOP Conference Series: Materials Science and Engineering* xx xx.
[22] Mursalim L O A M, Ruslan, Safitri R A, Azis T, Maulidiyah, Wibowo D, Nurdin M 2017 *IOP Conference Series: Materials Science and Engineering* xx xx.
[23] Nurhidayani, Muzakar M Z, Maulidiyah, Wibowo D, Nurdin M 2017 *IOP Conference Series: Materials Science and Engineering* xx xx.
[24] Zhou X-T, Ji H-B, Huang X-J 2012 *Molecules* **17** 1149.
[25] Maulidiyah, Ritonga H, Salamba R, Wibowo D, Nurdin M 2015 *Int. J. ChemTech Res.* **8** 645.
[26] Shao S, Wang W, Zhou K, Jiang F, Wu H, Koehn R 2017 *Materials Letter* **186** 193.

Acknowledgement
We acknowledge for financial support of the DRPM-Ministry of Research, Technology and Higher Education, the Republic of Indonesia.

Bond Strength Mechanism of Fly Ash Based Geopolymer Mortars: A Review

W W A Zailani[2,1, a], M M A B Abdullah[2,1, b], R A Razak[2,1, c], M R R M A Zainol[3, d], M F M Tahir[1, e]

[1] Center of Excellence Geopolymer and Green Technology (CEGeoGTech), School of Materials Engineering, Universiti Malaysia Perlis (UniMAP), P.O Box 77, D/A Pejabat Pos Besar, 01000 Kangar, Perlis, Malaysia.
[2] Faculty of Engineering Technology, Universiti Malaysia Perlis (UniMAP), P.O Box 77, D/A Pejabat Pos Besar, 01000 Kangar, Perlis, Malaysia.
[3] School of Civil Engineering, Universiti Sains Malaysia (USM), Engineering Campus, 14300 Nibong Tebal, P. Pinang, Malaysia

[a]warid.wazien92@yahoo.com, [b]mustafa_albakri@unimap.edu.my, [c]rafizarazak@unimap.edu.my, [d]ceremy@eng.usm.my, [e]faheem@unimap.edu.my

Abstract. Geopolymer possess many excellent properties such as high compressive and bond strength, long term durability, better acid resistance and also known as a "Sustainable Material" due to its low carbon emission and low energy consumption. Thus, it is a good opportunity to develop and explore not only for cement and concrete but also as geopolymeric repair materials. This reviews showed that good bonding properties between geopolymeric repair material and concrete substrate is important in order to acquire an enhanced resistance against penetration of harmful substances and avoiding respalling of the repair material by understanding the bonding behaviour. Bond strength depends to the properties of the repair materials itself and also the surface preparations of concrete substrate.

1. Introduction

Bonding strength is the major requirement of any successful repair material [1–3]. Thus, bond strength is one of the most important properties as concrete repair materials. Several studies have shown that the bond that exists between the repair material and concrete substrate is generally very weak in most of the repaired structures [4,5]. Several test methods such as pull off, splitting, flexural, and slant shear test have been proposed to evaluate the bond strength between repair materials and concrete substrate [6–8]. Thus, recent investigations in the field of geopolymers reveal a third category of mortars with high potential to be used in the field of concrete patch repair and slant shear test normally is used to study the bonding performance of repair material. Investigation on the utilization of high calcium fly ash geopolymer mortars containing OPC for use as repair material has been investigate and shows the strong bond at the interface zone.

The high calcium fly ash geopolymer mortar mixture not only gave sufficient high shear bond strength but also can be cured at room temperature because of the calcium in the system [9–13]. Thus, geopolymer becomes more practical for concrete repair and rehabilitation. The investigations have already been conducted in geopolymer field especially based on fly ash however, most of the

IOP Conf. Series: Materials Science and Engineering **267** (2017) 012008 doi:10.1088/1757-899X/267/1/012008

researchers covered the mixture development phase. Thus, it is a good opportunity to use some of their mixture formulation to develop this potential material as a new repair material. Therefore, the bonding mechanism between repair material and concrete substrate is reviewed in order to identify the factors that could improve the bond strength between geopolymer repair material and substrate.

2. Reaction Mechanism at Interfacial Transition Zone

Bond strength requires an understanding of chemical reaction between the concrete substrate and repair materials. Geopolymer repair material which is able to transform and hardened rapidly at low temperatures through geopolymerization process able to cure at ambient temperature [6–8], to be an alternative to epoxy resins for structural retrofitting [4], high adhesion strengths even at early ages [9], and the cheapest solution for repair and rehabilitation [7,8]. Geopolymer is formed using precursors containing alumina (Al_2O_3) and silica (SiO_2) such as fly ash (FA) [17], kaolin [18], granulated blast furnace slag (GBFS) [19], and palm oil fuel ash (POFA) [20] activated with alkali solutions. The Al_2O_3 and SiO_2 in the precursor will dissolve and form a three-dimensional amorphous aluminosilicate network [21].

However, the sialate network of geopolymer which composed of tetrahedral anions $[SiO_4]^{4-}$ and $[AlO_4]^{5-}$ need positive ions such as Na^+, K^+, Li^+, Ca^{++}, Na^+, Ba^{++}, NH^{4+}, H_3O^+ to compensate the electric charge of Al^{3+} in tetrahedral coordination as shown in Figure 1. When geopolymer is patch against the spalled of the reinforced concrete structure made from Ordinary Portland Cement (OPC), geopolymer is chemically bond to substrate due to the reaction products at the interface transition zone between OPC substrate [12,13].

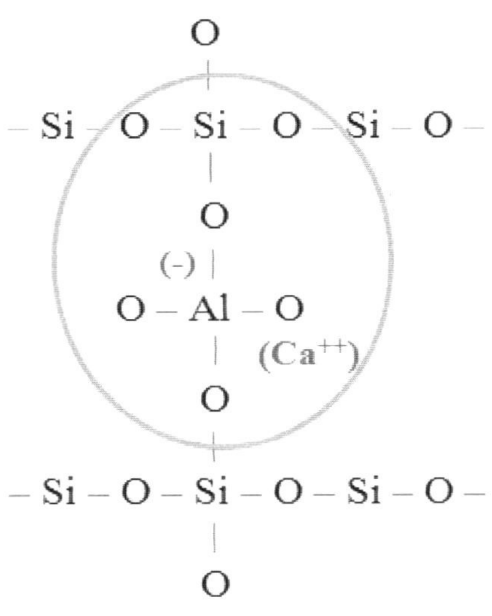

Figure 1. Schematic representation of the chemical bond at interface.

Calcium hydroxide or portlandite, which is available at the Ordinary Portland Cement (OPC) substrate surface, provides positive ion Ca^{++} and this positive ion is required for the geopolymerization process. Theoretically, the Ca^{++} will balance the negative charge of Al^{3+} ions in geopolymer thus, promoting to the bond strength between geopolymer repair material and substrate. Thus, interfacial transition zone through Scanning Electron Microstructure (SEM) shows that geopolymer made from Tungsten Mine

IOP Conf. Series: Materials Science and Engineering **267** (2017) 012008 doi:10.1088/1757-899X/267/1/012008

Waste Mud (TMWM) is chemically bond to the concrete substrate as shown in Figure 2. Moreover, the increase in Ca^{++} ion balanced the negative charge of Al^{3+} ions, which leading to a dense interface zone is shown in Figure 3. Therefore, the microstructure images give clear picture in order to evaluate the bond strengths of geopolymer repair materials. Geopolymer made with TMWM and fly ash are good candidates as an alternative bonding material for repair works.

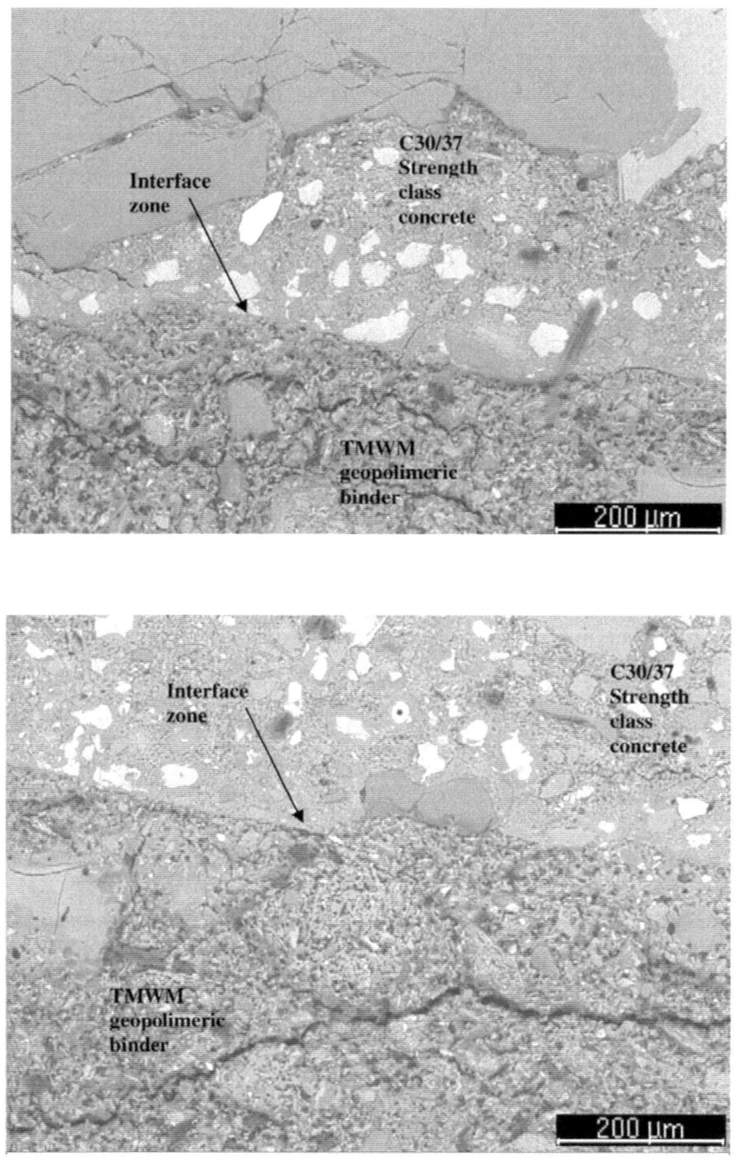

Figure 2. SEM micrographs of interfacial transition zone between geopolymer and substrate [8].

IOP Conf. Series: Materials Science and Engineering **267** (2017) 012008 doi:10.1088/1757-899X/267/1/012008

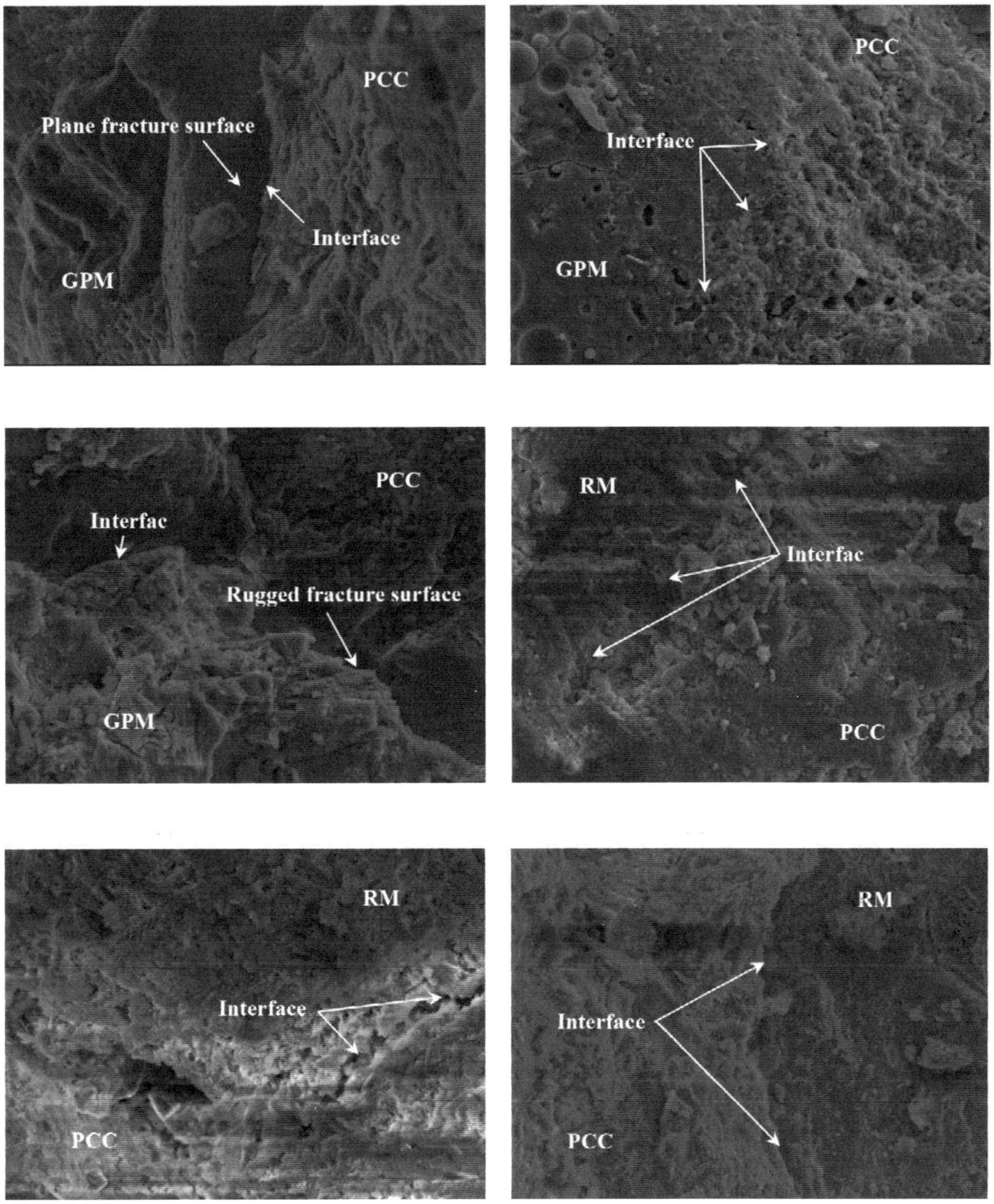

Figure 3. SEM micrographs of interfacial transition zone between concrete substrate (PCC) and geopolymeric binder (GPM) or repair material (RM) [9].

The microstructures image of these geopolymer (GPM), concrete substrate (PCC) and interface could be identified from unreacted fly ash, agglomeration of OPC and small line crossing along the interfacial, respectively. The negatively charged Al and tetrahedrally coordinated Al (III) atoms inside the network are charge-balanced by alkali metal cations such as Na^+ or K^+ (from alkaline activator) and Ca^{++} (from Portland cement grains react with water). Thus, geopolymer which is rich in Si^{4+} and

Al^{3+} ions can react with $Ca(OH)_2$ at the surface of substrate leading to bonding strength development at the contact zone.

3. Substrate Surface Treatment

Surface treatment or surface roughness is an important parameter to put into consideration in order to enhance the bond strength between repair material and concrete substrate as shown in Figure 4 such as sand blasting, wire brushing, drill holes, acid etching, against metallic formwork, and grooves. The surface of the concrete structure or damaged concrete surface area is treated by increasing the roughness surface and it is believed that it wills courage the bond strength between an old concrete and repair material [3]. Recently, many researchers tried different types of surface roughness and identify the best method in terms of practicality and strength [11,19]. This is because the bond strength depends not only on the characteristics of the repair material but also on the surface roughness of concrete substrate.

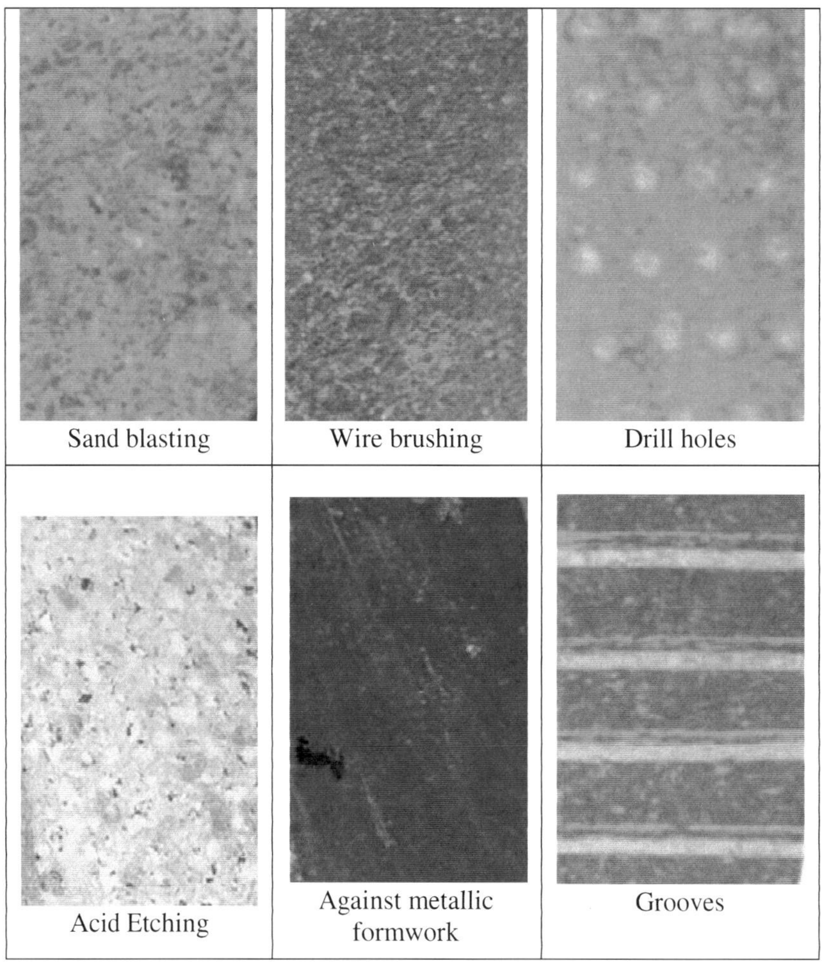

Figure 4. Different surface preparation of the concrete substrate.

Bond strength between the repair materials and substrate depends on the surface treatment of the substrate, as the surface treatment increases the bond strength increases [24]. Geopolymer possess very high bond strength however, some researchers believed that it is not affected by increasing the surface roughness of the concrete substrate [8]. The surface substrate that is rich in calcium hydroxide but lack exposed course aggregates which could contribute to improve bond strength due to silica dissolution from the aggregate surface. However, it is believed that surface treatment will have a significant impact to the bonding between repair material and concrete substrate although, the calcium hydroxide is reduced at the interface but this required for further analysis.

4. Conclusion

As conclusion, previous researchers proved that geopolymer can be used as repair materials and bond strength between repair materials and substrate is important to meet the requirement and effectiveness of the repair works. Therefore, the recommendation is to study the possibility of using the geopolymer as a repair material for a long term durability. Since one of the most critical factors affecting excellent repair material is the bond strength between a repair material and an existing concrete, a comprehensive analysis and study is required to understand the reaction and bond mechanism between geopolymer and concrete substrate.

5. Acknowledgments

The author would like to express his gratitude to the Center of Excellence Geopolymer & Green Technology (CEGeoGTech) for funding the research project.

6. References

[1] Sumesh M, Alengaram U J, Jumaat M Z, Mo K H, and Alnahhal M F 2017 *Construction and Building Materials* **148** 62–84
[2] Singh B, Ishwarya G, Gupta M, and Bhattacharyya S K 2015 *Construction and Building Materials* **85** 78–90
[3] Momayez A, Ehsani M R, Ramezanianpour A A, and Rajaie H 2005 *Cement and Concrete Research* **35** 748–757
[4] Pacheco-Torgal, Abdollahnejad Z, Miraldo S, Baklouti S, and Ding Y 2012 **36** 1053–1058
[5] Espeche A D and Lee J 2011 *Construction and Building Materials* **25** 1222–1235
[6] Momayez A, Ehsani M R, Ramezanianpour A A, and Rajaie H 2005 *Cement and Concrete Research* **35** 748–757
[7] Alanazi H, Yang M, Zhang D, and Gao Z 2016 *Cement and Concrete Composites* **65** 75–82
[8] Pacheco-Torgal F, Castro-Gomes J P, and Jalali S 2008 *Construction and Building Materials* **22** 154–161
[9] Phoo-Ngernkham T, Sata V, Hanjitsuwan S, Ridtirud C, Hatanaka S, and Chindaprasirt P 2015 *Construction and Building Materials* **98** 482–488
[10] Nie Q, Hu W, Ai T, Huang B, Shu X, and He Q 2016 *Construction and Building Materials* **125** 905–911
[11] Nath P and Sarker P K 2015 *Cement and Concrete Composites* **55** 205–214
[12] Somna K, Jaturapitakkul C, Kajitvichyanukul P, and Chindaprasirt P 2011 *Fuel* **90** 2118–2124
[13] Temuujin J, Riessen A, and Williams R 2009 *Journal of Hazardous Materials* **167** 82–88
[14] Khan M Z N, Nuruddin F A, Hao Y, and Hao H 2016 *Construction and Building Materials* **125** 809–820
[15] Nath P and Sarker P K 2014 *Construction and Building Materials* **66** 163–171
[16] Duan P, Yan C, and Luo W 2016 *Construction and Building Materials* **124** 69–73
[17] Al Bakri A M M, Kamarudin H, Bnhussain M, Nizar I K, Rafiza A R, and Zarina Y 2012 90–97
[18] Yun Ming L, Yong H C, Al Bakri M M, and Hussin K 2016 *Progress in Materials Science* **83** 595–629
[19] Islam A, Alengaram U J, Jumaat M Z, and Bashar I I 2014 *Materials and Design* **56** 833–841
[20] Zarina Y, Mustafa Al Bakri A M, Kamarudin H, Nizar I K, and Rafiza A R 2013 *Reviews on Advanced Materials Science* **34** 37–43
[21] Davidovits J 1994 *First International Conference on Alkaline Cements and Concretes* 131–149
[22] Phoo-ngernkham T, Sata V, Hanjitsuwan S, Ridtirud C, Hatanaka S, and Chindaprasirt P 2015 *Construction and Building Materials* **98** 482–488

IOP Conf. Series: Materials Science and Engineering **267** (2017) 012008 doi:10.1088/1757-899X/267/1/012008

[23] Tayeh B A, Abu Bakar B H, Megat Johari M A, and Voo Y L 2013 *Procedia Engineering* **54** 554–563
[24] Tayeh B A, Abu Bakar B H, Megat Johari M A, and Voo Y L 2012 *Construction and Building Materials* **36** 538–548

Mechanical Properties of Graphene-Rubber Nanocomposites

N. A. M. Anhar[1]*, M M. Ramli[1,2], N. A. M. A. Hambali[1], A. A. Aziz[3], S S Mat Isa[1,2], N. S. Danial[1,2], M. M. A. B. Abdullah[2]

[1] School of Microelectronic Engineering, Universiti Malaysia Perlis, Pauh Putra Campus, 02600 Arau, Perlis, Malaysia.

[2]Center of Excellence Geopolymer and Green Technology (CEGeoGTech), School of Materials Engineering, Universiti Malaysia Perlis (UniMAP), P.O. Box 77, D/A Pejabat Pos Besar, Kangar, Perlis, Malaysia

[3]High Value Added Rubber Products and Nanostructured Materials, Technology & Engineering Division, Rubber Research Institute of Malaysia, Malaysian Rubber Board, 47000 Sungai Buloh, Selangor, Malaysia.

E-mail: nur_asyikin1090@yahoo.com

Abstract. This research focused on development of wearable sensor device by using Prevulcanized Natural Rubber (PV) and Epoxidized Natural Rubber (ENR 50) latex incorporated with graphene oxide (GO), graphene paste, graphene powder and reduced graphene oxide (rGO) powder. The compounding formulation and calculation were based on phr (parts per hundred rubber) and all the samples were then tested for mechanical properties using Instron 5565 machine. It was found that the sonication effects on tensile strength may have better quality of tensile strength compared to non-sonicated GO. For PV incorporate GO, the optimum loading was best determined at loading 1.5 phr with or without sonication and similar result was recorded for PV/G. For ENR 50 incorporate graphene paste and rGO powder nanocomposite shows the best optimum was at 3.0 phr with 24 hours' sonication.

1. Introduction

The development of wearable device technology has widely increased and are demanded especially in healthcare industry. Recently, researchers are moving towards carbon based material as filler such as graphene for wearable sensor by reinforce with flexible polymer materials such as natural rubber to produce high flexibility and good conductivity of the device. Graphene is an atomic-scale honeycomb lattice made of carbon atoms [1] which possess extraordinary properties such as high specific area with approximately 2630 m^2/g [2], high Young's modulus (approximately 1100 GPa) [2] and high thermal conductivity (5000 W/m-K) [2] which is a promising material for biocompatible sensing as reported by [3,4]. J. H. Lee et. al. [4] discovered that the properties of graphene such as excellent electrical conductivity and high surface area allowed it to act as an "electron wire" between the redox centers of an enzyme or protein and an electrode's surface. The unique properties also possess the properties of high carrier mobility that can allows high signal-to-ratio and hence is a promising material for conductivity for electronic device applications.

IOP Conf. Series: Materials Science and Engineering **267** (2017) 012009 doi:10.1088/1757-899X/267/1/012009

In order to determine the best material to use as substrate and to have flexible and wearable device, this research will reveal the best nanocomposite material which used prevulcanized natural rubber incorporated graphene oxide (PV/GO) and graphene powder (PV/G) with and without sonication, and epoxidized natural rubber incorporate graphene paste (ENR 50/G) and reduced graphene oxide ENR 50/rGO).

2. Materials and Methods

2.1. Materials and sample preparation

Prevulcanized Natural Rubber (PV) and Epoxidized Natural Rubber (ENR 50) were obtained from Rubber Research Institute of Malaysia (RRIM). Graphene paste with 50% solid, water based material, graphene oxide (GO) and graphene powder were supplied individually by RRIM. Reduced graphene oxide (rGO) powder was supplied with purity of more than 99 wt. %. The compounded latex materials were categorized into 4 types of samples, A (control), B with loading 1.0 phr, C with loading 1.5 phr and D with loading 3.0 phr, respectively.

There were 3 groups of experiments conducted for this research as for group 1; PV incorporated GO paste with no sonication done labelled with PV/GOpa-ns, for group 2; PV incorporated GO paste with 24 hours sonication labelled with PV/GOpa-s and PV incorporated graphene powder with 24 hours sonication labelled with PV/Gpo-s, and finally, for group 3, ENR 50 incorporated graphene paste with 24 hours sonication labelled with ENR 50/Gpa-s and ENR 50 incorporated rGO powder with 24 hours sonication labelled with ENR 50/rGOpo-s.

Group 1 and 2 were done by redispersed graphene to 10 % (0.2 % of sodium dedocyl sulfate, SDS surfactant and 98 g of water) in aqueous medium and sonicated by using tip sonication Fisher Scientific Model 705 Sonic Dismembrator with 80 % amplitude for 24 hours (separated by a rest every 15 min). For group 3, the same method was applied but without any addition of surfactant.

2.2. Mixing and compounding latex

The compounding formulation was calculated based on phr (parts per hundred rubber) and summarized in Table 1 for group 1 and 2 and Table 2 for group 3. Both PV and ENR 50 latex was stirred by using IKA RW 20 Digital at 110 rpm for 1 hour to generate uniform latex. Then, the latex was sieved, weighed, compounded and stirred overnight. After 24 hours, the compounded latex was diluted to 30 % total solid content (TSC) by adding distilled water and continued stirring for 1 hour. The compounding latex was then casted on 12 cm x 12 cm plates and left to dry at room temperature for 2 to 3 days. The cast films were then vulcanized in an oven with 70 °C temperature for 2 hours directly on the plates. Figure 1 shows the physical state of mixing and compounding between PV and ENR 50 latex at different loading. It was observed that the higher the loading concentration, the darker the colour of the samples for PV and ENR 50 latex.

Table 1. Parts per hundred rubber formulation of PV latex and graphene for group 1 and 2.

	Samples			
Material	A (control)	B	C	D
phr[a]	-	1.0	1.5	3.0
PV latex (gm)	250.00	250.00	250.00	250.00
Graphene (gm)	0.00	15.13	22.69	45.38
Water to add (gm)	254.17	244.08	239.04	223.92

[a] Parts per hundred rubber.

Table 2. Parts per hundred rubber formulation of ENR 50 latex and graphene for group 3.

Material	Samples			
	A (control)	B	C	D
phr[a]	-	1.0	1.5	3.0
PV latex (gm)	250.00	250.00	250.00	250.00
Graphene (gm)	0.00	12.50	18.75	37.50
Water to add (gm)	166.67	158.33	154.16	141.67

[a] Parts per hundred rubber.

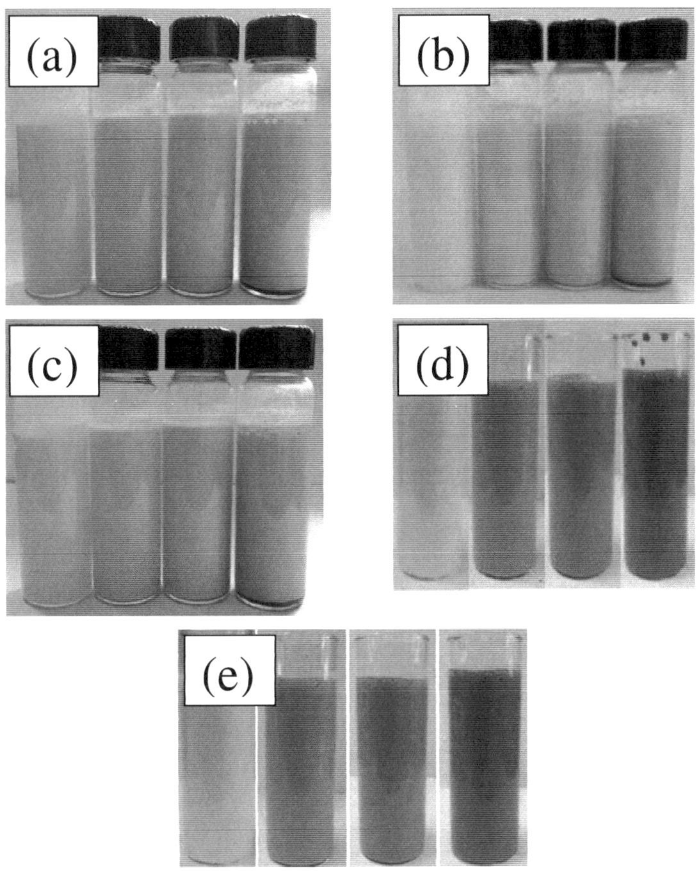

Figure 1. Samples of; (a) PV/GOpa-ns, (b) PV/GOpa-s, (c) PV/Gpo-s, (d) ENR 50/rGOpo-s, (e) ENR 50/Gpa-s, for loading from left; control, 1.0 phr, 1.5 phr, and 3.0 phr.

2.3. Measurement of tensile properties

Tensile properties for each sample were performed by using Instron machine (model 5565) equipped with a 10 N load cell at room temperature. The films were cut to dumbbell shape for 5 films every each samples and tested for every extension of 100 mm, 300 mm, 500 mm and 700 mm.

3. Results and discussions

3.1. Effects of GO paste sonication on tensile strength for PV latex

Figure 2 shows the effects of GO sonication on tensile strength of PV latex for group 1 (PV/GOpa-ns) and group 2 (PV/GOpa-s). For PV/GOpa-ns, it shows increasing of tensile strength from loading 1.0 to 1.5 phr with 16.6 MPa to 18.8 MPa. But, at loading 3.0 phr, the tensile strength decreased to 17.1 MPa. Compared to PV/GOpa-s, the tensile strength increased from 14 to 15 MPa for loading 1.0 to 1.5 phr. However, when loading further increased to 3.0 phr, the tensile strength drops to 13 MPa. This phenomenon was believed to be percolation phenomenon where the increasing of loading filler will subsequently increase the tensile strength but eventually it will level off at certain level (maximum level) as reported by A. Celzard *et.al* [5]. It was clearly observed that PV/GOpa-ns has better tensile strength compared to PV/GOpa-s. It can be pre-concluded that the best loading for PV incorporate GO paste was at 1.5 phr with or without sonication done. The effect GO sonication is not following the pattern as reported [6, 7]. It was expected that the longer the GO sonication time, the better the tensile strength of PV latex. However, this incident is still unclear as no finding was reported supporting this problem.

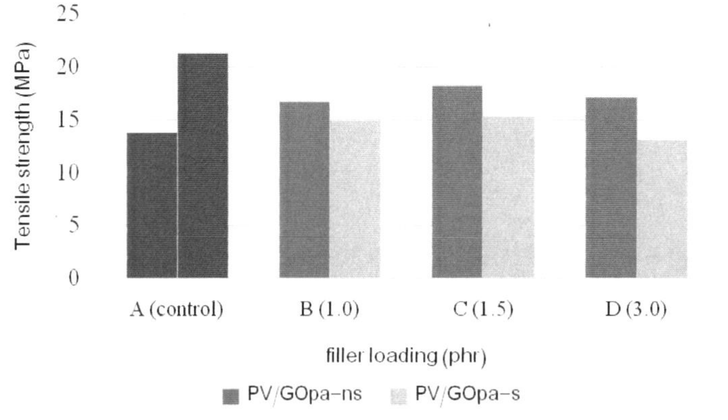

Figure 2. Effect of sonication of GO on tensile strength of PV/GOpa-ns and PV/GOpa-s at different loading.

3.2. Modulus at 100 % elongation (M_{100}) for PV/Gpa-ns and PV/Gpa-s

Figure 3 shows the modulus at 100 % elongation for PV/GOpa-ns and PV/GOpa-s. From this graph, it can be observed that the modulus at 100 % elongation (M_{100}) for PV/GOpa-s was directly proportional to tensile strength as M_{100} increased from 0.61 MPa to 0.69 MPa for loading 1.5 to 3.0 phr. It can be pre-concluded that the effect of sonication for GO paste may play its role on good dispersion in liquid media and strengthen the rubber network [8]. Subsequently, incorporation of GO has improved the mechanical properties of the PV latex. PV/GOpa-ns show a decreasing of tensile strength from 0.53 MPa to 0.42 from loading 1.0 to 3.0 phr. This incident mainly due to the inhomogeneous structures formed with the incompletely exfoliated graphene oxide flakes [9]. Even though the PV/GOpa-ns has higher tensile strength than PV/GOpa-s, it has very low fracture strength and toughness as reported by J.C. Feng et. al [9]. Hence, it can be concluded that the sonication of GO paste has improved the mechanical properties PV latex.

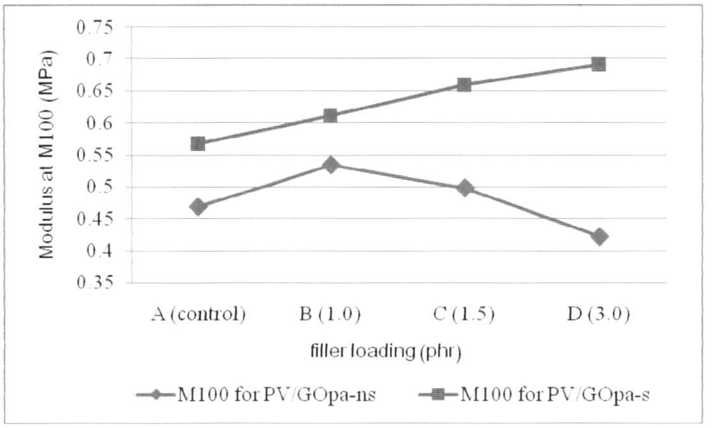

Figure 3. Effect of sonication of GO on modulus at 100 % elongation (M_{100}) for PV/GOpa-ns and PV/GOpa-s at different loading

3.3. Effect on tensile strength for PV/Gpo-s

Figure 4 shows the effect of PV latex tensile strength by the presence of graphene powder. It was observed that the increasing of tensile strength from 20.3 to 29.4 MPa for loading 1.0 to 1.5 phr. However, tensile strength decreased to 24.73 MPa at loading 3.0 phr. This finding is similar to group 2 (PV-GOpa-s) nanocomposite which shows the best loading was at 1.5 phr loading.

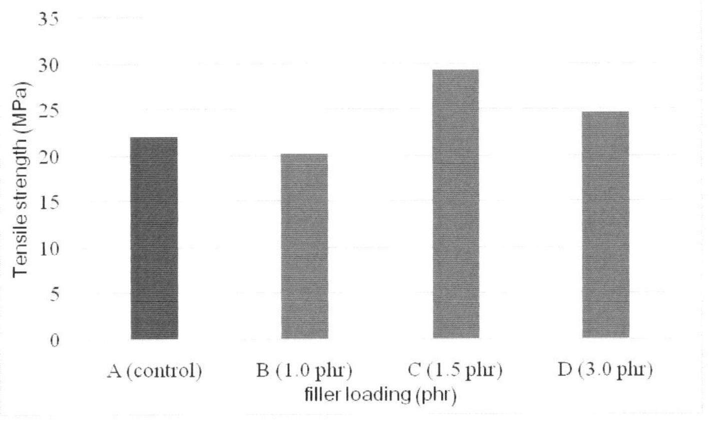

Figure 4. Effect of the presence of graphene powder on tensile strength of PV/Gpo-s at different loading

3.4. Modulus at 100 % elongation (M_{100}) for PV/Gpo-s

The modulus at 100 % elongation (M100) was investigated as shown in Figure 5. The presence of graphene in PV latex was observed to enhance the modulus at 100 % elongation for PV-graphene nanocomposite. The increasing of modulus was recorded from 0.67 to 0.81 MPa towards loading from 1.0 to 3.0 phr. It can be summarized that the good dispersion of graphene powder in PV latex solution has been achieved by sonicated the solution for 24 hours and subsequently increased the modulus at 100 % elongation.

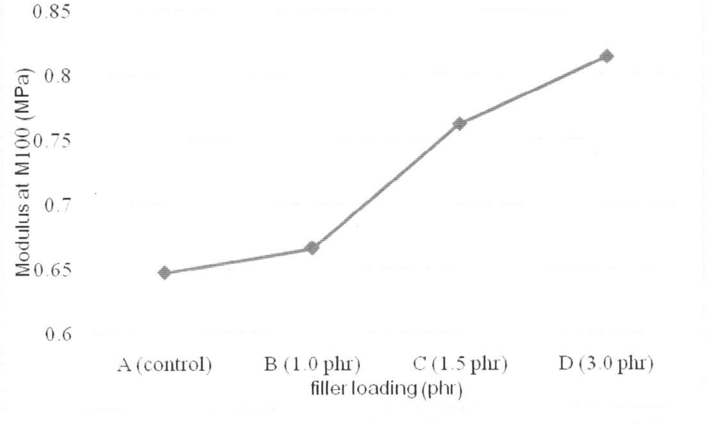

Figure 5. Effect of the presence of graphene powder on modulus at 100 % elongation (M_{100}) of PV/Gpo-s at different loading

3.5. Comparison of tensile strength at maximum load (MPa) between ENR 50/Gpa-s and ENR 50/rGOpo-s

The comparison of tensile strength at maximum load (MPa) between ENR 50/Gpa-s and ENR 50/rGOpo-s were carried out as shown in Figure 6. It was clearly observed that tensile strength of ENR 50/Gpa-s shows the dramatically drop from 1.3 MPa to 1.0 MPa by increasing the loading from 1.0 phr to 3.0 phr. It has been reported the percolation point of ENR latex incorporated graphene nanocomposites takes place at extremely low filler content which is less than 0.1 phr [10] and subsequently explain the reason on the decreasing of tensile strength of ENR 50/Gpa-s. For ENR 50/rGOpo-s shows the decreasing of tensile strength from 1.0 phr to 1.5 phr by 1.3 MPa to 1.2 MPa. But at loading of 3.0 phr, the tensile strength rises up to 1.3 MPa. This phenomenon led to the presence of epoxide group with adjacent double bond at ENR 50 rubber matrix [8]. It provided a strong intermolecular interaction and high cohesion energy which increase the tensile strength and toughness of the rubber [8].

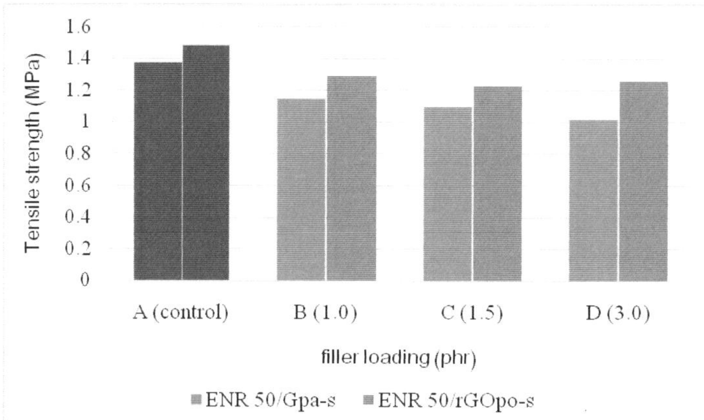

Figure 6. Comparison of tensile strength between ENR 50/Gpa-s and ENR 50/rGOpo at different loading

3.6. Modulus at 100 % elongation (M_{100}) between ENR 50/Gpa-s and ENR 50/rGOpo-s

Figure 7 shows the modulus at 100 % elongation (M_{100}) between ENR 50/Gpa-s and ENR 50/rGOpo-s. For ENR 50/Gpa-s, it shows that the tensile strength was increased by the increasing of loading which from 0.27 MPa to 0.4 MPa. This pattern was repeated for ENR 50/rGOpo-s from 1.0 to 3.0 phr by 0.35 MPa to 0.39 MPa. It was believed that the modulus at 100 % elongation (M_{100}) was directly proportional to tensile strength. Therefore, it can be concluded that the presence of graphene incorporated inside ENR 50 has greatly improved the tensile strength at 100 % modulus.

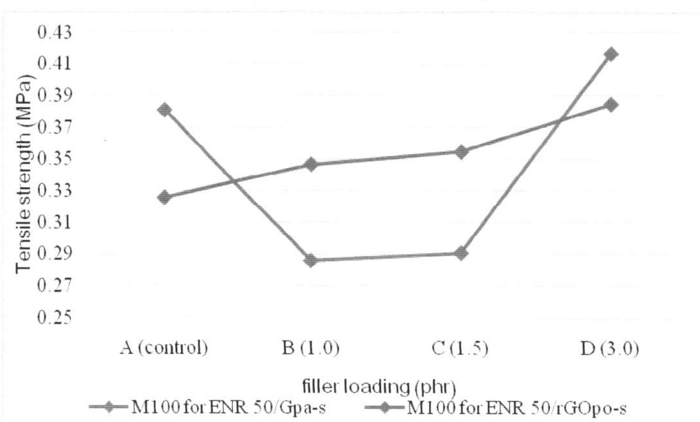

Figure 7. Comparison of modulus between ENR 50/Gpa-s and ENR 50/rGOpo-s at different loading

4. Conclusions

The effect of GO sonication into PV latex was successfully studied in term of its tensile strength. GO sonication gives the impact on increasing of modulus and fracture strength for PV latex. In general, the presence of sonicated GO in natural rubber has increased the tensile properties by 2 MPa. It was shown that the best loading for GO incorporate PV latex was at 1.5 phr with sonication done similar to PV/Gpo-s. The presence of graphene in PV latex enhanced the mechanical properties of the polymer. Hence, it has improved that graphene-PV latex nanocomposite is a promising material to enhance the mechanical properties of the stretchable polymer for future wearable device. For ENR 50 latex, percolation point had taken place at very low filler loading which less than 0.1 phr. Even though the tensile strength for this nanocomposite is not good as pure ENR latex, the presence of the carbon based material may slightly improve the modulus for the ENR nanocomposites. For ENR 50/Gpa-s and ENR 50/rGOpo-s, the best loading was recorded at 3.0 phr. The powder type rGO (ENR 50/rGOpo-s) can withstand higher stress for a longer time compared to the paste type of graphene (ENR 50/Gpa-s). The increasing of mechanical properties of ENR 50 such as modulus for graphene paste and rGO powder, is the result from the mobility of graphene that is grafted onto the surfaces of rubber [10]. Hence, it can be concluded that the presence of graphene may enhance the mechanical properties of the stretchable polymer such as natural rubber. Thus, this carbon based nanocomposite material is a promising material for the researchers to develop more wearable devices for future technology.

5. References

[1] Solov'yov I, Korol A and Solov'yov A 2017 Multiscale Modeling of Complex Molecular Structure and Dynamics with MBN Explorer (s.l: Springer-Verlag)

[2] Chen J, Bo Z and Lu G 2015 The Properties of Vertically-Oriented Graphene *Vertically Oriented Graphene* pp 11-18

[3] Ang P, Chen W, Wee A and Loh K 2008 Solution-Gated Epitaxial Graphene as pH Sensor *Journal of the American Chemical Society* **130** 14392-14393

[4] Kuila T, Bose S, Khanra P, Mishra A, Kim N and Lee J 2011 Recent advances in graphene-based biosensors *Biosensors and Bioelectronics* **26** 4637-4648

[5] Celzard A, McRae E, Deleuze C, Dufort M, Furdin G and Marêché J 1996 Critical concentration in percolating systems containing a high-aspect-ratio filler *Physical Review B* **53** 6209-6214

[6] Phua J, Teh P, Ghani S and Yeoh C 2016 Effect of Heat Assisted Bath Sonication on the Mechanical and Thermal Deformation Behaviours of Graphene Nanoplatelets Filled Epoxy Polymer Composites *International Journal of Polymer Science* **2016** 1-8

[7] Durge R, Kshirsagar R and Tambe P 2014 Effect of Sonication Energy on the Yield of Graphene Nanosheets by Liquid-phase Exfoliation of Graphite *Procedia Engineering* **97** 1457-1465

[8] Noriman N, Omar M, Sam S, Shamni N, Lokman Hakim I and Mustafa Al Bakri A 2016 *Advances in Environmental Biology* **7** 12

[9] Ye S and Feng J 2016 The effect of sonication treatment of graphene oxide on the mechanical properties of the assembled films *RSC Adv.* **6** 39681-39687

[10] Azira A, Kamal M and Rusop M 2016 *AIP Conference Proceedings* **1733** 020003

IOP Conf. Series: Materials Science and Engineering **267** (2017) 012010 doi:10.1088/1757-899X/267/1/012010

Numerical Analysis Study of Sarawak Barrage River Bed Erosion and Scouring by Using Smooth Particle Hydrodynamic (SPH)

M R R M A Zainol[1,4], M A Kamaruddin[2,4], M H Zawawi[3,4], and K A Wahab[1]

[1]School of Civil Engineering, Universiti Sains Malaysia, Penang, Malaysia
[2]School of Industrial Technology, Universiti Sains Malaysia, Penang, Malaysia
[3]College of Engineering, Universiti Tenaga Nasional, Bandar Baru Bangi, Malaysia
[4]Centre of Excellence Geopolymer and Green Technology (CEGeoGTech), Universiti Malaysia Perlis, Perlis, Malaysia

E-mail: ceremy@usm.my

Abstract. Smooth Particle Hydrodynamic is the three-dimensional (3D) model. In this research work, three cases and one validation have been simulate using DualSPHysics. Study area of this research work was at Sarawak Barrage. The cases have different water level at the downstream. This study actually to simulate riverbed erosion and scouring properties by using multi-phases cases which use sand as sediment and water. The velocity and the scouring profile have been recorded as the result and shown in the result chapter. The result of the validation is acceptable where the scouring profile and the velocity were slightly different between laboratory experiment and simulation. Hence, it can be concluded that the simulation by using SPH can be used as the alternative to simulate the real cases.

1. Introduction

Water flow has their own energy in the river. By that energy, the river changes their physical characteristics such as the shape which can be wider, deeper and longer. The material that has been transported from places to another places is caused by erosion of the river bed and river banks. Focusing on the river bed erosion, the changes of the river bed level can happen in both upstream and downstream of the river through the different of the head pressure. According to Chu (2014), river channel-bed erosion and deposition not only affects river sediment flux to the sea but also influences the river regime and associated management practices like in-channel navigation, flood control, and levee maintenance. These have been longstanding issues for many large rivers around the world, such as the Mississippi, the Amazon, the Yellow, the Yangtze and the Pearl.

[1] To whom any correspondence should be addressed.

Generally, at the barrage, the level of water at upstream is higher than the level of water at downstream. The greatest part of river bed to be eroded is at downstream of the river. When the barrage is opened, the highest head pressure of

water will flow to the lower head pressure and scouring or river bed erosion will occurred at those location. Based on Stefan et al. (2016) the residual flow energy will produce scour in an alluvial riverbed if no mitigation structure (such as a dissipation basin) is installed downstream of a grade-control structure. In the absence of a technical dissipation structure or a downstream apron, scour can occur in a loose riverbed downstream. Bhuiyan et al. (2007) investigated the effect of a W-weir (without upstream or downstream apron) on the downstream sediment transport processes, focusing on the scour-hole formation. The maximum scour depth was observed a short distance downstream of the weir, independent of the boundary conditions.

Therefore, by undergo the numerical simulation method the river bed erosion and scouring can be easily identified. The location of the phenomenon and other data can be obtained by using Smooth Particle Hydrodynamics (SPH). SPH could play a very useful role in determine the condition of the river bed at the barrage. The gathered information could lead to a new discovery and helps to improve the river condition.

1.1. Study area

This study actually is about the river bed erosion at downstream of the Sarawak Barrage which is over the Sarawak River connected to the sea. Sarawak River mouth is located near to the Kuching city. There is a barrage that also known as Kuching Barrage which has been constructed and completed in October 1997 which is located at the mouth of the Sarawak River as 'Figure 1'. The operation of the barrage started on January 1998 and the barrage has been built for the flood mitigation project. The barrage also act as a dam by providing fresh water for the water treatment plant. It also prevent or minimize saline intrusion to the water treatment plant.

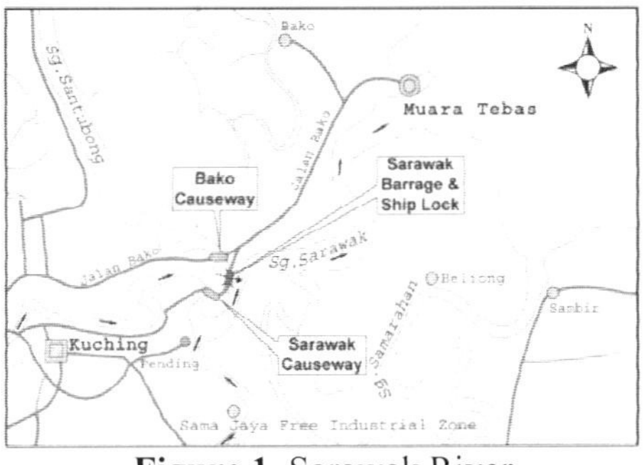

Figure 1. Sarawak River

The barrage has 5 movable gates that consist of radial and tidal gates and 1 ship lock with 2 radial gate which all the gates operated hydraulically. The barrage gates are 25 meters in width of each and a ship lock alongside of the barrage. The barrage also has a flyover on top of it with two causeways across Sarawak River at Pending and Santubong River at Bako

2. Numerical Method

For this study, the case that will be chosen in the software is Multi-phase case where it's liquid-sediment model solver.

The whole system is represented by a finite number of particles that carry individual mass and occupy individual space. Particle approximation aims to discretize the kernel approximation within the support domain. Fig. 2.1 provides a clearer picture on how the kernel function can be used to estimate a field variable of particle *i* within a radius *h*.

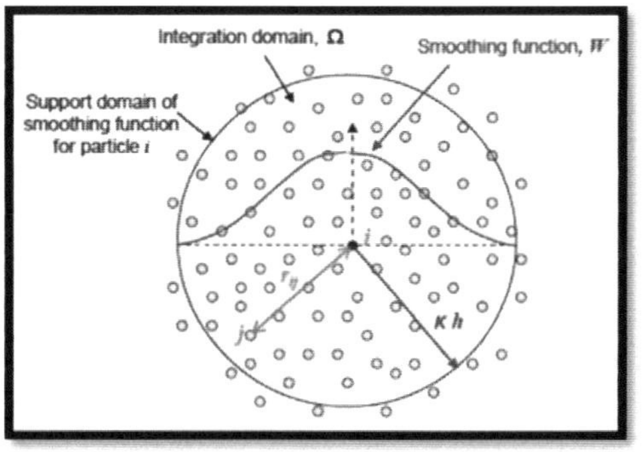

Figure 2. Support domain with a radius of κh and smoothing kernel function, W for particle i

The mass of the particles m_j is related to the finite volume of the particle, ΔV_j and its density of ρ_j, as follows:

$$m_j = \Delta V_j \rho_j \tag{1}$$

The integral form of the interpolation can be discretized as the summation of N neighboring particles (number of particles that interact with particle i and can be written as:

$$f(\boldsymbol{x}) = \int_{\Omega} f(\boldsymbol{x}')W(\boldsymbol{x}-\boldsymbol{x}')d\boldsymbol{x}' \cong \sum_{j=1}^{N} \frac{m_j}{\rho_j} f(\boldsymbol{x}_j) W(\boldsymbol{x}-\boldsymbol{x}_j,h) \tag{2}$$

It is noted that infinitesimal volume $dx'$is replaced by the finite volume of the particle ΔV_j. The final discretized form for the particle approximation function of particle i is:

$$< f(\boldsymbol{x}_i) >= \sum_{j=1}^{N} \frac{m_j}{\rho_j} f(\boldsymbol{x}_j) \cdot W_{ij} \tag{3}$$

where

$$W_{ij} = W(\boldsymbol{x}_i - \boldsymbol{x}_j,h) \tag{4}$$

Thus, the value of a variable at particle i is estimated as the sum of the values it assumes on the neighbouring points within the support domain weighted by the smoothing kernel function.

SPH involves the computation of discretized fluid into a set of particles, each particle is referred to a nodal point where physical quantities are computed as an interpolation of the values of the nearest particles. The integral approximation of any function $A(\vec{r})$ is

$$A(\vec{r}) = \int_{\Omega} A(\overrightarrow{r'})W(\vec{r} - \vec{r}',h)dr' \tag{5}$$

where h is the smoothing length and $W(r - r', h)$ is the weighting function or smoothing kernel.

The function $A(\vec{r})$ can be expressed in the discretized form and the approximation of the function at particle a is given by:

$$A(\vec{r}) = \sum_b m_b \frac{A_b}{\rho_b} W_{ab} \tag{6}$$

where $m_b$and $\rho_b$being the mass and density associated to the neighboring particle b, respectively. Besides, the term $W_{ab} = W(\vec{r_a} - \vec{r_b}, h)$ describes the weight function.

SPHysics provides four different kind of kernel definitions. For this particular research, the cubic spline kernel is selected and showed as follows:

$$W(r,h) \;=\alpha_D \begin{cases} 1 - \frac{3}{2}q^2 + \frac{3}{4}q^3 & 0 \le q \le 1 \\ \frac{1}{4}(2-q)^3 & 1 \le q \le 2 \\ 0 & q \ge 2 \end{cases} \tag{7}$$

where $\alpha_D = 1/(\pi h^3)$ in three-dimensional flow domain.

The momentum conservation equation in a continuum Eulerian field is shown as follows:

$$\frac{D\vec{v}}{Dt} = -\frac{1}{\rho}\vec{\nabla}P + \vec{g} + \vec{\Theta} \tag{8}$$

where $\vec{\Theta}$ denotes the diffusion term.

The momentum equation in SPH method can be described by different approaches based on various existing formulations of the diffusive terms. The momentum equation is implemented to determine the acceleration of a particle a as a result of the particle interaction with its neighbours, particles b:

$$\frac{d\vec{v}_a}{dt} = -\sum_b m_b \left(\frac{P_b}{{\rho_b}^2} + \frac{P_a}{{\rho_a}^2}\Pi_{ab}\right)\nabla_a W_{ab} + \vec{g} \tag{9}$$

Such that v being the velocity, P is the pressure, ρ is the density, m is the mass of particle, $\vec{g} = (0, 0, -9.81)\,\mathrm{m/s^2}$ is the three-dimensional gravitational acceleration vector and W_{ab} denotes the kernel function that dependent on the distance between particles a and b.

The mass of each particle is set to be constant, so that changes in fluid density are computed by solving the conservation of mass or continuity in SPH which is given as:

$$\frac{d\rho_a}{dt} = \sum_b m_b \vec{v}_{ab} \nabla_a W_{ab} \tag{10}$$

Pressure is calculated starting from density using Tait's equation of state as shown below:

$$P = B\left[\left(\frac{\rho}{\rho_0}\right)^{\gamma} - 1\right] \tag{11}$$

where the constants $B = {c_0}^2\rho_0/\gamma$ and $\gamma = 7$, for such that the reference density, $\rho_0 = 1000\,\mathrm{kg/m^3}$ and the corresponding speed of sound at this particular reference density,

$$c_0 = c(\rho = \rho_0) = \sqrt{(\partial P/\partial \rho)}\Big|_{\rho_0} \tag{12}$$

Particles are moved using the SPH variant. SPH ensures each particle movement is consistent with the average velocity of its neighboring particles on particle distribution. In other words, the purpose of SPH is to make fluid particle flows together with its neighboring fluid particles to make it more localized and stable.

3. Methodology

3.1. Data collection

Data collection are very important to obtain information in this study. Through the study site area, field data will be used in the SPH as the raw data. The data that gathered from site such as density of water, density of bed material, cohesion value of the bed material and also water depth will be as input to the program. Besides that, the details of the particle sizes and viscosity also important because it can give an effect to the movement of sediment particles in the simulation. All the parameters must have to be as an input in the SPH software.

Not only that, the characteristics and trend of the barrage also important such as the geometry. The geometry data are obtained through the drawing of Sarawak Barrage will be used for modelling during pre-processing process. To develop the model in the SPH program, the geometry of the barrage has been drawn in the SolidWork program in three-dimensional (3D) and then convert it to the STL file. Figure 3.7 shows the 3D drawing of Sarawak Barrage in scale by using SolidWork. All the geometry of the Sarawak Barrage need to be extracted and redraw back in AutoCADD every detail section to scale 1 to 25 before redraw back in SolidWork software as shown in Figure 3. Type of gates, size of the barrage, and shape of the barrage structure needs to draw clearly in the software. The dimension and criteria of the barrage should be proper managed.

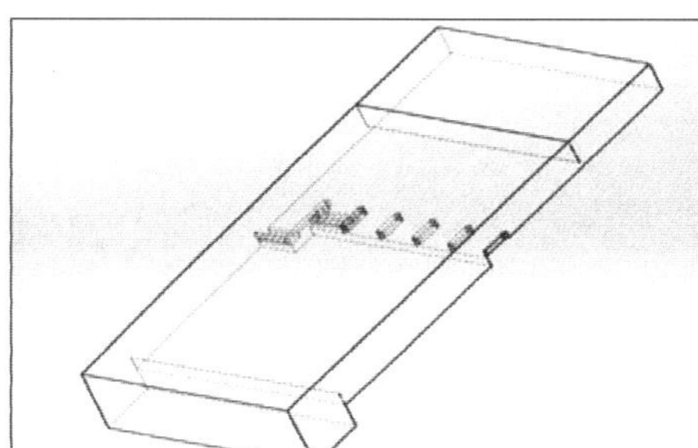

Figure 3. Sarawak Barrage model in 3D SolidWorks

3.2. Numerical analysis

The initial configuration of the simulation such as movement description and parameters are define in the GenCase file name which consider as Pre-processing process. All the input information contains in the XML format; Case_Def.xml. After running GenCase, two output files are created which are Case.xml and Case.bi4. All the parameters of the system configuration such as smoothing length, density, gravity, number of particles and moving boundaries are contained in Case.xml file. While for the Case.bi4 file contains the initial state of the particles is in BINX4(.bi4) format. Paraview software is used to visualize the particle geometries that create with GenCase (Case_All.vtk, Case_Bound.vtk and Case_Fluid.vtk).

Processing-process is where the case that has been create in pre-processing in the process and the SPH simulation performs via named DualSPHysics. All the input file XML file (Case.xml) and binary file (Case.bi4) that has been setup the parameters run through DualSPHysics code. The output files after run the software consist of binary format file that include the particle information of the simulation. This is where analysis of the numerical simulation in progress.

Post-processing process where the output of binary files of DualSPHysics will convert into different format by using PartVTK code that is can allow the output to be visualized and analysed. The results of the simulation can be viewed and analyse by using Paraview from the output files. Post processing is where the simulation and the result obtained from the numerical analysis DualSPHysics.

3.3. Validation

To study the effectiveness and the accuracy of the numerical simulation analysis DualSPHyics, the validation by experiment using the laboratory equipment should be made. For validation of the software, the experiment of the water flow through the sediment is by using the flume.

The experiment starting with by placing sand in the flumes as the sediment. The distance of the sand between the borders was set as 0.5m and the borders size is 0.05m x 0.05m x 0.3m each. The sediment height was set the same as the borders height which is 0.05m. Figure 4 show that the sediment setup.

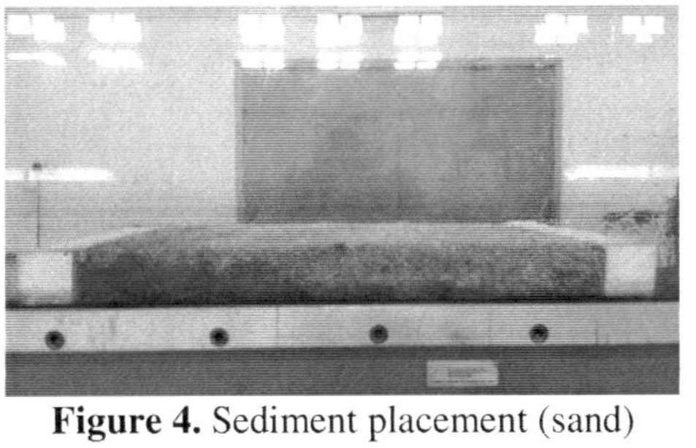

Figure 4. Sediment placement (sand)

Then the experiment was running within case $8.222\text{x}10^{-3}\text{m}^3/\text{s}$ flow rate by setting the valve for that flow rate. Then by using velocity meter, the velocity of the water flow is measure within five points as shown in the Figure 5. After that, measure the sediment high at several point within the depth of changing.

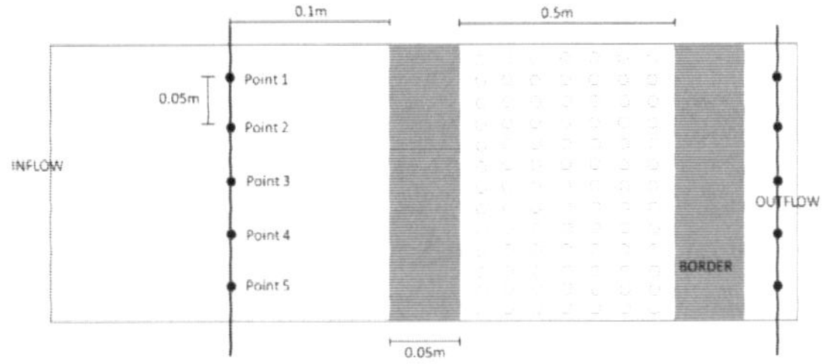

Figure 5. Plan view layout of flume with point measure velocity

4. Results and Discussion

4.1. Case 1

The water level for upstream for this case is 0.34m which has been scale from the real water depth 8.5m. While the water level for downstream has been set for 0.32m which has difference only 0.02m. To measure the scouring profile in the simulation, the distance of the location is 0.5m from the barrage pier. The scouring depth and changing is measured through the side view of the barrage and the point of location has been slice at the middle of each Gate 1 (A-A), Gate 2 (B-B), Gate 3 (C-C), and Gate 4 (D-D) as shown in Figure 6. The red colour of particle denotes sediment particle while the other denotes as liquid particle. Figure 7 shows the cut section of the sediment for the all gate and shape of the scouring profile.

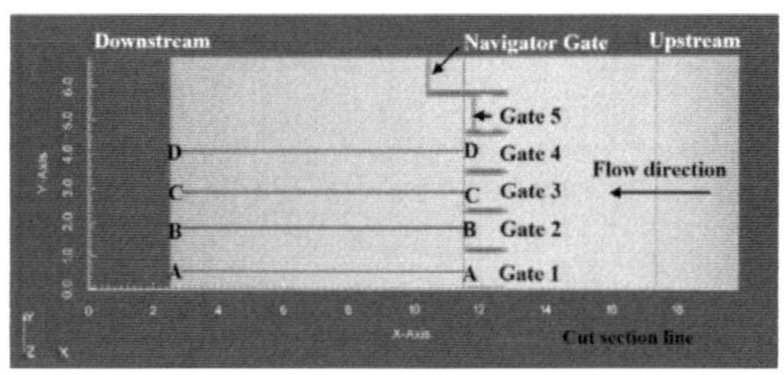

Figure 6. Plan view of barrage simulation

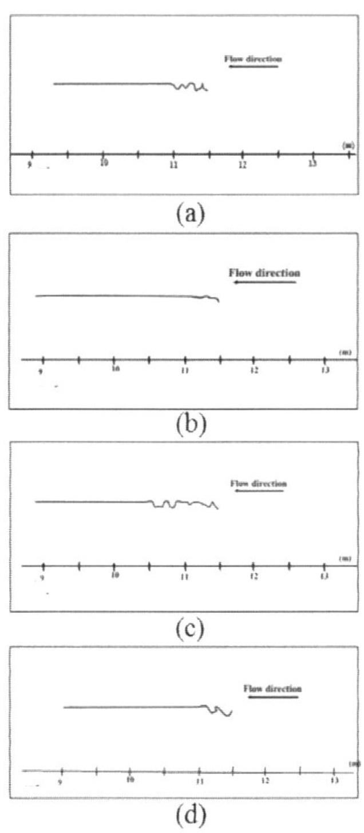

Figure 7. (a) Gate 1 (b) Gate 2 (c) Gate 3 (d) Gate 4 scouring profile

The sediment particle coordinates has been plotted in the Figure 8 with different gate. From that plot, we can see the scouring profile and the scouring depth very clearly. The maximum depth of the scouring is at the Gate 4 which approximately 0.09m. But for the other gate, the maximum scouring depth approximately 0.06m. Even though the scouring depth not very high, it still dangerous to the hydraulic structure because the scouring happen nearly to the apron of the Sarawak Barrage gate.

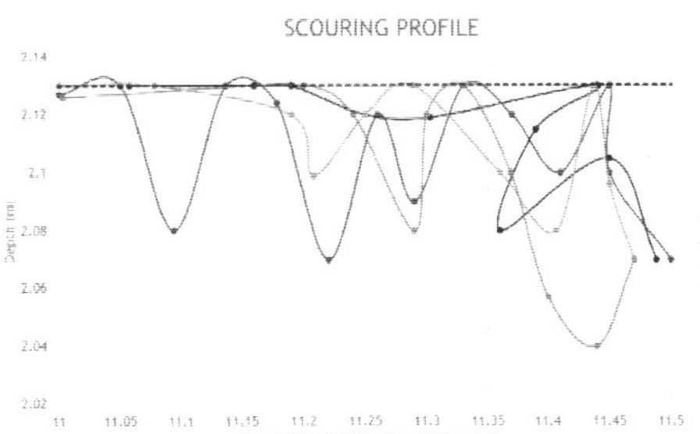

Figure 8. Scouring depth profile

The scouring pattern at the gates approximately similar. The scouring happen when the water passes the barrage apron and hit the water also sediment with different velocity value on each gate. The sediment was eroded by the water within 0.06m for the first phase after barrage apron and the scouring happened. All the sediment particle that has been eroded change their places from origin and moving to the direction of downstream. Figure 9 shows the distribution contour that has been cut to the section and view from the front for the four gate.

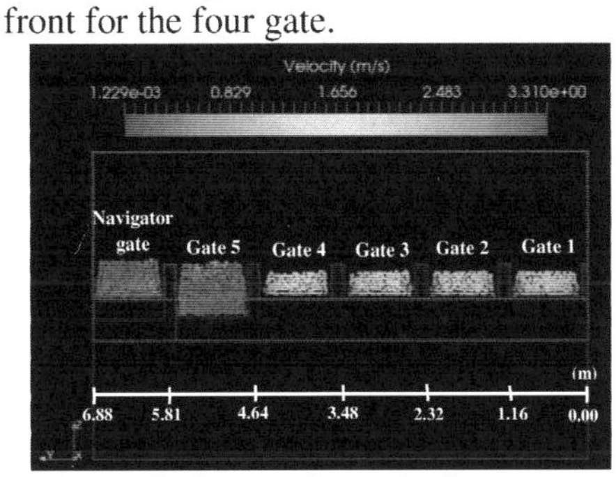

Figure 9. Cut section velocity contour

For the cut section of the Case 1, mostly the Gate 1, 2, 3 and 4 has the approximately same the average velocity which is 1.22m/s at the top. It is the highest velocity on that section. While at the middle of the liquid particle velocity is 1.1m/s which little bit less than the top velocity. But for the bottom velocity of the section is 0.6m/s. The deformation of the sediment particle was influenced by the velocity of the liquid. The flow of the top surfaces give the big impact of the stress for sediment particle which causes of the scouring and erosion.

4.2. Case 2

The water level at the downstream of the Sarawak Barrage is 0.26m from the bed level. While the water level at the upstream is same as first case 0.34m. The pattern of the scouring and the deformation of the river bed that has been eroded as shown in Figure 10. For this cases, the scouring depth less than the first case and more than the third case. The pattern of the scouring profiles are not very drastically eroded within the distance. The scouring profile almost the same for each gate.

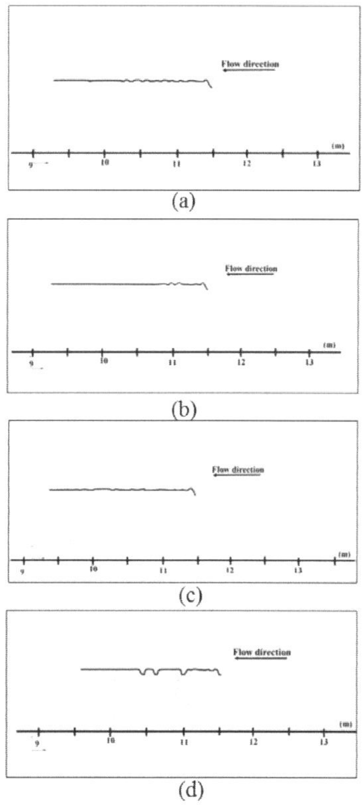

Figure 10. (a) Gate 1 (b) Gate 2 (c) Gate 3 (d) Gate 4 scouring profile

In Figure 11 the scouring profile of each gate shows that pattern and the scouring depth of the sediment within 0.5m from downstream barrage apron has been plotted by using the X-coordinates and the Z-coordinates. The result shown that the maximum scouring depth is approximately 0.05m and the value is same for each gate. This mean the velocity of the water flow have the same value during flow through the barrage.

Figure 11. Scouring depth profile

For the case of the 0.26m water level at the downstream, the velocity of the gate is different on each gates. The highest velocity value at the Gate 4 while the lowest is at Gate 1 with 0.73m/s and 0.48m/s, respectively. Figure 12 shows the velocity contour from the cut section of the barrage.

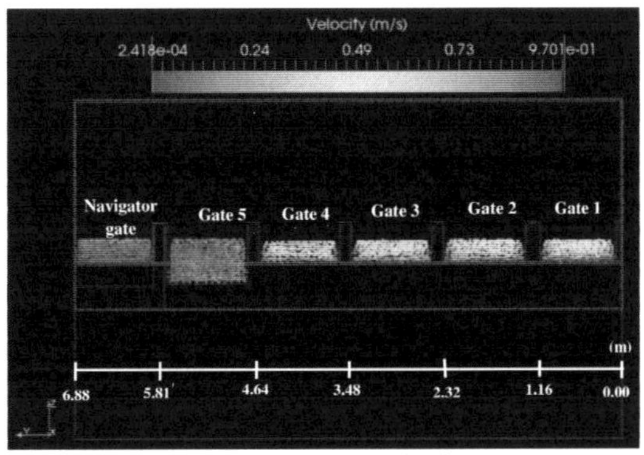

Figure 12. Cut section velocity contour

4.3. Case 3

For Case 3, the water level at downstream is 0.22m while at the upstream is still same. After 5 second simulation, the result has been taken same as the previous case location. Only the water level at downstream has a different. The sediment particle change can been seen clearly as shown in Figure 13. From the sediment particle view from the side of the barrage, the profile of the scouring can be seen clearly.

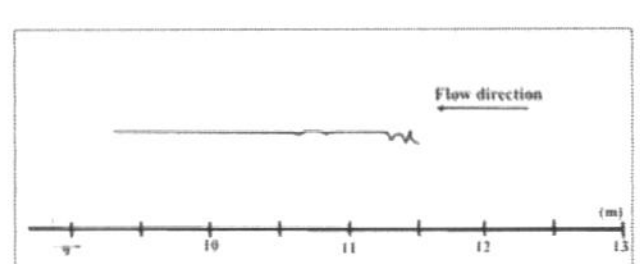

(a)

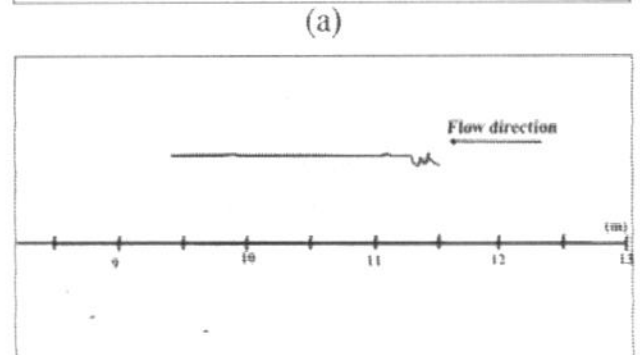

(b)

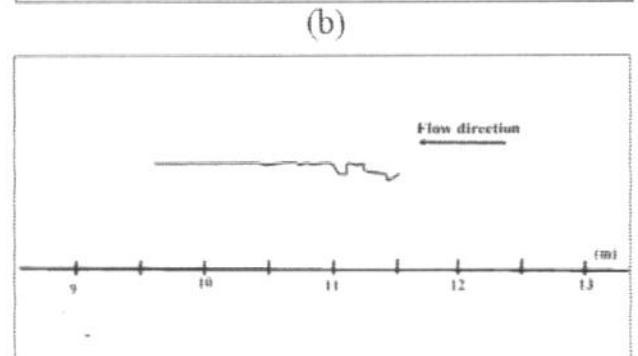

(c)

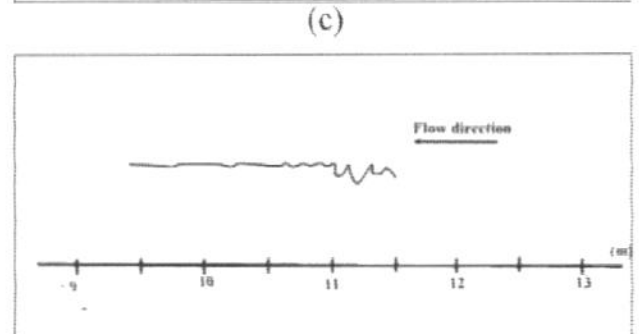

(d)

Figure 13. (a) Gate 1 (b) Gate 2 (c) Gate 3 (d) Gate 4 scouring profile

The scouring depth and the scouring profile plotted in Figure 3.9. The result shown that the highest depth of the scouring is in front Gate 3 with the depth approximately 0.13m. The scouring depth for this case is critical where all sediment particle has been eroded very significantly. The pattern of the scouring also can be seen clearly in Figure 14. The pattern of the scour profile for this case has large deformation from the original profile as compared to the other cases.

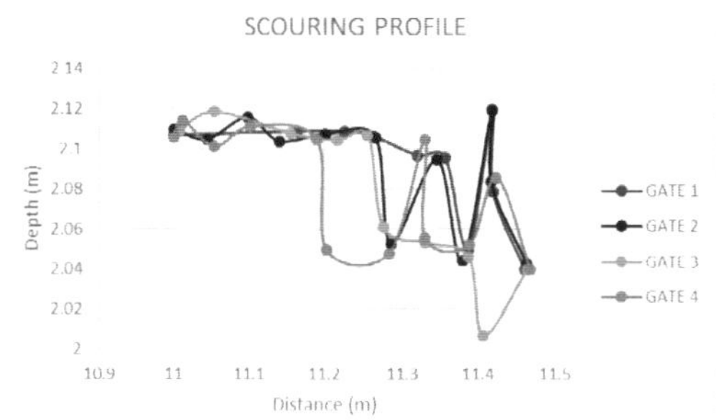

Figure 14. Scouring depth profile

The highest average velocity of this case at the top of the Gate 4 which is 0.9m/s. The water level for this simulation cases is the lowest which is 0.22m at the downstream of the barrage. Because of that water is much lower than the water level upstream, the water particle is moving very fast due to the less of stress to interact with the other particle. It is also due the head different of the water level that affect to the water energy. The contour of the velocity as shown in Figure 15.

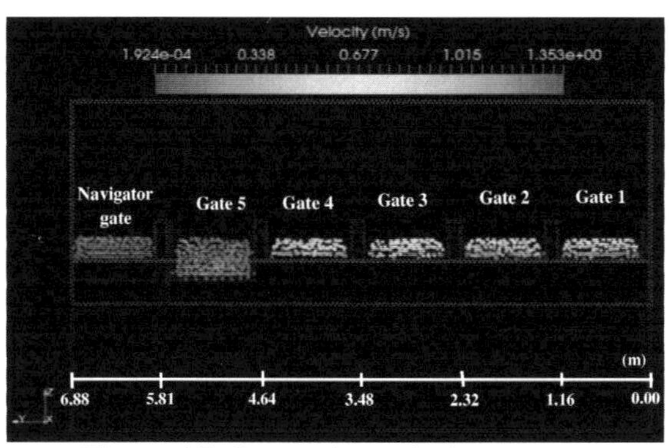

Figure 15. Cut section velocity contour

The result obtains also show the same phenomenon of the velocity where at the top level is the highest, while the bottom level velocity is the lowest and the middle velocity is in intermediate between top and bottom. The lowest velocity is at the bottom of Gate 1. But the value does not differ much with the bottom Gate 2. While for the Gate 3 and 4 at the bottom, the average velocity is same. That's mean the water flow with the constant rate and at normal flow. The contour of the velocity on each gate can be seen clearly in Figure 15.

4.4. Validation

To make sure the results obtained from the simulation program were accurate, the validation of the software need to be investigated between the experiment and the numerical analysis simulation. The validation has been made based on the velocity of the water flow in the experiment and the numerical simulation. By measuring the velocity in the experiment at the middle of water depth by using Nixon Streamflo velocity meter, Table 1 shows the result obtained from the experiment.

Table 1. Velocity obtained by Nixon Streamflo meter

Point	1	2	3	4	5
Velocity (m/s)	0.219	0.223	0.225	0.28	0.221

The average velocity in the experiment is 0.223m/s. From the simulation by using SPH, the point to measure the velocity are same as the laboratory experiment. But the raw velocity from SPH is in the velocity vector. After change the vector to the magnitude velocity, the value get the real velocity. Table 2 shows the result obtained from the simulation.

Table 2. Velocity obtained by numerical simulation

Point	1	2	3	4	5
Velocity (m/s)	0.23	0.26	0.27	0.3	0.22

As can be seen in Figure 16 Coefficient of Determination (R2), comparison between observation and simulation has been made. The R2 value between the observation in the laboratory experiment with simulation is approximately 0.9 which results the simulation value are acceptable. That is mean the difference of the velocity are still in the range of acceptable.

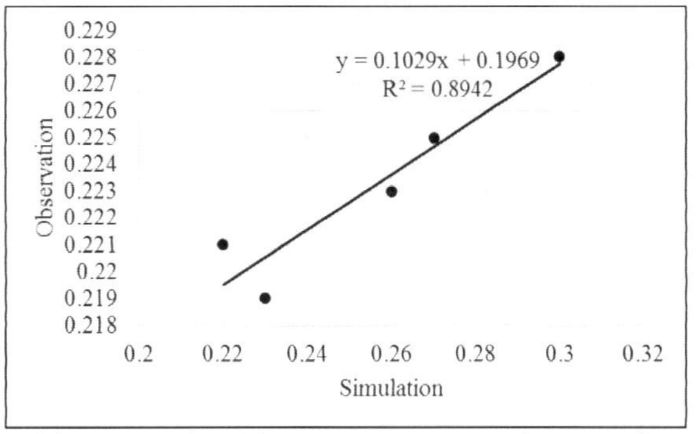

Figure 16. R2 relationship between observation and simulation

The shape of the scouring or degradation of the sediment particle can be used to validate the result between the experiment and simulation. The result obtains from the laboratory experiment by measuring the depth of the scouring as shown in Figure. 17 and the depth of the scouring for the simulation in Figure 18. Figure 19 show scouring shape pattern.

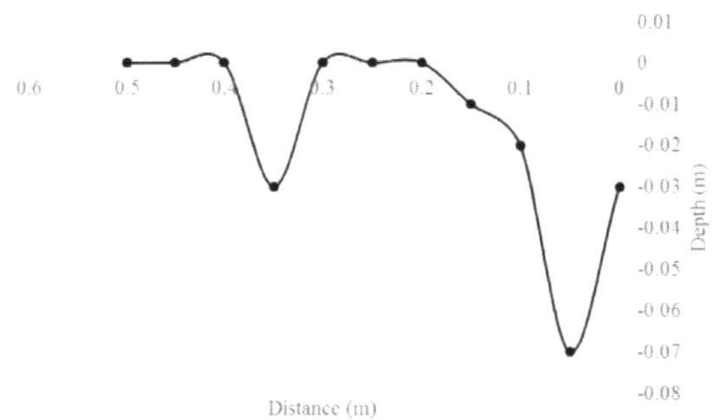

Figure 17. Experimental scouring profile

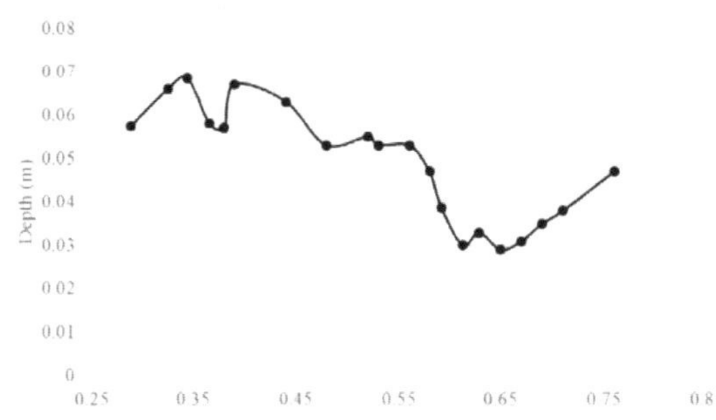

Figure 18. Simulation scouring profile

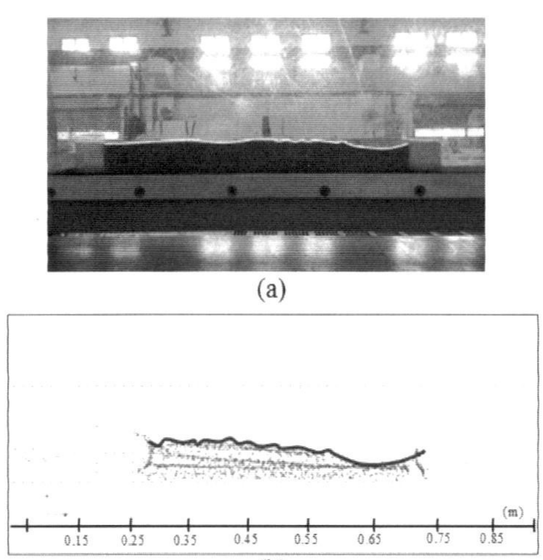

(a)

(b)

Figure 19. (a) Experimental (b) simulation scouring shape pattern

As a result, the pattern of the scouring profile explicit similar pattern between the laboratory experiment and numerical analysis simulation. The maximum scouring depth of the experiment is

0.07m and for the simulation is 0.03m. From the result as shown, the profile of the scouring by SPH simulation is acceptable and can be used to simulate the real cases.

5. Conclusion

This research described the usefulness of the DualSPHysics (SPH) three-dimensional (3D) to solving the hydraulic problems by using numerical analysis simulation of the Sarawak Barrage. The simulation of the cases is to get the velocity of the flow and the scouring profiles. Three cases have been implemented in this research study of the Sarawak Barrage river bed erosion which is Case 1, Case 2 and Case 3. The different water level at downstream of the barrage between Case 1, Case 2 and Case 3 were 0.32m, 0.26m and 0.22m respectively. While the upstream water level of the barrage for the three cases remain same as 0.34m. To prove the result from the simulation, the validation of one case has conducted to ensure the accuracy of the result and also acceptable.

The validation process was conducted by comparing the velocity and scouring profile between the laboratory experiment and the numerical analysis simulation SPH. As the result, numerical analysis of SPH program can be used to simulate the real cases because it was proven by the validation in this research. However, high performance computer need to be used for running the real cases because the software denotes the fluid as the particle. As a final conclusion, the objectives of the study have been achieved.

The result of the simulation has been validated with laboratory experiment and the data obtained was accepted. The result obtained showed that it has slightly different between the experiment and simulation. The average velocity of the experiment was 0.223m/s while the numerical simulation was 0.256m/s. Through the chart coefficient of determination between experimental observation and simulation velocity, the R2 value was approximately 0.9. Thus, the simulation value are acceptable. Based on the case that has been proved, it can be concluded that the SPH numerical simulation can be used in this study.

Sarawak Barrage has been simulated for three cases in which different water level at the downstream have been implemented to get the velocity of the water flow and scouring profile. The maximum scouring was 0.13m depth which simulated for Case 3. The minimum scouring depth among the three cases was 0.05m which is for Case 2. From the scouring profile, the different level of water between upstream and downstream influence the scouring depth of the sediment. Besides that, flow velocity of water also affected due to differ of the water level at upstream and downstream. The maximum velocity of water at the barrage gate was 0.9m/s for case 3 while the lowest velocity was Case 2 which is 0.73m/s. Hence, it can be concluded that the higher the velocity of water from upstream, the higher the scouring depth at downstream when there was much difference in water level between upstream and downstream. The result obtained from this study was logically can be accepted due to the normal phenomenon. From the simulation result, improvement can be made to the site to mitigate all the problems.

6. References

[1] Chu, Z., 2014. The dramatic changes and anthropogenic causes of erosion and deposition in the lower Yellow (Huanghe) River since 1952. Geomorphology, 216, pp.171–179.

[2] Fourtakas, G., 2014. Modelling multi-phase flows in Nuclear Decommissioning using SPH A thesis is submitted to The University of Manchester for the degree Georgios Fourtakas School of Mechanical, Aerospace and Civil Engineering.

[3] Jüstrich, S. et al., 2012. Mobile Riverbed Scour Downstream of a Piano Key Weir. Journal of Hydraulic Engineering, 142(11), pp.1–12.

[4] Fourtakas, G. & Rogers, B.D., 2016. Modelling multi-phase liquid-sediment scour and resuspension induced by rapid flows using Smoothed Particle Hydrodynamics (SPH) accelerated with a Graphics Processing Unit (GPU). Advances in Water Resources, 92, pp.186–199.

[5] Jilani, A.N. & Hashemi, S.U., 2013. Numerical investigations on bed load sediment transportation using SPH method. Scientia Iranica, 20(2), pp.294–299.

[6] Law, I.N., Oon, Y.W., Law, P.L., Kho, F.W.L., 2011. Impacts of Barrage Flushing and Flooding-in Operations on Saline Intrusion Upstream. UNIMAS e-Journal of Civil Engineering, 2(March), pp.18–24.

[7] National Geographic Society (2017), Erosion. Available from :http://www.nationalgeographic.org/encyclopedia/erosion/

Potential use of Plastic Waste as Construction Materials: Recent Progress and Future Prospect

M A Kamaruddin[1,2], M M A Abdullah[2], M H Zawawi[3] and M R R A Zainol[4]
[1]Environmental Technology Division, School of Industrial Technology, Universiti Sains Malaysia, 11800 Penang, Malaysia
[2]Center of Excellence Geopolymer & Green Technology (CEGeoGTech), Universiti Malaysia Perlis, Perlis, Malaysia
[3]College of Engineering, Universiti Tenaga Nasional, Jalan Serdang, Seksyen 11, Selangor, 43650 Bandar Baru Bangi, Malaysia
[4]School of Civil Engineering, Universiti Sains Malaysia, Engineering Campus, 14300 Nibong Tebal, Penang,

E-mail: anuarkamaruddin@usm.my

Abstract. Plastic associates products based have been considered as the world most consumer packaging solution. However, substantial quantities of plastic consumption have led to exponential increase of plastic derived waste. Recycling of plastic waste as valued added product such as concrete appears as one of promising solution for alternative use of plastic waste. This paper summarized recent progress on the development of concrete mixture which incorporates plastic wastes as partial aggregate replacement during concrete manufacturing. A collection of data from previous studies that have been researched which employed plastic waste in concrete mixtures were evaluated and conclusions are drawn based on the laboratory results of all the mentioned research papers studied.

1. Introduction

According to Rochman, Browne [1], in the year 2012 alone, it was estimated that about 280 million tonnes of plastic has been produced worldwide. From that amount, about 130 million tonnes of the plastics were landfilled or recycled. Of the remaining 150 million tonnes, plastic will find their place in daily lives of human being. Meanwhile, the rest of the plastic fraction find their final way as litters in the oceans or landfilled. Consequently, the plastic waste brings serious environmental threat to modern society because it is made up from several toxic chemicals, and therefore plastic pollutes soil, air and water if not properly managed or treated [2, 3]. Accordingly, when most of the available plastic today is made up from non-biodegradable sources, land-filling by using plastic would mean burying the harmful material for over a period until it naturally degrades. In their original condition, any plastic materials would increase the waste volume during landfilling. However, their degradation rate and bulky in nature creates environmental risks tremendously. Besides, the plastic waste mass may hinder the ground water movement [4]. Plastic waste may usually in the form of film and hard plastic may contains harmful metal based elements such heavy metal, at which when mixed up with water or rain

[1] anuarkamaruddin@usm.my

water can impede soil and receive water. Besides, plastic garbage can impede the rate of percolation and in turns would deteriorate the soil fertility if it is mixed with soil [4]. Moreover, plastic waste which is mostly disposed into surface water, public drain, river or sea water can create imbalance of the water and aquatic life [5].

Aquatic life such as plants and animals might have entangled with plastic waste, which can damage their health. According to The Star newspaper article [6], Malaysians, in general generate about 30,000 metric tonnes of solid waste each day. In particular, plastic waste makes up about 13% of solid waste composition, which means that about 4,000 metric tonnes of plastic waste are generated across Malaysia daily. Until now, Malaysia has yet to fully utilize on the relentless advantages of recycled plastics, which can be applied in road construction or pavement to improve strength and increase the road durability, as insulator or conduit in building construction, as raw material or fibre for textiles manufacturing, as fasteners for bulk items tying, low strength plastic furniture and other post-consumer goods. As far as the authors are concern, recycling plastics is a feasible option as an alternative way from common waste stream destination. Because plastic waste is derived from hydrocarbon-based material, its exert relatively high calorific value which can be used for incineration or boiler. However, burning of plastics at lower temperature may releases toxic and poisonous chemical gases into the air, including dioxins which is harmful to the human being. Plastic waste can also be used to produce new plastic based products after submitting to reprocessing line [7]. However, this method is seen uneconomical because the recycled plastic degrades in quality over time and thus there is growing demand for new plastic for the original product.

In terms of civil engineering application, recycling of plastic waste as cementitious based materials, such as cement mixtures or concrete mixture appear as better option for alternative plastic waste disposal. This is due to its economic and ecological advantages which can substitute or replace certain portion of aggregate in concrete mix. Besides, some alteration method of plastic waste could be ideal candidate as lightweight concrete pavement with low strength application. As such, there are already a lot of published reports that have been studied concerning the ability of plastic waste mixed together with cementitious based mixtures such as polyethylene terephthalate (PET) bottle [8, 9], polyvinyl chloride (PVC) pipe [10], high density polyethylene (HDPE) [11], spent plastic waste [12], expanded polystyrene foam (EPS) [13], glass reinforced plastic (GRP) [14], polycarbonate [15], thermoplastic recycled polystyrene [16], polypropylene fiber [17] as an aggregate, or mixture in the manufacturing of concrete.

In general, plastic is lightweight, water retainer and resistant, expandable, strong, and very cheap to produce. These are the attractive qualities that contribute to over-consumption of plastic based goods. Instead, if plastic is 100% made from hydrocarbon intermediates, it is very durable and lead to slow degradation. According to Plastic-Pollution Organization, plastic materials that are used in our daily consumption has become attractive that initiate an undeniable behavioural needs which led to over-consuming. This behaviour simultaneously pollutes the environment seriously. Previous review on application of plastic waste in concrete manufacturing has been reported previously [2, 18]. However, there is a very limited information on the post-consumer plastic utilization for their detail physical and chemical properties. This review aims to provide summary on recent progress of plastic derived waste utilization as partial aggregate replacement for concrete mixture and their future prospects for the efficient plastic waste diversion.

1.1. Current Scenario of Plastic Waste

In general, plastics are mostly used in diverse types of application daily. From industrial to family consumption, plastic is considered as universal material that can resolve wide range of problems. However, it has been found that and plastic items end up in the waste stream even after a single use only within a short period of time after purchasing especially for packaging purpose [19]. According to [18] plastic waste can be treated either by landfilled or incineration of recycled back based on municipal solid waste hierarchy. Due to technological advancement, landfilling of plastic waste is considered the least preferable method because it requires a large space, reduce the landfill lifespan and causes persistent pollution problems. Incineration process is adopted in some developed countries because of high combustible value of the plastic material and exhibit low moisture content. Thus, complete elimination of this waste and their successful rate is high as compared to Asia countries. This is because, most of the plastic especially plastic film used by most of Asia's community are rich with moisture. These properties impede the ability of incinerator to completely burn off these waste that may produce a great amount of dioxins if the temperature is lower than 800 °C. Until now, only EU has documented data on plastic production made available to the public but not for Asia countries. Figure 1 shows recent statistic on plastic production worldwide. The trend shows that the production of plastic throughout the world is increasing over times. The rest of the world has been produced more than 1,986 million tonnes of plastic since 2005. This staggering increase of plastic production worldwide shows that there is no clear indicator when the transformation from plastic based production will shift into alternative type of plastic substitute because the end destination of the plastic, if not properly managed and disposed will find their way in landfills, rivers, or sea.

1.2 Alternative Use of Plastic Waste

There are a lot of studies have been reported and published concerning the alternative use of plastic waste into different type of products [20-24]. Plastic waste formed for high grade resins have been recycled from used and spoilt plastics including: automotive parts, home appliances, cloths (textiles), mulches, and films. Accordingly, plastic waste treatment and reprocessing techniques could be divided into four major categories which are re-cycling, mechanical, chemical and energy recovery as reported by Al-Salem, Lettieri [19]. Meanwhile, Ishaiba [25] described types of plastic and their potential recycling techniques as shown in Table 1. Based on the table, common types of plastics that are mostly reprocessed including polyvinyl chloride (PVC), polystyrene (PS), polyethylene terephthalate (PET), polypropylene (PP) and polyethylene (PE).

Table 1. Plastic types and their potential recycling method [25].

Plastic source	Characteristics	Common use of virgin plastic	Common use for recycled plastic
Polyethylene Terephthalate (PET)	Clear hard plastic, suitable for fibre	Soft drink and mineral water bottles	Clear and soft film for Packaging and wrapping, rug fibers, rain coats
Low density polyethylene (LDPE)	Soft, flexible plastic, milky white, unless a pigment is added	Lids of food containers, garbage bags, and rubbish bins	Soft film, wrapping industry, plant packaging and nurseries bags
High density Polyethylene (HDPE)	Commonly used plastic in white or coloured	Puckered shopping bags, milk storage bags (freeze)	Compost bins, detergent bottles, crates, and mobile rubbish bins
Unplasticised Polyvinyl chloride (UPVC)	Hard rigid plastic, clear type	Sanitary piping, plumbing pipes and fittings	Dishwasher bottles, toiletries detergent bottles, tiles, and plumbing pipe fittings
Plasticized Polyvinyl chloride (PPVC)	Flexible, clear, elastic Plastic	Garden hose, shoe soles, blood bags and tubing	Hose inner core, and industrial flooring
Polypropylene (PP)	Hard, but flexible plastic	Ice-cream containers, potato crisp bags,	Compost bins, kerb side recycling crates,

		stools and chairs	and worm factories
Polystyrene (PS)	Stiff but brittle plastic. Clear in nature and glossy	Cheap, transparent kitchen ware, light fittings, bottles, toys, and food containers	Laundry pegs, coat hangers, and video/CD boxes
Polyester (EPS)	Foamed, lightweight, energy absorbing, and thermal insulation	Hot drink cups, and takeaway food containers	spools, rulers, and video/CD boxes
Polyamides (PA)	Nylons	fibers, toothbrush bristles, and fishing lines	

Even though there are diverse types of recycling techniques for plastic waste as mentioned earlier, the reuse of plastic waste for construction materials can be considered as a promising method to maximize this waste. Through this method, plastic waste can be revitalized directly it disposed by substitute the plastic waste for partial aggregate replacement could enhance the environment sustainability or construction materials.

2. Plastic Aggregates and Plastic Fiber in Concrete Mix

Concrete is made up from coarse and fine aggregates, cement and water. Concrete is the most prevalent construction materials due to the fact that the raw materials are easily available and relatively low cost [26]. It also provides better fire resistance than any other building materials. Traditionally, concrete contains numerous weakness and flaws if no appropriate preconditioner is in place. For example, propagation of micro cracks of concrete under uniform concentrated applied load can be contributed to low tensile strength of concrete. Therefore, it is expected better performance of concrete structure that can withstand higher tensile strength as well as the flexural strength which could be obtained by introducing closely spaced fibres. As a matter of fact, concrete can withstand higher compression stress but low in tensile strength. In normal concrete mix, aggregate typically accounts for 65 to 85% of the mass concrete volume. In addition, aggregate plays a significant role in concrete strength development which can be characterised based on their slump value, compressive strength, dimensional stability, and durability [8]. Therefore, by replacing partial aggregate utilization in concrete mix preparation will provide alternative solution to the other potential use of plastic wastes.

Briefly, there are two forms of plastics waste which are plastic aggregate (PA) and plastic fibre (PF) commonly employed for building materials as described by Gu and Ozbakkaloglu [18]. PAs are employed to replace coarse aggregates (CA) and fine aggregates (FA). Normally, the PA possess lower bulk density than granite, limestone or basalt. Therefore, they are preferably being employed for lightweight concrete. PAs can be obtained by applying mechanical recycling method. In contrast, plastic fiber (PF) are used as reinforcement which can replace common steel fibre that can improve mechanical and strength durability [26]. The major drawbacks of common steel fibre as concrete reinforcement are their susceptible to corrosion especially on concrete surface when exposed to marine or saline water without having appropriate protection. Table simplifies the characteristics of PFs used in the concrete manufacturing. The tables also describe the source and method use to obscure PA and PF that have been reported in literature elsewhere.

2.1 Application of Plastic Waste in Concrete Mix

There are a lot of important factors need to be considered when applying plastic waste ad partial aggregate replacement. As for the present work, authors attempt to summarizes all the reported literatures mentioning concrete properties based on selective work. Physical properties of concrete, may include density, slump value, mechanical properties, covering splitting tensile strength, compressive strength, Young modulus, flexural strength, abrasion resistance, impact resistance and pulse velocity, durability properties including change in strength, chloride attacks, absorption, creep, shrinkage, carbonization, sulfide attack and several others physical and chemical properties are discussed. Table 3 simplifies some of the requirement when conducting test on the prepared concrete samples.

Table 2. Characteristics of plastic aggregate and plastic fiber and their characteristics.

Parameter	Plastic source	Reference
Plastic type	PET bottle ground into pieces and sieve according to sieve size	[2, 27-29]
	High density polyethylene (HDPE)	[30]
	Expanded polystyrene (EPS) crushing and form bead	[31]
	PVC pipe crushed to aggregate	[32]
	PET fibers from melting process	[33]
	Glassfiber reinforced plastic (GRP) fiber	[34]
	PET plastic bottle shredded into fiber	[35]
	Virgin plastic as partial fine aggregate	[36]
	Glass reinforce plastic fiber	[37]
	Polyurethane (PUR) foam waste as coarse aggregates	[38]
Size (Typical)	2-11 mm (Coarse aggregate) <2.36 mm (Fine aggregate)	[26]
	0.02 μm to 600 μm	[34]
	Microplastic - diameter ranges from 5 to 100 mm and length is 5–30 mm	[39]
	Macro plastic 30–60 mm and cross section of 0.6–1 mm^2	[40]
	Coarse aggregate (8/20 mm)	[38]
Density (kg/m^3)	Concrete aggregate (220-240) Fine aggregate (310-340)	[41]
	PET lightweight (844)	[42]
	38 113 225	[37]

337

Table 3. Test requirement for concrete mix preparation form plastic waste.

Test	Factors to be considered	Possible contribution to strength development
Slump	Water/cement ratio Substitution level of plastic aggregate/fiber Shape of plastic aggregate /fiber	Generally, increasing the amount of PA/PF would reduce the slump value Due to non-uniform shape of PA/PF Low fluidity Absorption capacity Aggregate with different specific gravity
Unit weight and density	Amount of PA or PF Substitution level of plastic aggregate/fiber Shape of plastic aggregate /fiber	Generally, increasing the level of substitution reduce the density for PA Similar observation, if using PF, little changes of density
Air content	Amount of air content	Incorporation of PA and PF increase the air content Due to plastic and natural aggregate could not bind together thus increase porosity
Compressive strength	Water/cement ratio Substitution level of plastic aggregate/fiber Shape of plastic aggregate /fiber Aspect ratio and geometry of fibers	Increasing the Water/cement ratio or substitution level of plastic aggregate/fiber led to reduction in compressive strength Due low elastic modulus Due to low bond strength between surface of plastic and aggregate If using high ultimate tensile strength fiber would increase compressive strength (i.e: polypropylene) than PET fiber
Splitting tensile	Substitution level of plastic	Increasing Substitution level of

strength	aggregate/fiber Shape of plastic aggregate /fiber	plastic aggregate would reduce tensile strength Due to non-uniformity shape of PA/PF Due to change in modulus elasticity However, not the case for PF Increasing amount of PF increased tensile strength (i.e PET, PP)
Elastic modulus (E_c)	Substitution level of plastic aggregate/fiber Shape of plastic aggregate /fiber Type of waste plastic Porosity of aggregates Transition zone characteristics	Ec of PA concrete much lower than conventional concrete with the same w/c Significant lower of Ec can be observed if the shape of PA become more irregular For the case of PF, not much different in Ec as in conventional concrete

3. Future prospect
This paper presents a critical review of the recent published reports on the for plastic waste based materials for aggregate replacement in concrete mix. Many studies reported in recent years on these materials indicate that the use of recycled plastic aggregates and fibers as partial aggregate replacement is gaining significant interests from many researchers. The findings of the review also show that the use of these materials can improve concrete properties under appropriate mix composition with the main motivation is to find alternative destination of the plastic wastes rather than direct disposal at the landfill. This is because, if optimum solution of plastic waste can be realized, it is estimated that about 30% of total waste disposed from solid waste could be reduced. As far as the authors are concern, most of the plastic derived based polymeric are difficult to be degraded for decades, even for centuries. By utilizing this waste as partial aggregate replacement, it could be inferred that these materials could be preserved inside concrete structures for ages. Although no studies have yet to forecast the service life of concrete structures containing plastic waste materials, the use of plastic waste in concrete can contribute meaningfully toward a more sustainable and holistic construction industry. Thus, long-term performance of plastic waste in concrete, and their environmental impact after its service life are recommended to be explored further.

4. References
[1] Rochman, C.M., et al., *Policy: Classify plastic waste as hazardous.* Nature, 2013. **494**(7436): p. 169-171.

[2] Saikia, N. and J. de Brito, *Use of plastic waste as aggregate in cement mortar and concrete preparation: A review.* Construction and Building Materials, 2012. **34**: p. 385-401.

[3] Kamaruddin, M.A., et al. *Resource recovery from municipal solid waste by mechanical heat treatment: An opportunity.* in *AIP Conference Proceedings*. 2017. AIP Publishing.

[4] Silva, R., J. De Brito, and R. Dhir, *Properties and composition of recycled aggregates from construction and demolition waste suitable for concrete production.* Construction and Building Materials, 2014. **65**: p. 201-217.
[5] Kamaruddin, M.A., M.S. Yusoff, and A.H. Ahmad. *Waste to Wealth: Recovery of Recyclable Items and Biodegradable Wastes from Night Markets at Seberang Perai Selatan, Penang*. in *Applied Mechanics and Materials*. 2015. Trans Tech Publications.
[6] Star, T., *Saturday, 21 January 2017 Solving the plastic problem.* http://www.thestar.com.my/opinion/letters/2017/01/21/solving-the-plastic-problem/, 2017.
[7] Bogas, J.A., J. de Brito, and J.M. Figueiredo, *Mechanical characterization of concrete produced with recycled lightweight expanded clay aggregate concrete.* Journal of Cleaner Production, 2015. **89**: p. 187-195.
[8] Saikia, N. and J. de Brito, *Mechanical properties and abrasion behaviour of concrete containing shredded PET bottle waste as a partial substitution of natural aggregate.* Construction and building materials, 2014. **52**: p. 236-244.
[9] Iucolano, F., et al., *Recycled plastic aggregate in mortars composition: Effect on physical and mechanical properties.* Materials & Design, 2013. **52**: p. 916-922.
[10] Binici, H. and O. Aksogan, *Eco-friendly insulation material production with waste olive seeds, ground PVC and wood chips.* Journal of Building Engineering, 2016. **5**: p. 260-266.
[11] Shanmugapriya, M. and H. Santhi, *Strength and Chloride Permeable Properties of Concrete with High Density Polyethylene Wastes.* International Journal of Chemical Sciences, 2017. **15**(1).
[12] Jha, J., et al., *Behavior of plastic waste fiber-reinforced industrial wastes in pavement applications.* International Journal of Geotechnical Engineering, 2014. **8**(3): p. 277-286.
[13] Sayadi, A.A., et al., *Effects of expanded polystyrene (EPS) particles on fire resistance, thermal conductivity and compressive strength of foamed concrete.* Construction and Building Materials, 2016. **112**: p. 716-724.
[14] Pastor, J., et al., *Glass reinforced concrete panels containing recycled tyres: Evaluation of the acoustic properties of for their use as sound barriers.* Construction and Building Materials, 2014. **54**: p. 541-549.
[15] Wang, J., et al., *Morphologies and properties of polycarbonate/polyethylene in situ microfibrillar composites prepared through multistage stretching extrusion.* Journal of Applied Polymer Science, 2014. **131**(7).
[16] Dalhat, M. and H. Al-Abdul Wahhab, *Properties of Recycled Polystyrene and Polypropylene Bounded Concretes Compared to Conventional Concretes.* Journal of Materials in Civil Engineering, 2017. **29**(9): p. 04017120.
[17] Yang, S., et al., *Properties of self-compacting lightweight concrete containing recycled plastic particles.* Construction and Building Materials, 2015. **84**: p. 444-453.
[18] Gu, L. and T. Ozbakkaloglu, *Use of recycled plastics in concrete: A critical review.* Waste Management, 2016. **51**: p. 19-42.
[19] Al-Salem, S.M., P. Lettieri, and J. Baeyens, *Recycling and recovery routes of plastic solid waste (PSW): A review.* Waste Management, 2009. **29**(10): p. 2625-2643.
[20] Rajaeifar, M.A., R. Abdi, and M. Tabatabaei, *Expanded polystyrene waste application for improving biodiesel environmental performance parameters from life cycle assessment point of view.* Renewable and Sustainable Energy Reviews, 2017. **74**: p. 278-298.
[21] Gu, F., P. Hall, and N. Miles, *Development of composites based on recycled polypropylene for injection moulding automobile parts using hierarchical clustering analysis and principal component estimate.* Journal of Cleaner Production, 2016. **137**: p. 632-643.
[22] Al-Hadithi, A.I. and N.N. Hilal, *The possibility of enhancing some properties of self-compacting concrete by adding waste plastic fibers.* Journal of Building Engineering, 2016. **8**: p. 20-28.

[23] Muise, I., et al., *Attitudes to the recovery and recycling of agricultural plastics waste: A case study of Nova Scotia, Canada.* Resources, Conservation and Recycling, 2016. **109**: p. 137-145.
[24] Kreiger, M., et al., *Life cycle analysis of distributed recycling of post-consumer high density polyethylene for 3-D printing filament.* Journal of Cleaner Production, 2014. **70**: p. 90-96.
[25] Ishaiba, A.A., *Mechanical Properties of Concrete Using Recycled Plastic*. 2015, The Islamic University-Gaza.
[26] Yin, S., et al., *Use of macro plastic fibres in concrete: a review.* Construction and Building Materials, 2015. **93**: p. 180-188.
[27] Frigione, M., *Recycling of PET bottles as fine aggregate in concrete.* Waste management, 2010. **30**(6): p. 1101-1106.
[28] Ling, T.-C. and C.-S. Poon, *A comparative study on the feasible use of recycled beverage and CRT funnel glass as fine aggregate in cement mortar.* Journal of cleaner production, 2012. **29**: p. 46-52.
[29] Ferreira, L., J. de Brito, and N. Saikia, *Influence of curing conditions on the mechanical performance of concrete containing recycled plastic aggregate.* Construction and Building Materials, 2012. **36**: p. 196-204.
[30] Vanitha, S., V. Natarajan, and M. Praba, *Utilisation of waste plastics as a partial replacement of coarse aggregate in concrete blocks.* Indian Journal of Science and Technology, 2015. **8**(12).
[31] Tang, W., Y. Lo, and A. Nadeem, *Mechanical and drying shrinkage properties of structural-graded polystyrene aggregate concrete.* Cement and Concrete Composites, 2008. **30**(5): p. 403-409.
[32] Alhozaimy, A. and M. Shannag, *Performance of concretes reinforced with recycled plastic fibres.* Magazine of Concrete Research, 2009. **61**(4): p. 293-298.
[33] Kim, S.B., et al., *Material and structural performance evaluation of recycled PET fiber reinforced concrete.* Cement and concrete composites, 2010. **32**(3): p. 232-240.
[34] Asokan, P., M. Osmani, and A.D. Price, *Improvement of the mechanical properties of glass fibre reinforced plastic waste powder filled concrete.* Construction and Building Materials, 2010. **24**(4): p. 448-460.
[35] Ramadevi, K. and R. Manju, *Experimental investigation on the properties of concrete with plastic PET (bottle) fibres as fine aggregates.* International journal of emerging technology and advanced engineering, 2012. **2**(6): p. 42-46.
[36] Rai, B., et al., *Study of waste plastic mix concrete with plasticizer.* ISRN Civil Engineering, 2012. **2012**.
[37] Osmani, M. and A. Pappu, *An assessment of the compressive strength of glass reinforced plastic waste filled concrete for potential applications in construction.* Concrete Research Letters, 2010. **1**(1): p. 1-5.
[38] Ben Fraj, A., M. Kismi, and P. Mounanga, *Valorization of coarse rigid polyurethane foam waste in lightweight aggregate concrete.* Construction and Building Materials, 2010. **24**(6): p. 1069-1077.
[39] Nili, M. and V. Afroughsabet, *The effects of silica fume and polypropylene fibers on the impact resistance and mechanical properties of concrete.* Construction and Building Materials, 2010. **24**(6): p. 927-933.
[40] Yin, S., et al., *Fiber preparation and mechanical properties of recycled polypropylene for reinforcing concrete.* Journal of Applied Polymer Science, 2015. **132**(16).
[41] Kan, A. and R. Demirboğa, *A new technique of processing for waste-expanded polystyrene foams as aggregates.* Journal of Materials Processing Technology, 2009. **209**(6): p. 2994-3000.

[42] Choi, Y.W., et al., *Characteristics of mortar and concrete containing fine aggregate manufactured from recycled waste polyethylene terephthalate bottles.* Construction and Building Materials, 2009. **23**(8): p. 2829-2835.

Acknowledgments
This research is supported by Solid Waste Management Cluster (SWAM), Grant No. 1001/CKT/870023 administered by Science and Engineering Research Center (SERC) Universiti Sains Malaysia. Authors gratefully acknowledge contribution in this paper whom directly or indirectly towards the outcome of the present research

Durability of Geopolymer Lightweight Concrete Infilled LECA in Seawater Exposure

R A Razak [1, 2, a] M M A B Abdullah [1, 2, b] Z Yahya[1,2, c] and M S A Hamid[2, d]
[1] Center of Excellence Geopolymer and Green Technology, School of Materials Engineering, Universiti Malaysia Perlis (UniMAP), 01000, P.O. Box 77, D/A Pejabat Pos Besar, Kangar, Perlis, Malaysia.
[2] Faculty of Engineering Technology, Universiti Malaysia Perlis (UniMAP), Kampus Uniciti Alam, Padang Besar, Perlis, Malaysia.

[a] Corresponding author: rafizarazak@unimap.edu.my
[b] mustafa_albakri@unimap.edu.my
[c] zarinayahya@unimap.edu.my
[d] sarifhamid4545@gmail.com

Abstract. This paper describes a development of lightweight concrete using lightweight expanded clay aggregate (LECA) in fly ash (FA) based geopolymer immersed in seawater. The objective of this research is to compare the performance of geopolymer concrete (GPC) with ordinary Portland cement (OPC) concrete infilled lightweight expanded clay aggregate (LECA) in seawater exposure. Geopolymer concrete is produced by using alkaline activator to activate the raw material, FA. The highest compressive strength of this study is 42.0 MPa at 28 days and 49.8 MPa at 60 days. The density for this concrete is in the range of 1580 kg/m^3 to 1660 kg/m^3. The result for water absorption is in the range of 6.82% to 14.72%. However, the test results of weight loss is in the range between 0.30% to 0.43%.

1. Introduction

A better knowledge for the production of geopolymer cements based on natural materials (calcined kaolin, volcanic scoriae) and waste materials (slag, fly ash) becomes more crucial to improve the local economy. Ordinary Portland cement (OPC) producing high carbon dioxides (CO_2) emissions that cause global warming, hence geopolymer has been introduced to solve this problem. The utilization of geopolymer material as substituent to OPC, the mixture used like fly ash, silica fume and blast furnace slag sand can reduce the average of five times CO_2 compared to OPC [1]. The use of geopolymer has advantages like corrosion resistance, fire resistance, high compressive strength, tensile strength and rapid strength which enables geopolymer becomes as alternative material to OPC [2].

Lightweight aggregate concrete is defined as concrete of low density using lightweight aggregate such as expanded clay [3, 4]. LECA has small hole that could be absorbing and maintain environmental pollution [5, 6]. LECA is generated in kiln rotate horizontal about 1200°C by wet process, by utilizing bloated clay. LECA will be uses to replace coarse aggregate for the production of lightweight concrete [7].

Ordinary concrete typically contains about 12%, 8% mixing water, and 80% aggregate by mass. This means that in addition to 1.5 billion tons of cement, the concrete industry is consuming annually 9 billion tons of sand and rock together with one billion ton of mixing water [8].

LECA is used to replace the coarse aggregate and fly ash is used for the synthesis of green cement as alternative materials to produce a lightweight concrete in seawater exposure. Past study showed less study on the geopolymer concrete against seawater exposure. So this study is carried out to examine the performance of geopolymer lightweight concrete against seawater exposure.

Performance of fly ash based geopolymer incorporating LECA for lightweight concrete will be determined in term of compressive strength, water absorption, density, and weight loss compared to OPC lightweight concrete.

2. Materials and experimental methods

2.1 Mix Design

For this research, NaOH molarity of 12M and Na_2SiO_3/NaOH ratio of 2.5 are fixed [9]. The mass of fly ash, NaOH and Na_2SiO_3 are fixed for all samples. Table 1 shows the details of mix design for geopolymer and OPC.

TABLE 1. The details of concrete mix design.

Description (Geopolymer)	**Mix design**	**Description (OPC)**	**Mix Design**
Solid/liquid (Fly Ash/ Alkaline activator) ratio	2.0	Solid/Liquid (cement)	2.0
Na_2SiO_3/NaOH ratio	2.5	Water	2.5
Mass of Fly Ash (g)	798.0	Mass of Cement	798.0
Mass of NaOH (g)	114.29	Mass of Water	997.5
Mass of Na_2SiO_3 (g)	285.71		

The initial mix design for the aggregate sample is used to find the high compressive strength between the geopolymer with the proportion as shown in Table 2.

TABLE 2. The proportion materials used for GPC and OPC concrete.

GPC	Fly Ash (g)	Alkaline Activator		Aggregate	
		NaOH (g)	Na_2SiO_3(g)	LECA (g)	Sand (g)
	798.0	114.29	285.71	266	798
OPC	OPC (g)	Water (g)		Aggregate	
				LECA (g)	Sand (g)
	798.0	997.5		266	798

2.2 Experimental Procedure

The aggregate and the fly ash is mix together in the mixer. The mix is first dry mix to evenly distribute the fly ash and the aggregate. Then the solution of NaOH and Na_2SiO_3 are mixed to form alkaline activator. The activator is pour into the mixer to mix with the dry materials. After the mixing is homogeneous, the geopolymer paste is casted into 100mm x 100mm x 100mm mould. The samples is demoulded after 24 hours. Then the initial sample is weighted, then immerse in seawater according to the curing days of 3,7,14, 28 and 60 days. All samples will be tested for weight loss to measure the erosion of the cube, compressive strength, density, and water absorption. The test for water absorption is done according to BS 1881-122 (1983) [10]. The compressive strength test is conducted according to BS 1881-116 (1983) [11].

For weight loss testing, the sample cubes are weighed after removed from seawater, then the cube has been scraped off using a metal brush, and after being scraped, the cube sample is weighed a second time to get the difference result caused by erosion.

3. Result and discussion

3.1 Density

The density of concrete can be categorized either as normal, lightweight or heavyweight concrete. According to the result, the density of the OPC lightweight concrete and geopolymer lightweight concrete (GPC) are varied from 1580 to 1660 kg/m^3. The densities for both types of concrete are below than 1800 kg/m^3, which can be classified in the group of lightweight concrete.

3.2 Compressive Strength

Figure 1 shows the graph of compressive strength for geopolymer lightweight concrete (GPC) and ordinary Portland lightweight concrete (OPC) infilled LECA at various ages. The highest strength of GPC lightweight concrete is 49.8 MPa at 60 days and the second highest is 40.7 MPa at 28 days. Meanwhile for OPC lightweight concrete, the highest strength is 24.5 MPa at 60 days and 22.5 MPa at 28 days. The strength of GPC is two times (103%) higher than the strength of the OPC due to the chemical bonds of Si-O-Si and Si-O-Al produced through geopolimerization process which affect in increasing the compressive strength of GPC. The early strength of GPC at 3 days shows higher than OPC. During the laboratory works, the GPC samples were hardened rapidly than OPC. This shows that the GPC also applicable on site works and the time to complete the works can be reduced while the early high strength can be achieved [3, 4].

The results of compressive strength for GPC lightweight concrete shows high strength even using the LECA as a replacement of coarse aggregate and exposed in seawater exposure . The bond strength between geopolymer paste and LECA shows better bond strength and stable in seawater immersion compared to coarse aggregate.

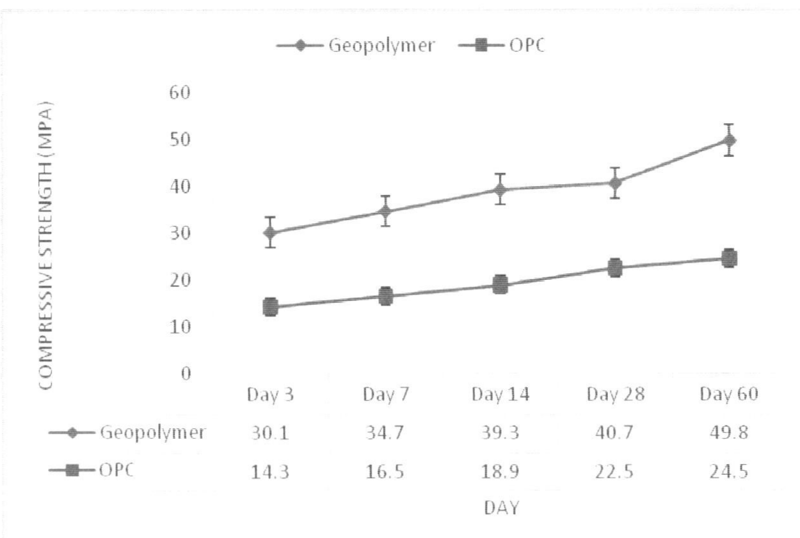

FIGURE 1. Compressive strength of GPC and OPC lightweight concrete.

3.3 Water Absorption

The water absorption for OPC concrete is in the range of 10.41% to 14.59% and the water absorption for geopolymer is in the range of 6.82% to 14.72%. Figure 2 shows the graph of water absorption of GPC and OPC lightweight concrete. Both results show not much difference for water absorption for seawater exposure. However, GPC shows slightly lower than OPC due to geopolymer gel produced in the GPC lightweight concrete which acts as a boundary to prevent the absorption of water in the concrete. The water absorption value was decreasing with the increasing of ages. The higher the curing days, the higher the matrix bonds presented in the samples and consume to high strength as proved in the compressive strength results.

IOP Conf. Series: Materials Science and Engineering **267** (2017) 012012 doi:10.1088/1757-899X/267/1/012012

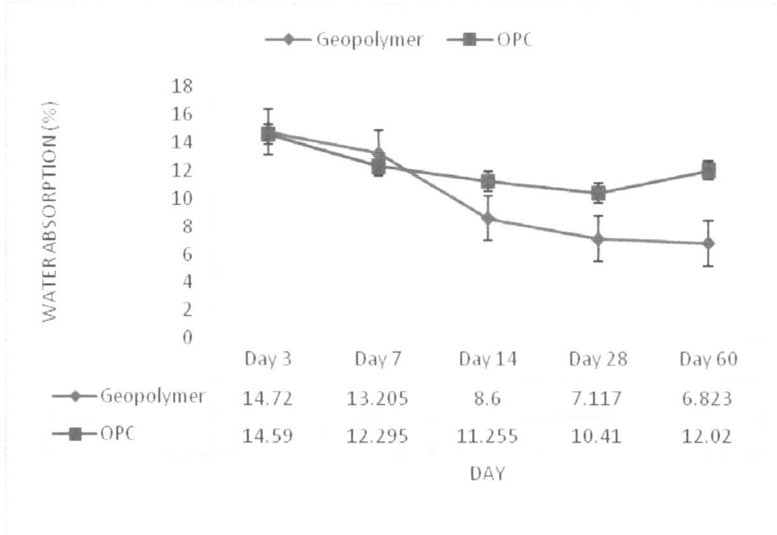

FIGURE 2. Water absorption results for GPC and OPC lightweight concrete.

3.4 Weight Loss

Weight loss test was performed to investigate the erosion of concrete when exposed to seawater. Figure 3 shows the graph of percent weight loss in the concrete at various ages. The lowest weight loss for OPC lightweight concrete is 0.135% at 3 days and the highest is 0.7% at 60 days. Then the lowest weight loss for GPC lightweight concrete is 0.303% at 3 days and the highest is 0.435% at 60 days. It shows that GPC lightweight concrete has lower weight loss at longer curing period in seawater compared to OPC lightweight concrete. This shows that GPC lightweight concrete have more durability in immersion of seawater. Further research should be done for longer curing period in seawater to investigate the weight loss of concrete.

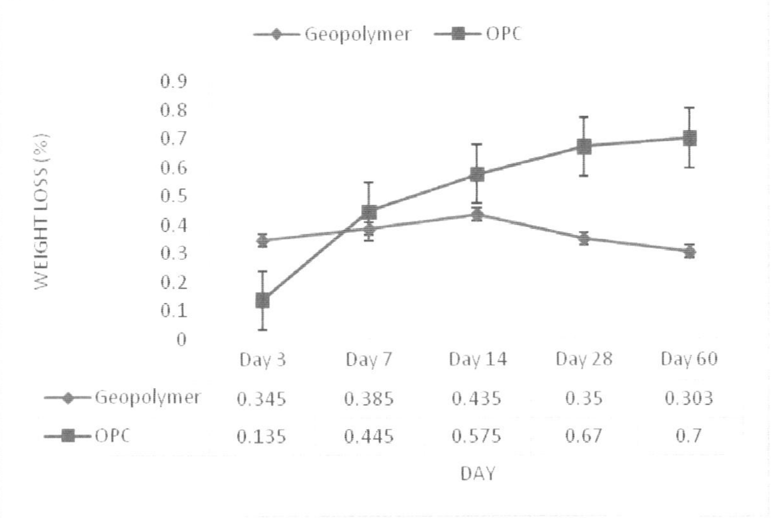

FIGURE 3. Percentage of weight for GPC and OPC lightweight concrete at various ages.

4. Conclusions

Based on the experimental result of this study, it can be concluded that:

1. In geopolymer concrete, the density of geopolymer concrete is below 1800 kg /m^3 which can be classified as lightweight concrete to be used as structure.
2. The highest compressive strength of GPC lightweight concrete is 49.8MPa at 60 days which is two times higher than OPC lightweight concrete infilled LECA and immersed in seawater.
3. In addition, the water absorption of GPC lightweight concrete is in the range of 6.82% to 14.72% which is merely difference with OPC lightweight concrete.
4. GPC lightweight concrete has lower weight loss at longer curing period in seawater compared to OPC lightweight concrete. The durability of GPC lightweight concrete is better than OPC concrete for longer curing period when immersed in seawater.

5. References

1. A. M. M. Al Bakri, H. Kamarudin, O. Abdulkareem, C. M. Ruzaidi, A. R. Rafiza, and M. N. Norazian. Applied Mechanics and Materials **110-116**, 734–739 (2011).
2. S. Lee, H. T. Jou, A. Van Riessen, W. D. A. Rickard, C. M. Chon, and N. H. Kang, N. H. Construction and Building Materials **52**, 221–226 (2014).
3. R. Arellano Aguilar, O. Burciaga, and J. I. Escalante. Construction and Building Materials, **24(7)**, 1166-1175 (2010).
4. S. Bazzaz Bonabi, J. Kahani Khabushan, R. Kahani, and A. Honarbakhsh Raouf. Materials and Design **64**, 310-315 (2014).
5. M. N. Sepehr, H. Kazemian, E. Ghahramani, A. Amrane, V. Sivasankar, and M. Zarrabi. Journal of the Taiwan Institute of Chemical Engineers, **45(4)**, 1821–1834 (2014).
6. A. Ardakani, and M. Yazdani. (2014). Applied Clay Science **93-94**, 28-34 (2014).
7. E. D. Murat, N. A. Hakan, and S. Cengiz. (2015). Construction and Building Materials **101**, 260–267 (2015).
8. G. Satheeshkumar and B. Selvam. International Journal of Emerging Technology in Computer Science & Electronics (IJETCSE), **13(1)**, 536–541 (2015).

9. A. M. Mustafa Al Bakri, H. Kamarudin, M. Bnhussain, A. R. Rafiza and Y. Zarina, ACI Mater. Journal, **109 (5)**, 503-508 (2012).
10. British Standard BS 1881-122. Testing concrete, Method for determination of water absorption (1983).
11. British Standard BS1881-116. Testing concrete, Part 116: Method for determination of compressive strength of concrete cubes (1983).

The Influence of Sintering Method on Kaolin-Based Geopolymer Ceramics with Addition of Ultra High Molecular Weight Polyethylene as Binder.

A Romisuhani[1*], M M AlBakri [1,2] , H Kamarudin [1,2] , S V Andrei [3]

[1]Center of Excellence Geopolymer and Green Technology (CEGeoGTech), School of Materials Engineering, Universiti Malaysia Perlis (UniMAP), P.O Box 77, D/A Pejabat Pos Besar, 01000 Kangar, Perlis, Malaysia.
[2] Faculty of Engineering Technology, Universiti Malaysia Perlis (UniMAP), P.O Box 77, D/A Pejabat Pos Besar, 01000 Kangar, Perlis, Malaysia.
[3] Gheorghe Asachi Technical University of lasi, Faculty of Materials Science and Engineering, Blvd. D. Mangeron 71, 700050, lasi, Romania

[1*]romisuhani@studentmail.unimap.edu.my, [1,2]mustafa_albakri@unimap.edu.my

Abstract. The influence of sintering method on kaolin-based geopolymer ceramics with addition of Ultra High Molecular Weight Polyethylene as binder were studied. Geopolymer were formed at room temperature from kaolin and sodium silicate in a highly alkaline medium, followed by curing and drying at 80 °C. 12 M of sodium hydroxide solution were mixed with sodium silicate at a ratio of 0.24 to form alkaline activator. Powder metallurgy technique were used in order to produce kaolin geopolymer ceramics with addition of Ultra High Molecular Weight Polyethylene. The samples were heated at temperature of 1200 °C with two different sintering method which are conventional method and two-step sintering method. The strength and density were tested.

1. Introduction

Inorganic aluminosilicates produced at low temperatures, also known as geopolymers, have potential applications as matrix in cements, coatings, and ceramics [1]–[3]. Synthetic alkali aluminosilicate material called geopolymer are produced through a reaction between aluminosilicate materials with alkali or alkali silicates under highly alkaline conditions [4][5][6]. Chemically, geopolymers contain of cross-linked units of AlO_4 and and SiO_4 tetrahedra, where charge-balancing cations are provided by alkali metal cations such as Li^+, Na^+, K^+, and Cs^+ [4].

The characteristics of geopolymer includes high strength, fast or low setting, low shrinkage, acid resistance, low thermal conductivity and fire resistance. For the reason, geopolymers are considered to be suitable for a broad range of engineering materials applications [7][8][9]. It can be used as sealants, concretes, fire and heat resistant fibre composites, ceramics, etc., depending on the chemical composition of the source material and the activators [10]. Furthermore, geopolymer technology provides a new route to produce advanced ceramics with beneficially short fabrication time with the proper processing procedure and low cost. The geopolymer with amorphous to semi-crystalline behavior will transform into crystalline ceramic phases on heating [11] [12] [13]. Geopolymer also can

[1] Faculty of Engineering Technology, Universiti Malaysia Perlis (UniMAP), P.O Box 77, D/A Pejabat Pos Besar, 01000 Kangar, Perlis, Malaysia : romisuhani@studentmail.unimap.edu.my

be directly transformed into final structural ceramic part of interest and also convenient to design the chemical composition of the final product [14].

Porous ceramics are significantly outstanding permeability and have a large surface area. Their greater properties can be extensively used in diverse fields such as ceramic filters, bioceramics, fuel-cell electrodes and membrane reactors [15]. In the spite of widen the applications, these ceramic should not have only large interconnections and high porosity but also high in strength and dense networks. Therefore, the addition of binder in production of geopolymer ceramics can be one of the solution to achieve all those excellent properties. The binder were used purposely to deform the plasticity between ceramic particles hence transmit an adequate strength to the green compact [16].

The sintering method will allow densification to occur without stimulating the grain growth which is suitable to microstructural refinement lead to produce dense ceramics. A great deal of different methodologies has been developed with a view to improve the mechanical properties through powder consolidation. Other than conventional method that were used by many researchers on their study of ceramic fabrication, the two-step sintering is also among the favourable method found to effectively control the undesirable grain coarsening during densification[17]. Furthermore, the removal of binder usually take place in oxidising atmosphere at moderately elevated temperatures, and generally become complete at around 600 °C to 700 °C for polymers [16]. Thus, sintering method plays an important role in order to produce ceramic materials with the desired properties.

This paper presents a study on the effect of sintering method of the kaolin-based geopolymer ceramics with addition of UHMWPE as binder. The addition of polymer (UHMWPE) into geopolymer ceramic materials prepared by using powder metallurgy method were sintered at 1200 °C. The effect of two different methods which are conventional method and two-step sintering method on flexural strength and density were studied.

2. Experimental

2.1 Materials

Kaolin was provided by Associated Kaolin Industries Malaysia as Si-Al sources materials. The summarization of chemical composition of kaolin has been analysed by X-ray fluorescence (XRF) (Table 1). Malvern particle size analyser were used to analyse the particle size distribution of aluminosilicate materials. Sodium hydroxide (NaOH) powder used was of caustic soda micropearls, 99% purity with brand name of Formosoda-P made in Taiwan. Sodium silicate (Na_2SiO_3) chemical composition are 9.4% of Na_2O, 30.1% of SiO_2, and 60.5% of H_2O (modulus, SiO_2/Na_2O = 3.2), specific gravity at 20 °C = 1.4 g/cm^3 and viscosity at 20 °C = 0.4 Pa s. The UHMWPE in powder form, with a molecular weight of 5 X 106 g/mol and density of 0.94 g/ml was used as binder in kaolin-based geopolymer ceramics and supplied from Ticona Engineering Polymer, China.

Table 1. Chemical composition of kaolin.

Composition	SiO_2	Al_2O_3	Fe_2O_3	K_2O	TiO_2	MnO_2	ZrO_2	LOI
Percentage (%)	54.0	31.7	4.89	6.05	1.41	0.11	0.10	1.74

2.2 Samples preparation

The 12 M of NaOH solution were mixed with Na_2SiO_3 at a ratio of 0.24 to form an alkaline activator solution. In order to dissolve the silica completely, the solution was then allowed to mature for 24 h. Kaolin powder were mixed with alkaline activator with the solid-to-liquid ratio of 1.0. A mechanical

mixer was used to stir the mixture for few minutes to obtain a homogeneous mixture. The details of the mixtures involved were shown in Table 2. The ratio was selected based on our previous work, as it allowed an optimal physical and mechanical properties. The kaolin-based geopolymer samples were crushed using mechanical crusher after the samples cured in the oven at 80 °C for 24 h. Then the samples were sieved by using 150 μ siever to obtained fine powder.
The UHMWPE (4 wt. %) were added into kaolin-based geopolymer powder by using planetary mill in dry condition method. The mixture was mixed for 4 minutes at 100 rpm with reverse direction until achieve their homogeneity. After that, the mixture was compacted using a 12-mm-diameter cylindrical stainless steel die at 5 tons for 2 min. The final green body was sintered in a high temperature Carbolite furnace 1200 °C with two different sintering method which are conventional and two-step sintering method as shown in Figure 1 with a cooling and heating rate of 5 °C/ min.

Table 2. Mix design details.

S/ L	NaOH molarity	Na_2SiO_3/ NaOH	Total molar ratio		
			Na_2O/ SiO_2	SiO_2/ Al_2O_3	H_2O/ Na_2O
1	12	0.24	0.3950	3.2022	11.5730

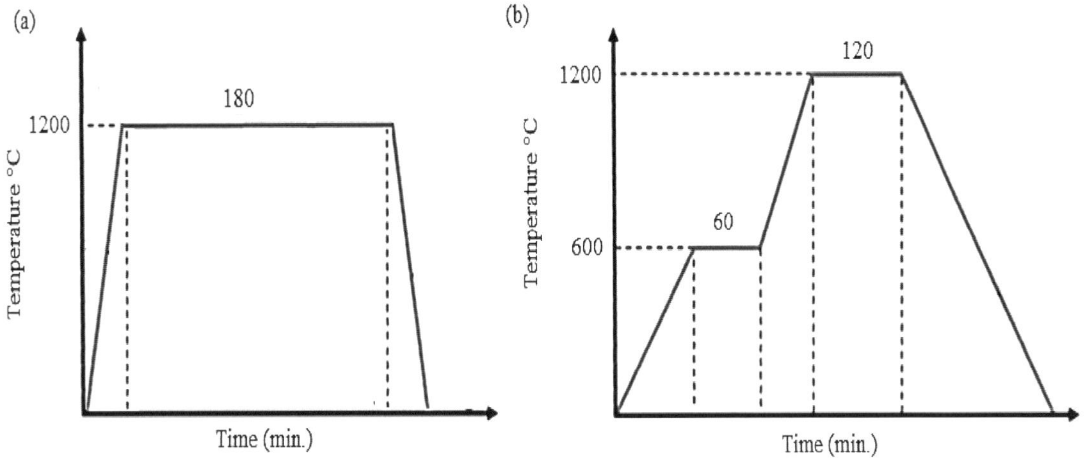

Figure 1. Sintering method used in fabrication of kaolin-based geopolymer ceramic with addition of UHMWPE (a) conventional method and (b) two-step sintering method.

2.3 Testing

2.3.1 Flexural Strength

Flexural strength test was conducted on specimens (7 mm x 5 mm x 52 mm) by using a three-point-bending fixture on an instron-500 tester. With a span length of 30mm at a crosshead speed of 0.5 mm/min, five samples for each group were subjected to flexural strength testing in accordance to ASTM C1161.

2.3.2 Density

Pycnometry (AccuPyc II 1340 He pycnometer, Micromeritics, GA, USA) were used to analyzed density of the geopolymer ceramic. A cylinder shaped sample with diameter of 12 mm and length of 4 mm were used in this experiment.

3. Results and Discussion

3.1 Flexural Strength

The effect of sintering method on the kaolin-based geopolymer ceramics heated at 1200 °C is shown in Figure 2. The flexural strength of kaolin-based geopolymer ceramics sintered using two-steps sintering method is higher compared to the kaolin-based geopolymer ceramics sintered using conventional method. The pre-treatment heating in two-steps sintering method increased the fracture toughness of the grain boundaries, reduced the subcritical growth rate of sintering flaws hence resulting in increasing of the strength [18]. The addition of 4 wt. % of UHMWPE to the kaolin-based geopolymer ceramics shows the highest strength of 94.32 MPa. The effect of binder on kaolin-based geopolymer ceramics is obvious with an increasing of strength greater than the kaolin-based geopolymer ceramics without addition of UHMWPE. This contribution to the strength were due to the advantageous of the carbonaceous residue for the resistance against crystallization thus produces a complex microstructure for carbon-rich of geopolymer ceramics [19].

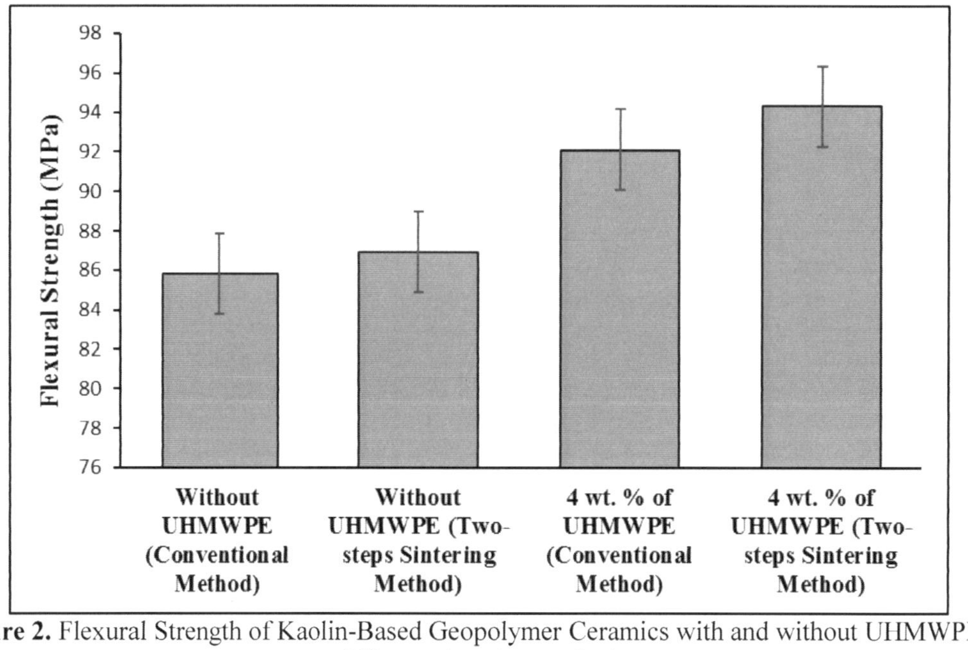

Figure 2. Flexural Strength of Kaolin-Based Geopolymer Ceramics with and without UHMWPE at different sintering method

3.2 Density Analysis

Figure 3 shows the density of the kaolin-based geopolymer ceramics with and without addition of UHMWPE at different sintering method. Kaolin-based geopolymer ceramics with and without addition of UHMWPE that sintered with two-steps sintering method shows a lower density compared to the kaolin geopolymer ceramics sintered with conventional method. The kaolin-based geopolymer ceramics with addition of 4 wt. % of UHMWPE sintered with two-steps method achieved the lowest density of 2.095 g/cm^3. It was clear that by having pre-treatment heating at lower temperature consequently reduce the densification rate. According to the Maglia et al (2013), the neck formation among particles during pre-heating leads to the more compaction, produce a slighter distribution in pore size and promoting the eliminating of fine particle thus becoming a factor in decreasing the density [20]. The decreasing of density in terms of addition of UHMWPE as a binder were contributed by the thermal decomposition of UHMWPE that resulting in formation of pores hence affected the density of the green part [21].

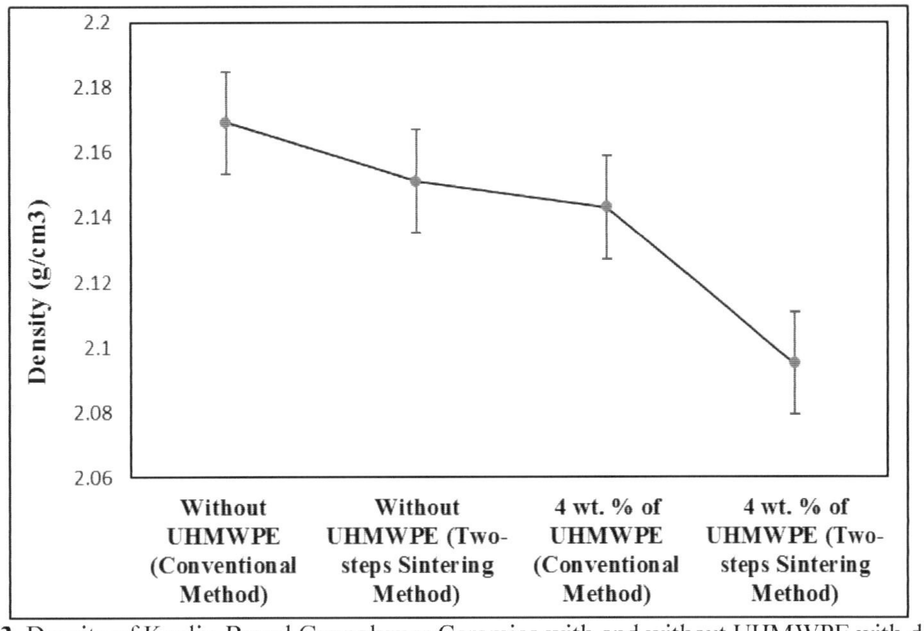

Figure 3. Density of Kaolin-Based Geopolymer Ceramics with and without UHMWPE with different sintering method

4. Conclusion

This paper is purposely to study the influence of sintering method on kaolin-based geopolymer ceramics with addition of Ultra High Molecular Weight Polyethylene (UHMWPE) as a binder. Results determined that the sintering method used affected the strength of the kaolin-based geopolymer ceramics. The flexural strength of kaolin-based geopolymer ceramics sintered by using two-steps sintering method is higher compared with using conventional method. Kaolin-based geopolymer ceramics with addition of 4 wt. % UHMWPE give the highest strength which is 94.32 Mpa. The density of the kaolin-based geopolymer ceramics with addition of UHMWPE using two-steps sintering method give a lowest value (2.095 g/cm^3) and decreasing in density were observed for the addition of UHMWPE for both method of sintering.

5. Acknowledgments

The authors of the present work wish to dedicate their great thanks to Centre of Excellence Geopolymer System Research @UniMAP.

6. References

[1] Davidovits J 1991 *J. Therm. Anal.* **37**(8) 1633
[2] Davidovits J and Davidovics M 1991 *Geopolymer: Ultra-High Temperature Tooling Material for The Manufacture of Advanced Composites*," *How Concept Becomes Reality* vol 2 p 1939
[3] He P, Jia D, Lin T, Wang M, and Zhou Y 2010 *Ceram. Inter.* **36**(4) 1447
[4] Heah C Y, Kamarudin H, Mustafa Al Bakri A M, Bnhussain M, Luqman M, Khairul N I, Ruzaidi C M and Liew Y M 2012 *Constr. Build. Mater.* **35** 912
[5] Barbosa V F F, Mackenzie K J D and Thaumaturgo C 2000 *Int J. Inorg. Mater.* **2** 309
[6] Mustafa Al Bakri A M, Kamarudin H, Bnhussain M, Nizar I K, Rafiza A R, and Izzat A M

2011 *J. Appl. Sci. Res.* **7**(7) 1199
[7] Duxson P, Fernandez-Jimenez A, Provis J L, Lukey G C, Palomo A, and Van Deventer J S J 2007 *J. Mater. Sci.* **42**(9) 2917
[8] Cheng T W and Chiu J P 2003 *Miner. Eng.* **16**(3) 205
[9] Cheng T W, Lee M L, Ko M S, Ueng T H, and Yang S F 2012 *Appl. Clay Sci.* **56** 90
[10] van Jaarsveld J G, van Deventer J S, and Lukey G 2002 *J.Chem. Eng.* **89**(1) 63
[11] Duxson P, Lukey G C, and van Deventer J S J 2006 *J. Non-Crystal. Sol.*, **352** 5541
[12] Barbosa V F F, and MacKenzie K J D 2003 *Mater. Res. Bull.* **38**(2) 319
[13] Peigang H, Dechang J, Meirong W, and Yu Z 2011*Ceram. Int.* **37**(1) 59
[14] Peigang H, Dechang J, Meirong W, and Yu Z 2010*Ceram. Int.* **36** 2395
[15] Boaro M, Vohs J M, Gorte R J, and Boaro M 2003 *J. Am. Ceram. Soc.* **86**(3) 395
[16] Baklouti S, Bouaziz J, Chartier T, and Baumard J F 2001 *J. Eur. Ceram. Soc.* **21**(8) 1087
[17] Ramesh S, Gan C J, Bang LT, Niakan A, Tan C Y, Purbolaksono J, Hari Chandran, Ramesh S, Yap B K, and Teng W D 2015 *J. Ceram. Process. Res.* **16**(6) 683
[18] Lóh N J, Simão L, Faller C A, De Noni A, and Montedo O R K 2016 *Ceram. Int.* **42**(11)12556
[19] Mera G, Riedel R, Poli F, and Müller K 2009 *J. Eur. Ceram. Soc.* **29**(13) 2873
[20] Maglia F, Tredici I G, and Anselmi-Tamburini U 2013 *J. Eur. Ceram. Soc.* **33**(6)1045
[21] Lu P K, and Lannutti J J 2000 *J. Am. Ceram. Soc.* **83**(10) 2536

IOP Conf. Series: Materials Science and Engineering **267** (2017) 012014 doi:10.1088/1757-899X/267/1/012014

Study on The Geopolymer Concrete Properties Reinforced with Hooked Steel Fiber

M M A B Abdullah[1,2,a], M F M Tahir[1,b], M A F M A Tajudin[1,c] J J Ekaputri[3,1,d], R Bayuaji[3,1,e] and N A M Khatim[1,f]

[1] Center of Excellence Geopolymer and Green Technology (CEGeoGTech), School of Materials Engineering, Universiti Malaysia Perlis (UniMAP), Perlis, Malaysia.

[2] Faculty of Engineering Technology, Universiti Malaysia Perlis (UniMAP), Perlis, Malaysia.

[3]Civil Engineering Department, Faculty of Civil Engineering and Planning, Institut Teknologi Sepuluh Nopember (ITS), Surabaya, Indonesia

E-mail: mustafa_albakri@unimap.edu.my

Abstract. In this research, Class F fly ash and a mixture of alkaline activators and different amount of hooked steel fiber were used for preparing geopolymer concrete. In order to analyses the effect of hooked steel fiber on the geopolymer concrete, the analysis such as chemical composition of fly ash, workability of fresh geopolymer, water absorption, density, compressive strength of hardened geopolymer concrete have been carried out. Mixtures were prepared with fly ash to alkaline liquid ratio of 2.0 with hooked steel fibers were added to the mix with different amounts which are 1%, 3%, 5% and 7% by the weight of the concrete. Experimental results showed that the compressive strength of geopolymer concrete increases as the hooked steel fibers increases. The optimum compressive strength obtained was up to 87.83 MPa on the 14th day. The density of geopolymer concrete are in the range between 2466 kg/m^3 to 2501 kg/m^3. In addition, the workability value of geopolymer without hooked steel fibers is 100 mm while the workability value of geopolymer with hooked steel fibers are between 60 mm to 30 mm.

1. Introduction

Concrete is the most often used material in construction besides wood. Concrete which is a main building material is broadly used in the construction of infrastructures such as buildings, bridges, highways, dams, and others facilities. The increasing of the human population leads to the increasing the demand for the building constructions. One of the materials that is usually used as a binder in the manufacture of concrete is the ordinary Portland cement (OPC) [1].

At the same time, the amount of by-products such as fly ash produced by energy and mineral industries are increasing day by day that are currently disposed as a waste material. Fly ash is one of the coal combustion products which is composed that are driven out of the boiler with the flue gases. It is generally captured with scrubbers, electrostatic precipitators, or filter fabric baghouses in power plants, then sluiced to settling basins (wet) prior to disposal or stored in silos (dry) for disposal or sale. Fortunately, many fly ash are diverted for a good purpose in order to reduce its amount. One of the resolution to reduce the amount of fly ash and to yield more environmentally friendly concrete is to replace OPC with these by product materials. In this respect, a new type of inorganic cementless binder called geopolymer have been discovered as an OPC replacement. Generally, many researchers suggest that Class F fly ash is more suitable to be applied for geopolymer concrete [1-3]. The most significant composition in the fly ash with regards to geopolymer concrete are silicon (Si) and aluminium (Al). The major consequence on the fly ash chemical composition comes from the type of coal. Fly ash that have higher content of calcium oxide (CaO) is claimed to produce higher compressive strength due to the development of calcium-aluminate-hydrate and other calcium compounds particularly in the early ages [4].

Geopolymer is a new inorganic polymeric material that has undergone a marked development in the past years. The geopolymer term was first used by Davidovits in its research which represents the activation of sodium hydroxide produced alumina-silicate components which was found to be as an alternative binder to the OPC [5]. Geopolymer is formed from the reaction of a source material that is rich in silica and alumina with alkaline liquid [2]. Geopolymerisation involves a heterogeneous chemical reaction between alkali metal silicate solutions and solid aluminosilicate oxides at highly alkaline conditions and mild temperatures yielding amorphous to semi crystalline polymeric structures, which consist of Si–O–Si and Si–O–Al bonds [6]. If geopolymer is compared with OPC, geopolymers show many advantages. It generally consists of covalent bonds and has such advantages as being excellent in volume stability, light in weight, fire resistant, and low in density, all of which arouse much attention from researchers [7].

Besides that, the combination of sodium silicate and sodium hydroxide solution are suitable to be used for the activation of fly ash based geopolymer concrete [8]. Hardjito stated that the most frequent alkaline liquids used in geopolymerisation is a mixture of sodium hydroxide and sodium silicate solution. The addition of sodium silicate solution to the sodium hydroxide solution as the alkaline liquids will enhanced the reaction between base materials and the solution [4]. Alkaline concentration is a huge element for controlling the leaching of alumina and silica from fly ash particles, geopolymerisation process and mechanical properties of solidified geopolymer [9]. Previous researcher also affirmed that the presence of NaOH in the alkaline activators solution can make the reaction that takes place more briskly and the gel is less smooth [10]. The most common sodium hydroxide concentration used by many researchers is 12 M. Suriya proved that the suitable molarity used for mixing of NaOH is 12 M [11].

Fiber reinforced concrete (FRC) is a composite materials made with Portland cement, aggregates and incorporating discrete discontinuous fibers. Plain and unreinforced concrete is known as a brittle material with a low tensile strength and a low strain capacity [12]. The bond between the concrete and the reinforcing steel fibers is a significant mechanism for the performance of reinforced geopolymer concrete as a composite material [13]. Several experimental done by previous researchers pointed that the utilization of fibers especially steel fibers enhanced the impact resistance of the geopolymer concrete [14].

In order to improve the properties of geopolymer concrete such as durability, ductility, flexural strength and fracture toughness, the concept of using fibers as reinforcement in geopolymer concrete is introduced. Steel, glass and synthetic fibers such as polypropylene fibers were used as a reinforcement in concrete since 1960 and the research of fibers reinforced concrete continues until today [15]. In this research, geopolymer concrete was produced by Class F fly ash as the base material and reinforced with hooked steel fibers. A mixture of sodium silicate solution and sodium hydroxide solution were used to react with aluminium and silicon in the fly ash to form the paste that bound the aggregates and steel fibers in the mixture to yield the geopolymer concrete.

2. Experimental Method

2.1. Raw Materials

The fly ash used is attained from Manjung Power Plant, Lumut, Perak which is of low calcium (Class F) is used as the base material and equivalent to ASTM C618. The chemical composition of fly ash obtained from the analysis as tabulated in Table 1 showed that the calcium oxide (CaO) content is less than 20% and the most abundant oxides in this fly ash are silica oxide (SiO_2) as the main constituents with 38.80% followed by aluminium oxide (Al_2O_3) and iron (III) oxide (Fe_2O_3) with 14.70% and 19.48% respectively.

Table 1. XRF analysis data of fly ash composition

Component	SiO_2	Al_2O_3	Fe_2O_3	TiO_3	CaO	MgO	SO_3	K_2O	MnO	BaO	SrO
Percentage (%)	38.80	14.70	19.48	1.02	18.10	3.30	1.50	1.79	0.16	0.27	0.11

Based on these results, the fly ash used can be classified as Class F type according to the ASTM C618. It is mainly pozzolanic where its silicon dioxide content reacts with the calcium hydroxide from the cement hydration process to form the calcium silicate hydrate (CSH) gel and produced cementitious compounds which is suitable to be used in the geopolymer [16]. Besides that, this fly ash has the higher CaO content which can develop the higher compressive strength in the early ages. Other researcher also claimed that fly ash that have higher CaO content can increase the compressive strength of the concrete. It is also stated that the calcium content present in fly ash played a significance role in developing the compressive strength. The presence of calcium ions produced a quick reactivity. Thus, the geopolymer will yields quickly hardening in shorter curing time [17].

2.2. Mix Design and Process

In preparation of 12 M NaOH solution, 960 g of NaOH pellets are dissolved in distilled water in a volumetric flask for getting 12 M solution. The NaOH solution is left for settling down for 24 hours. After 24 hours, 250 g of NaOH solution and 100 g of Na_2SiO_3 solution are mixed. When both are mixed and stirred gradually, an exothermic reaction takes place and extreme amount of heat is evolved. The mixture is left for settling down about 1 hour. Hence, for safety factor, hand gloves are used. For geopolymer concrete samples, the fly ash and aggregates are dry mix. Then, alkaline activators is added to the dry mix and wet mixing is done for about 3 to 4 minutes. Lastly, hooked steel fibers is added to the wet mix in different weight such as 28 g, 84 g, 140 g and 196 g. Table 2 shows the details of mix proportions of geopolymer concrete reinforced with hooked steel fibers.

Table 2. Mix proportions of geopolymer concrete reinforced with hooked steel fibers

Ingredients	Fly ash	Fine aggregates	Coarse aggregates	Hooked steel fibers	NaOH	Na_2SiO_3
Quantity (g)	700	700	1050	28	250	100
				84		
				140		
				196		

2.3. Curing and Moulding Process

Fresh geopolymer with or without hooked steel fibers are poured into (100x100x100) mm size cubes and (100x100x500) mm size beams steel moulds. The samples are left for 24 hours in room temperature and then they are demoulded. After that, the samples are then left in the laboratory ambient conditions until the day of testing. After 14 days, weight of samples were taken to determine density, water absorption and tested in compressive strength testing machine.

IOP Conf. Series: Materials Science and Engineering **267** (2017) 012014 doi:10.1088/1757-899X/267/1/012014

3. Results and Discussion

3.1. Workability

Figure 1 shows the slump test result of fresh geopolymer mix with and without hooked steel fibers. For geopolymer mix without hooked steel fibers (GPC 0), the workability value is 100 mm. For geopolymer mix with 1% hooked steel fibers (GPC 1), the workability value is 60 mm. The workability value for geopolymer mix with 3% hooked steel fibers (GPC 3) is 49 mm. While the workability value for geopolymer mix with 5% hooked steel fibers (GPC 5) and geopolymer mix with 7% hooked steel fibers (GPC 7) are 39.5 mm and 30 mm respectively. The trend of workability in Figure 1 shows the workability values are gradually decreased as the increasing amount of hooked steel fibers from GPC 0 to GPC 7. This may be cause by the hindrance which is provided by the hooked steel fibers to the free flow. As stated by Sunilkumar, the slump value are between 65 mm and 100 mm for the fresh geopolymer concrete while for geopolymer concrete with steel fibers, the slump value are between 28.2 mm and 27.5 mm [3]. To be conclude, as the amount of hooked steel fibers increases, the workability will be decreases.

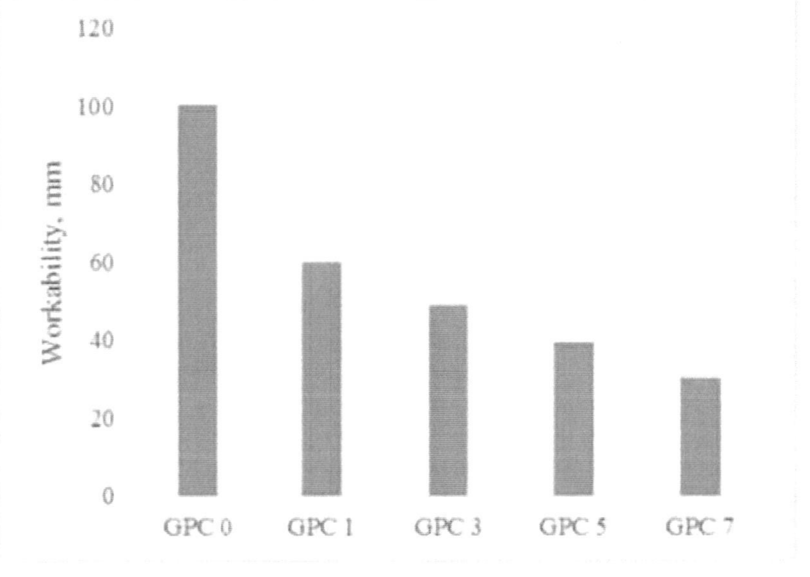

Figure 1. Workability graph for different mix of geopolymer concrete

3.2. Water Absorption

Water absorption of all mixtures are measured at the age of 14 days and the results are presented in Figure 2. According to the Figure 2, the percent of water absorption are reduced where the average absorption of GPC 0, GPC 1, GPC 3, GPC 5 and GPC 7 are 3.83%, 3.81%, 3.35%, 3.33% and 2.94% respectively. The water absorption for GPC 0 until GPC 3 declined slightly while for GPC 5 and GPC 7 decreases abruptly with the increasing of hooked steel fibers.

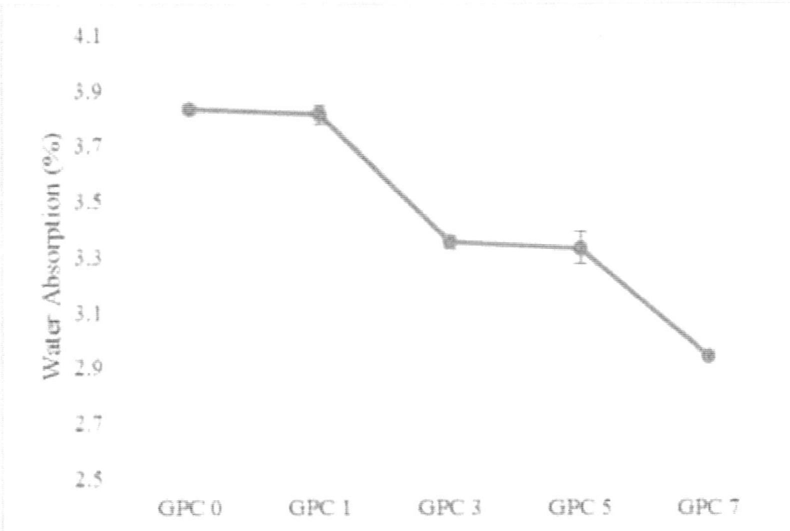

Figure 2. The relationship between water absorption and different mix of geopolymer concrete

This shows that the concrete become less permeability because the pore in the concrete have been closed by the addition of hooked steel fibers in geopolymer concrete. The geopolymer concrete with higher water absorption having the lowest compressive strength [17]. It is proved that GPC 0 have the maximum water absorption and having the minimum compressive strength. While GPC 7 have the minimum water absorption and having the maximum compressive strength.

3.3. Density

The density of geopolymer concrete after 7 days is presented in Figure 3. Density value for GPC 0 , GPC 1, GPC 3, GPC 5 and GPC 7 are 2466 kg/m^3, 2477 kg/m^3, 2487 kg/m^3, 2491 kg/m^3 and 2501 kg/m^3 respectively. The density of GPC 7 with 7% hooked steel fiber is higher than GPC 0 without hooked steel fiber. It is verified that as the loading of hooked steel fibers increased, the density also shows an increment.

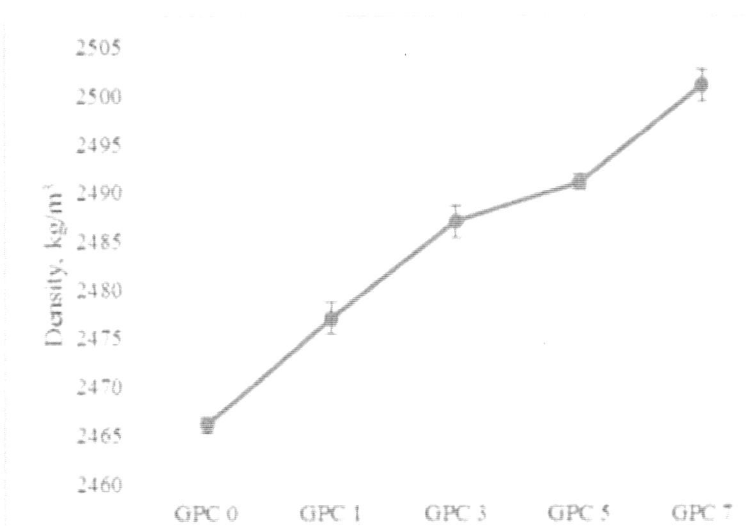

Figure 3. Density with different mix of geopolymer concrete

IOP Conf. Series: Materials Science and Engineering **267** (2017) 012014 doi:10.1088/1757-899X/267/1/012014

3.4. Compressive Strength

Figure 4 demonstrates the effect of addition of hooked steel fibers on the compressive strength with the increasing of age of concrete. From Figure 4, it can be observed that at the age of 7 days, the trend show that the strength is slightly increase from GPC 0 until GPC 7. The compressive strength of GPC 0, GPC 1, GPC 3, GPC 5 and GPC 7 for 14 days are found to be 12.45 MPa, 18.25 MPa, 20.87 MPa, 22.35 MPa and 26.55 MPa respectively. The increase in compressive strength are about 46.6%, 67.6%, 79.5% and 113.3% for GPC 1, GPC 3, GPC 5 and GPC 7 respectively with respect to GPC 0 concrete. While at the age of 14 days, the trend show that the strength is increase rapidly from GPC 0 until GPC 7. The compressive strength of GPC 0, GPC 1, GPC 3, GPC 5 and GPC 7 for 14 days are found to be 51.30 MPa, 59.47 MPa, 63.17 MPa, 70.23 MPa and 87.83 MPa respectively. The increase in compressive strength are about 15.9%, 23.1%, 36.9% and 71.2% for GPC 1, GPC 3, GPC 5 and GPC 7 respectively with respect to GPC 0 concrete.

In terms of inclusion of hooked steel fibers, the compressive strength also increased. It is proved that the compressive strength of GPC 1, GPC 3, GPC 5 and GPC 7 are increased enormously than GPC as shown in Figure 4. The optimum strength is obtained in GPC 7 which have the higher amount of hooked steel fibers. In addition, compressive strength can also increase as the age of the concrete increases. The compressive strength for 14 days showed a big increase in strength compared to 7 days. As the age of concrete increases, the compressive strength will also increases for all concrete. The longer the age of concrete improves the polymerisation process producing higher compressive strength of concrete [18]. Other researchers also claimed that geopolymer concrete with steel fibers will reach its maximum strength in less than 20 days.

There is an explicit relationship between compressive strength and density of geopolymer concrete with hooked steel fibers. For each geopolymer mix, it shows the increment of density as the compressive strength increase. Steel fibers are firstly used to control and prevent the drying shrinkage while in the concrete [19]. So, it is proved that as the density increase, the geopolymer concrete with hooked steel fibers become more durable compared to geopolymer without hooked steel fibers.

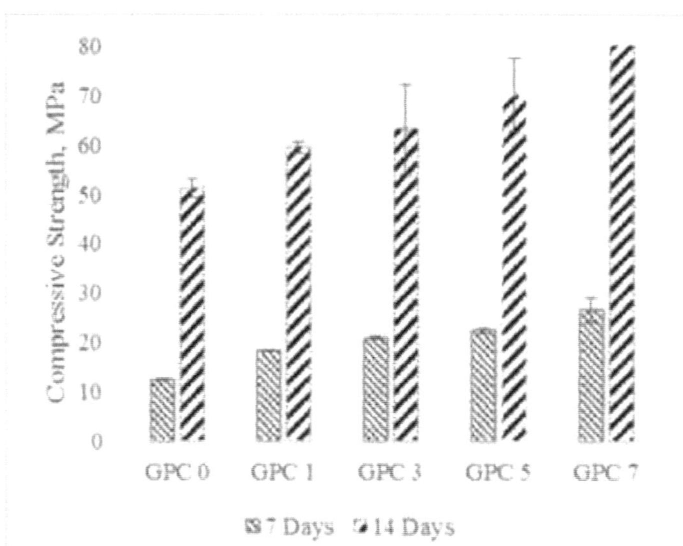

Figure 4. Effect of inclusion of hooked steel fibers on compressive strength

4. Conclusion

The purpose of this paper is to investigate the geopolymer concrete properties reinforced with hooked steel fibers. The compressive strength of the geopolymer concrete is highly influenced by the amount of CaO content in the fly ash. Besides that, the compressive strength also increases as the amount of the hooked steel fibers increases. The highest compressive strength is found to be 87.83 MPa at 14 days. Geopolymer concrete with 7% of hooked steel fibers give a higher grip strength to the concrete. Besides that, the highest density is found to be 2501 kg/m^3 at 14 days for GPC 7. While for workability and water absorption, GPC 7 have the lowest value which is 30 mm and 2.94% respectively. This can conclude that, as the amount of hooked steel fibers increases, the workability and water absorp tion will be decreased and the density will increased.

5. Acknowledgements

The author would like to extent our appreciation to the Center of Excellence Geopolymer & Green Technology (CEGeoGTech) for funding and ensuring the research project successfully completed.

6. References

[1] Izzat, A.M., Al Bakri, A.M.M., Kamarudin, H., Moga, L.M., Ruzaidi, G.C.M., Faheem, M.T.M., Sandu, A.V. 2013. *Materiale Plastice,* Volume **50**, Issue 3, September 2013, Pages 171-174.
[2] I. Arya A. & Matthews M.P. 2014. *International Journal of Engineering Research & Technology,* Vol. **3**, pp. 825-829.
[3] Sunilkumar S.P. & Amol A.P. 2015. *International Journal of Research in Advent Technology,* Vol. **3**, pp. 70-74.
[4] Hardjito D. & Rangan B.V. 2005. *Research Report*, pp. 1-103.
[5] Sivakumar A. & Srinivasan K. 2014. *International Journal of Civil, Environmental, Structural, Construction and Architectural Engineering,* Vol. **8**, pp. 1275-1278.
[6] Hardjito D., Wallah S.E., Sumajouw D.M.J. & Rangan B.V. 2004. *Invited Paper*, pp. 1-10.
[7] Haoyang S., Jinyu X. & Weibo R. 2016. *Ceramics International,* Vol. **42**, pp. 3888–3898.
[8] Subhash V.P., Yuwaraj M.G. & Sanjay S.J. 2015. *Advances in Structural Engineering,* pp. 1619-1634.
[9] Ramkumar G. Sundarkumar S. & Sivakumar A. 2015. *International Journal of Advance Research in Science and Engineering*, Vol. **4**, pp. 1717-1725.
[10] Duchesne J., Duong L., Bostrom T. & Frost R. 2010. *Waste and Biomass Valorization*, Vol. **1**, pp. 367-377.
[11] Suriya A.P. & Senthil G.K. 2015. *International Journal of Engineering Trends and Technology,* Vol. **21**, pp. 396-399.
[12] Nguyen V.C. 2004. *Seminar Material* Ho Chi Minh: Faculty of Civil Engineering, Ho Chi Minh City University of Technology. pp. 108-116.
[13] Prabir K.S. 2011. *Journal of Materials and Structures*, Vol. **44**, pp. 1021-1030.
[14] Murali G., Santhi A.S. & Ganesh G.M. 2014. *Journal of Applied Science and Engineering*, Vol. **17**, pp. 259-266.
[15] Vijai K., Kumutha R. & Vishnuram B.G. 2010. *International Journal of the Physical Sciences,* Vol. **5**, pp. 1419-1423.
[16] Atteshamuddin S.S. & Subhash V.P. 2013. *Indian Journal of Materials Science*, Vol. **2**, pp. 1-8.
[17] Mustafa Al Bakri, A.M., Kamarudin, H., Bnhussain, M., Nizar, K., Rafiza, A.R., Zarina, Y. 2012. *Reviews on Advanced Materials Science*, **30** (1), pp. 90-97
[18] Vijai K., Kumutha R., & Vishnuram B.G. 2011. *Asian Journal of Civil Engineering*, pp. 377-385.
[19] Mustaqqim A.R., Zuhayr M.G., Azizi A.M., Fazdliel A.I., Norlia M.I. & Izzati M.N. 2015. *Applied Mechanics and Materials*, Vol. **754-755**, pp. 373-377.

IOP Conf. Series: Materials Science and Engineering **267** (2017) 012015 doi:10.1088/1757-899X/267/1/012015

The High Teperature Influence on Geopolymer Fly Ash Mixture's Compressisive Strength with Insudtrial Waste Material Substitution.

R Bayuaji[1], B Wibowo[1], S Subekti[1], S E Santoso[1], E Hardiyanto[1], Y Kaelani[1], L L Mallu[2]
Department of Civil Infrastructure Engineering, Institut Teknologi Sepuluh Nopember, Indonesia (1)
Faculty of Civil Engineering and Planning, Institut Teknologi Adhi Tama Surabaya (2)
Email: bayuaji@ce.its.ac.id

Abstract
This research aimed to figure out the influence of fly ash mixture from the industrial waste at the temperatures of 150°C, 450°C, 750°C viewed from the strength and resistance of geopolymer paste. As a result, cement will be substituted by industrial waste like fly ash. This experimental research was conducted on the mix design of geopolymer concrete which was made by dimension with 2.5 cm in diameter and 5 cm in height from four mixture composition of fly ash and industrial waste i.e. 100% fly ash, 50% fly ash+50% bottom ash, 50% fly ash+50% sandblast, and 50% fly ash+50% carbide waste. Each mixture was tested in terms of porosity and compressive strength.
In conclusion, in the mixture of 50% fly ash+50% Sandblast and 50% fly ash+50% bottom ash in 12 molars, 1.5 activator comparison can be used to substitute fly ash at high temperature. Meanwhile, the mixture of 50% fly ash+50% carbide waste in 8 molars, 0.5 activator comparison has very small strength remaining if it is compared to the mixture of fly ash and other industrial waste (Bottom ash and Sandblast). The performance of mixture paste of 50% fly ash+50% carbide waste was very vulnerable after being burnt. Consequently, it cannot be used as the main structure at high temperature.

Keywords: geopolymer, fly ash, industrial waste, high temperature.

1. Introduction

Fire can strike at any type of buildings, ranging from home residences to high rise buildings. Considering fire strike based on the building type, home residence was the number one to have a high risk of fire strike with the total of 65,6% occurrences followed by 13,8% of industrial building and 20,7% of general building respectively [1]. The building's high temperature from the fire strike could decrease the concrete's compressive strength [2]. This situation could lead into various different structural damage, with the worst case of concrete cracking [3] which make it visible to see its steel confinement. In addition, the catastrophe damage caused by the fire strike could also showed the peeling effect in which we could see the segregation of cement paste and aggregate, making the concrete lost its cohesiveness with binders and aggregates [4]. During this time, cement paste was compressed while the aggregate will swell, causing crack on concrete which can lower the concrete quality. Another reason to conduct this study was to eliminate the use of Portland cement. Recently, there is a dispute among the environment experts about the use of Portland cement as construction material. The environment experts showed a serious concern on the carbon foot print produced during the Portland cement production which could worsen the climate change. During the cement production, the kiln need to reach 1400°C in order to produce 1-ton cement, during the process it emit 0.55 ton of CO_2[6]. This fact reminded us to continue to find other alternative for Portland cement considering its danger to the nature. One of the alternative could be found by making geopolymer.[7-12]. Geopolymer concrete is a modern concrete which used natural binder rich in silica and alumina [13]. The silica and alumina content need to be dissolved with alkali solution in order to produce chemical reaction which later on used as the binder for the aggregate. In addition, there are other materials from coal burning waste material to use for geopolymer concrete called fly ash. There are also numerous of other resource of waste material to substitute the use of Portland cement in geopolymer concrete. Fly ash is a coal burning waste material. Each fly ash has its own distinctive content and varied in each place. This might happen due to the different resource and process used during the production. Therefore, it is hard to generalize all fly ash quality. This study will explain more on the influence of geopolymer pasta made from industrial waste in reducing the use of burned fly ash. Therefore, a new alternative will be found to substitute the use of Portland cement from industrial waste like fly ash. The fly ash was aimed to substitute the use of Portland cement in conventional concrete.

2. Research Method

The research method used in this study was based on the laboratory experimental. The source material of fly ash and bottom ash was collected from steamed power plant (PLTU) Paiton, Probolinggo. While the sandbasting (silica fume) waste was collected from PT. Swadaya Graha, Gresik. The carbide waste materials were collected from PT. Z.

During the research, SEM-EDX, XRD (X-Ray Diffraction) and XRF (X-Ray Fluorescence) were used to determine the chemical composition of each material. While the alkali solution activator used in this study was natrium silicate (Na_2SiO_3) and natrium hydroxide (NaOH) with molarity of 12M for 100% fly ash mixture, 50% fly ash+ 50% bottom ash, and 50%fly ash + 50% sandblast. In addition, we also use 8M for the mixture of 50% fly ash + 50% carbide waste.

The sample used in this study was made from a cylinder with diameter of 2,5 cm x 5 cm. The curing sample age was determined as 28 days and 56 days which was burned at the temperature of 150°C, 450°C and 750°C for 2 hours. The compressive strength test for the cylinder was done at ITS structure lab Surabaya.

Table 1. The Composition of Materials And Sample Code

Code	Notes	Composition	
		74% of Materials	26% of Activator
FA 12-1.5	100% *fly ash*	74%	26%
BF 12-1.5	50% *fly ash*+50% *bottom ash*	50% dari 74%	
SF 12-1.5	50% *fly ash*+50% *sandblast*	50% dari 74%	
LKF 8-0.5	50% *fly ash*+50% *limbah karbit*	50% dari 74%	

3. Result and Discussion

3.1 Result and Discussion

Some of the material test results can be seen in Figure 1,2,3, and 4.

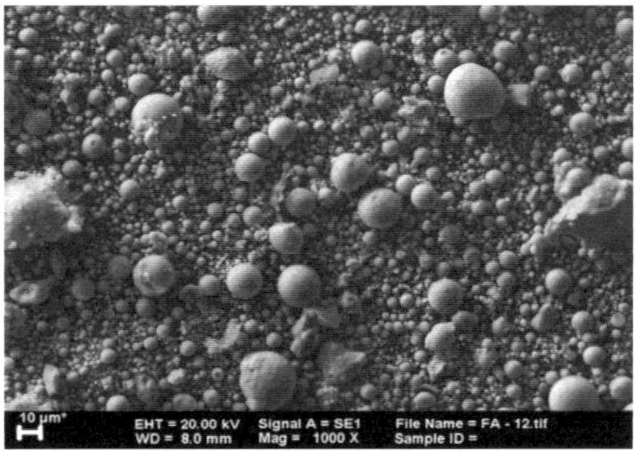

Figure 1. Fly Ash 1000x Zoom

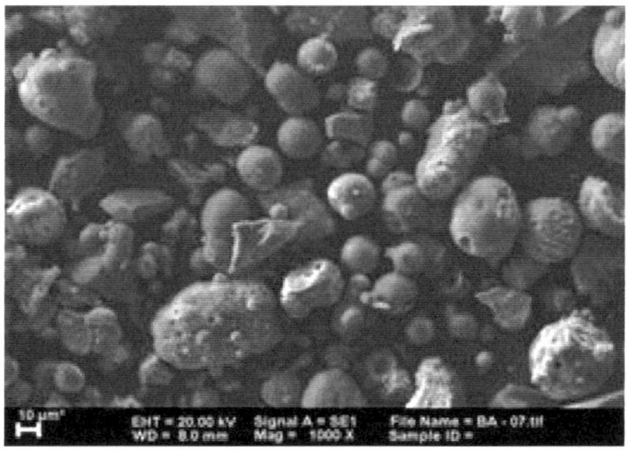

Figure 2. Bottom Ash 1000x Zoom

Figure 3. Sandblasting Waste Material 1000x Zoom

Figure 4. Carbide Waste 1000x Zoom

If we look at the content test from Scanning Electron Microscopy (SEM) above, we could see that fly ash tends to have a round and solid shape. While bottom ash showed that it has more oval, bigger shape and has more pores compare to fly ash. This fact has made bottom ash has less compressive strength compare to fly ash and sandblast even though bottom ash has more SiO_2 content. Sandblast material has a polygonal and solid shape. While carbide waste material has a fragile-look and abstract shape. Every particle shape on each material will affect the cohesiveness and influence the inter-locking force.

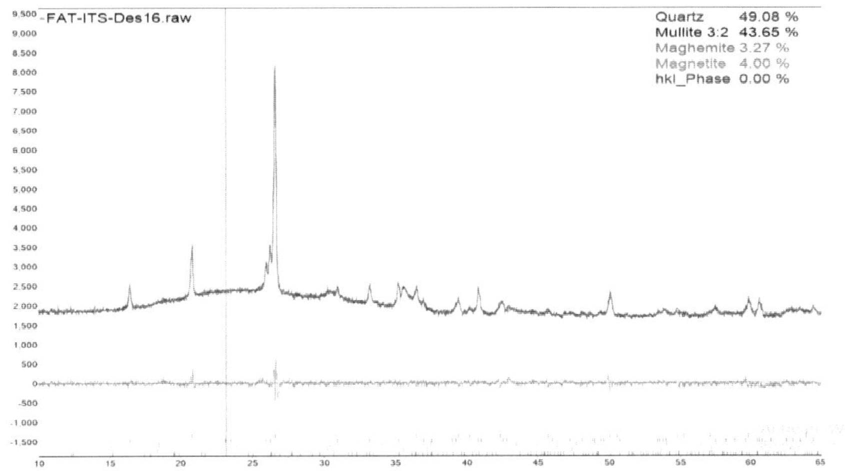

Figure 5. XRD Test Result-Fly Ash

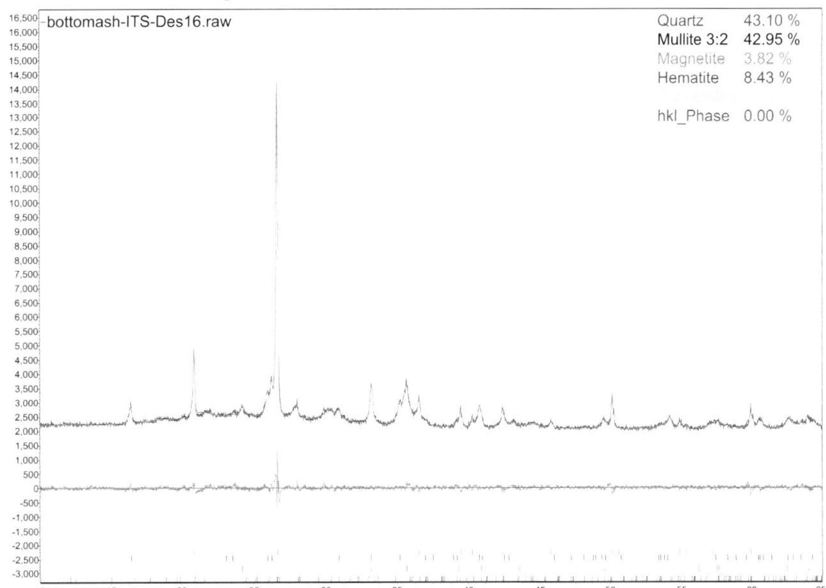

Figure 6. XRD Test Result -Bottom Ash

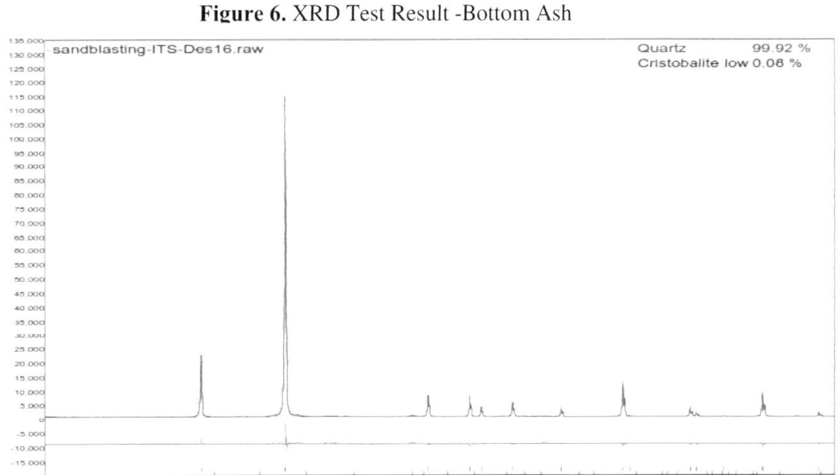

Figure 7. XRD Test Result – Sandblasting Waste Materials

Table 2. XRF Material Test Result

NO	Chemical Oxide	Fly Ash	Carbide Waste	Bottom Ash	Sandblasting
1	Al_2O_3	24.25	2.01	2.93	22.01
2	SiO_2	47.10	3.91	89.91	49.53
3	Fe_2O_3	16.07	0.87	4.71	18.80
4	CaO	5.83	91.48	0.73	4.56
5	MgO	2.62	0.23	0.16	1.98
6	K2O	1.64	0.07	0.68	1.12
7	TiO2	1.16	0.14	0.28	1.15
8	Na2O	0.65	0.72	0.08	0.39
9	P2O5	0.19	0.02	0.02	0.13
10	SO3	0.21	0.27	0.03	0.07
11	Others	0.29	0.27	0.46	0.26
Total		100.00	100.00	100.00	100.00

According to XRF test, it is noted that the fly ash used in this study has more than 70% content of Si + Al + Fe and less that 10% content of CaO (5,83%). Therefore, the used fly ash was considered as F class fly ash based on ASTM C 618-84. Considering the chemical content on each material (Table 4.7), fly ash and sandblast has SiO_2 content twice more as Al_2O_3. While bottom ash has more SiO2 content than Al_2O_3.Carbide waste material has the most CaO among other with the total percentage of 91,48 %. In addition, Carbide waste material also has the least compressive strength among other materials. These facts have supported the theory of geopolymer bond whereby Al and Si combined with NaOH as activator to form a strong geopolymer bond
.

4. Analysis and Result of Compressive Strength

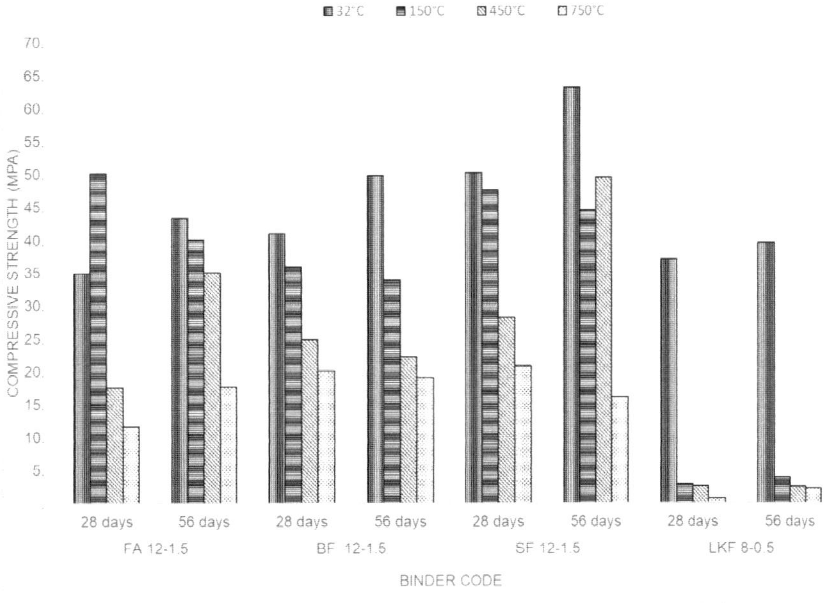

Figure 8. Result of Compressive Strength

According to Figure 8, the highest compressive strength gained at the 100% fly ash composition, reaching its strength at 50,11 MPa under the heating temperature of 150°C and followed by 28 days of setting time. While the least compressive strength gained at the combined carbide waste material with fly ash composition, reaching its strength at 0.75 MPa under the heating temperature of 750°C and followed by 28 days of setting time. The heating temperature observation at 150°C and 28 days of setting time showed a better compressive test result that those for 56 days for all geopolymer pasta composition except the composition of 100% Fly Ash + 100% carbide waste material. While the heating temperature observation at 450°C showed that only 100% Fly Ash + 100% Bottom Ash and 100% Fly Ash + 100% carbide waste material compositions have more compressive strength at 28 days of setting time than those in 56 days. The heating temperature observation at 750°C showed that only 100% Fly ash + 100% Sandblast and 100% Fly ash + 100% Bottom Ash compositions have a good compressive strength at 28 days of setting time

5. Conclusion

Based on the conducted research, it could be concluded that:

1. The higher the temperature set, the less compressive strength gained and the bigger porosity for geopolymer pasta. Unless the composition changed into 50% Fly ash + 50% Sandblast with 56 days of setting time heated at the temperature of 450°C. this composition will resulted into a higher compressive strength than 150°C heating temperature.

2. The highest compressive strength was obtained at the composition of:
    - 50% fly ash + 50% sandblast, heated at the temperature of 150°C. resulted in 47.57 MPa compressive strength.
    - 50% fly ash + 50% sandblast heated at the temperature of 450°C resulted in 49.50 MPa compressive strength.
    - 50% fly ash + 50% sandblast, heated at the temperature of 750°C resulted in 20.81 MPa compressive strength.
3. Sandblast and fly ash materials have twice more SiO_2 content as Al_2O_3 and has a good compressive strength at high temperature. These facts have supported the theory of geopolymer bond whereby Al and Si combined with NaOH as activator to form a strong geopolymer bond
4. Based on SEM test, it is noted that fly ash particle has smaller rounded and more solid shape compare to bottom ash. While sand blast seems to have a polygonal and solid shape. These facts will determine each material's compressive strength.

5. The advantage and disadvantage of each composition based on the compressive strength and porosity are:
    - 50% Fly ash + 50 % Sandblast composition produce the best compressive strength and the least porosity at each heating temperature. Yet, it has inconsistency of compressive strength towards the vary heating temperature.
    - 50% Fly ash + 50 % Bottom ash has a consistency fall in compressive strength due to the fall in heating temperature. In addition, it also has a fair compressive strength compared to the other sandblast substitution material.
    - 50% Fly ash + 50 % carbide waste material have no advantage compare to other materials substitutes due to its bad compressive strength consistency and porosity.
6. Sandblast and bottom ash materials could be used to substitute fly ash for geopolymer pasta which resisted in high temperature.

6. Reference

1. Surabaya, D.K., *Kejadian Kebakaran Tahun 2016. Surabaya.* 2016.
2. Kodur, V. and A. Agrawal, *Effect of temperature induced bond degradation on fire response of reinforced concrete beams.* Engineering Structures, 2017. **142**: p. 98-109.
3. Abid, M., et al., *High temperature and residual properties of reactive powder concrete–A review.* Construction and Building Materials, 2017. **147**: p. 339-351.
4. Song, T.-Y., et al., *Bond Behavior of Concrete-Filled Steel Tubes at Elevated Temperatures.* Journal of Structural Engineering, 2017. **143**(11): p. 04017147.
5. Li, L., et al., *Effects of cement dosage and cooling regimes on the compressive strength of concrete after post-fire-curing from 800° C.* Construction and Building Materials, 2017. **142**: p. 208-220.
6. Davidovits, J., *Environmentally Driven Geopolymer Cement Applications.*, in *Geopolymer 2002 Conference.* . 2002: Melbourne, Australia,.
7. Ridho Bayuaji, M.S.D., Boedi Wibowo, Nur Ahmad Husin, Srie Subekti, Januarti Jaya Ekaputri, *The Influence of Chloride Environment on Compressive Strength*

of Geopolymer Concrete with Fly Ash Using Taguchi Approach. Applied Mechanics and Materials, 2015. **754-755**: p. 400-405
8. MS Darmawan, et al., *The Effect of Chloride Environment on Mechanical Properties Geopolymer Binder with Fly Ash.* Key Engineering Materials, 2014. **594**: p. 648-655.
9. Bayuaji, R., et al. *Mechanical Properties of MIRHA-Fly Ash Geopolymer Concrete*. in *Materials Science Forum*. 2015. Trans Tech Publ.
10. Bayuaji, R., et al. *The Effect of High Temperature on Compression Strength of Geopolymer Paste*. in *MATEC Web of Conferences*. 2017. EDP Sciences.
11. Bayuaji, R., et al. *The Effect of Corrosive Environment on Geopolymer Concrete Tensile Strength*. in *MATEC Web of Conferences*. 2017. EDP Sciences.
12. Bayuaji, R., et al., *Utilization of High Calcium Content Fly Ash: Flexural Strength of Geopolymer Concrete Beams in Sea Water Environment.* The Open Civil Engineering Journal, 2016. **10**(1).
13. Davidovits, J., *Global Warming Impact on the Cement and Agregate Industries.* Word Resource Review, 1994. **Vol.6, No.2**: p. 263-278.

Compressive Strength and Modulus of Elasticity of Concrete with Cubed Waste Tire Rubbers as Coarse Aggregates

Y Haryanto[1], N I S Hermanto[1], G Pamudji[1] and K P Wardana[1]
[1]Department of Civil Engineering, Faculty of Engineering, Jenderal Soedirman University, Jl. Mayjend Sungkono Km 5, Purbalingga, Indonesia

E-mail: yanuar_haryanto@yahoo.com

Abstract. One feasible solution to overcome the issue of tire disposal waste is the use of waste tire rubber to replace aggregate in concrete. We have conducted an experimental investigation on the effect of rubber tire waste aggregate in cuboid form on the compressive strength and modulus of elasticity of concrete. The test was performed on 72 cylindrical specimens with the height of 300 mm and diameter of 150 mm. We found that the workability of concrete with waste tire rubber aggregate has increased. The concrete density with waste tire rubber aggregate was decreased, and so was the compressive strength. The decrease of compressive strength is up to 64.34%. If the content of waste tire rubber aggregate is more than 40%, then the resulting concrete cannot be categorized as structural concrete. The modulus of elasticity decreased to 59.77%. The theoretical equation developed to determine the modulus of elasticity of concrete with rubber tire waste aggregate has an accuracy of 84.27%.

1. Introduction

One industry which has the potential for rapid development in Indonesia is the tire industry. The Indonesian tire industry experienced a 5% increase of tire sales, from 96.57 million units in 2012 to 101.43 million in 2013. This figure comprises car tire sales of 47.26 million units, and motorcycle tire sales of 54.16 million units [1]. This number is predicted to increase continuously and rapidly [2], while only a small portion of waste tire is regularly discarded. Meanwhile, more than 50% of tire waste is disposed of without special treatment [3]. Tire waste disposal has become a major environmental issue in major cities around the world. The easiest and cheapest way to dispose the waste is by burning, but the pollution generated makes it unacceptable and it is prohibited by law in some countries. It is estimated that approximately one billion tires are withdrawn from global use annually [4-6].

One feasible solution to overcome those problems is the use of waste tire rubber to replace aggregate in concrete. Several researchers who have investigated the use of waste tire rubber as an aggregate replacement in concrete reported that it reduces compressive strength [7-9] due to poor adhesion of the waste rubber tire aggregate to cement paste [3]. Treatment with a different method to increase the adhesion of the tire rubber waste aggregate has also been studied [10, 11]. In another study, it was suggested that concrete with waste tire rubber aggregate of a certain size may reach a compressive strength of 20 MPa at a water–cement ratio of 0.35 [12] and it also showed improved

[1] To whom any correspondence should be addressed.

post-cracking behaviour, although its bond strength didn't differ significantly from normal concrete in pull-out test results [13].

As a heterogeneous multiphase material, concrete is a mixture of cement paste and fine and coarse aggregates, with a range of sizes and shapes. With regard to its mechanical behaviour, concrete is often considered to be a three-phase composite structure, consisting of aggregate particles, the cement paste matrix in which they are dispersed, and the interfacial transition zone (ITZ) around the aggregate particles and cement paste [14]. In some applications of concrete, it is demanded that concrete should have low unit weight, high strength, high toughness and high impact resistance. Although concrete is the most commonly used construction material, it does not always fulfil these requirements [15]. One of the ways to improve these properties might be the addition of the rubber into concrete as an aggregate. The objectives of this work are to complement previous studies by discussing the effect of waste tire rubber aggregate in cuboid form on the compressive strength and concrete elasticity modulus.

2. Experimental Work

The materials used for the concrete mixture were cement type 1, fine aggregate such as sand from Mount Merapi in Yogyakarta, and coarse aggregate from local stone crushers in Purbalingga. Fine aggregate should have the specific gravity of 2.62 with a fine modulus of 3.07. The coarse aggregate had a specific gravity of 2.69 with a 20 mm maximum size of aggregate. The waste tire rubber as a replacement for coarse aggregates was made in cuboid form (Figure 1), and had a specific gravity of 1.00 with a size variation of 5 mm, 10 mm, 15 mm and 20 mm. The substitution percentage in each specimen varied between 0%, 20%, 40%, 60%, 80% and 100%. The composition of the concrete mixture was named according to the content of waste tire rubber aggregate, as presented in Table 1.

Figure 1. Waste tire rubber aggregate.

Table 1. Composition of concrete mixture.

Code	Aggregate variation	Material Requirement				
		Cement (kg)	Water (kg)	Gravel (kg)	Sand (kg)	Rubber (kg)
A	0%R, 100%C	405	210	1059	706	0
B	20%R, 80%C	405	210	847	706	79
C	40%R, 60%C	405	210	635	706	159
D	60%R, 40%C	405	210	423	706	238
E	80%R, 20%C	405	210	211	706	318
F	100%R, 0%C	405	210	0	706	397

%R: Rubber content, %C: Course aggregate content

To check the concrete-mixture workability, a slump test was used as shown in Figure 2. The slump test was conducted according to SNI 1972: 2008 [16]. The compressive strength test and modulus of elasticity test were conducted at 28 days on 72 cylindrical test specimens with the height of 300 mm and diameter of 150 mm in accordance with SNI 1974: 2011 and SNI 03-4169-1996 [17, 18]. The concrete compressive strength was determined with an experiment using (1), while the modulus of elasticity of concrete was determined with an experiment and empirical test [19] based on (2) and (3).

$$f_c^{'} = \frac{P}{A} \tag{1}$$

$$E = \frac{S_2 - S_1}{\varepsilon_s - 0.00005} \tag{2}$$

$$E = W_c^{1.5}\ x\ 0.043\sqrt{f_c^{'}} \tag{3}$$

in which:

$f_c^{'}$ = Compressive strength (MPa)
P = Maximum load (N)
A = Area of specimen (mm^2)
E = Modulus of elasticity (MPa)
S_1 = Stress when the load reaches 40% of the maximum load (MPa)
S_2 = Stress when the contraction reaches 0.00005
ε_1 = Strain corresponding to the stress S_1 (MPa)
W_c = Unit weight of concrete (kg/m^3)

Figure 2. Slump test.

3. Results and Discussion

3.1. Workability

The workability property of the fresh concrete produced from the slump test is shown in Figure 3. Waste tire rubber aggregate has a poor interlocking action with cement paste [20] and causes the concrete mix to become more diluted, hence the slump value for each content of waste tire rubber aggregate increased. The slump values for all the concrete mixtures ranged between 125–151 mm.

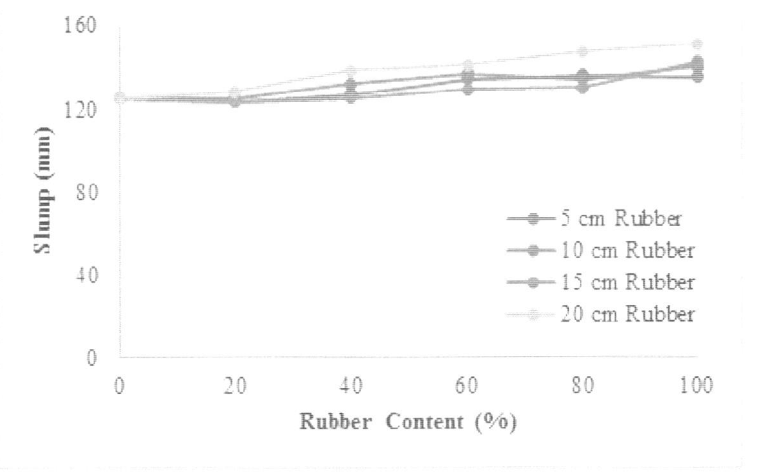

Figure 3. Workability.

3.2. Density
Figure 4 shows the average density at 28 days of the specimens that were prepared for the compressive strength test and modulus of elasticity. As shown in the below figure, the density decreased with the addition of tire rubber waste aggregate. The decrease in density was caused by the low specific gravity of waste tire rubber aggregate [9]. A decrease in density can be expected in the utilization of concrete with waste tire rubber aggregate for architectural applications such as interior construction [21].

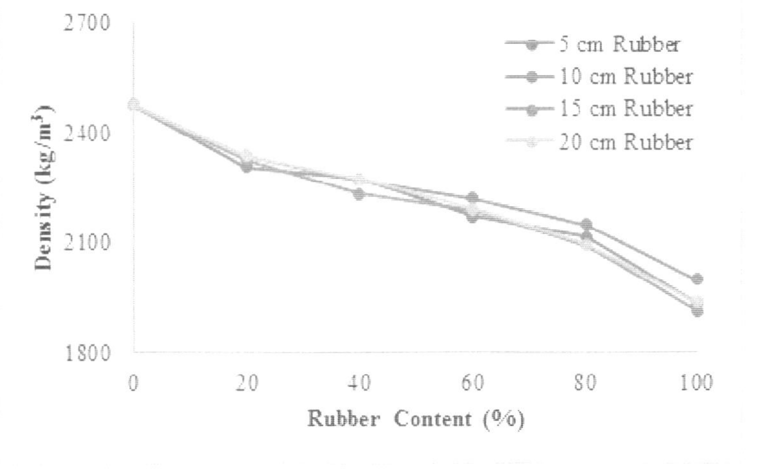

Figure 4. Density.

3.3. Compressive Strength
The increased content of waste tire rubber aggregate caused the compressive strength to decrease as shown in Figure 5. The rubber particles were much softer than the cement matrix around the rubber, so the stress was centered on the part that led to cracks at the rubber and cement interface at the time of loading [10]. The cracks that occurred subsequently spread and caused a loss of strength and eventual failure. Furthermore, the use of softer waste tire rubber aggregate led to decrease strength because the concrete compressive strength was strongly influenced by the aggregate properties. If the content of

waste tire rubber aggregate is more than 40%, then the resulting concrete cannot be categorized as structural concrete. The biggest decrease of compressive strength occurred in the concrete with 100% rubber, with the size of 20 mm, i.e. equal to 64.34%.

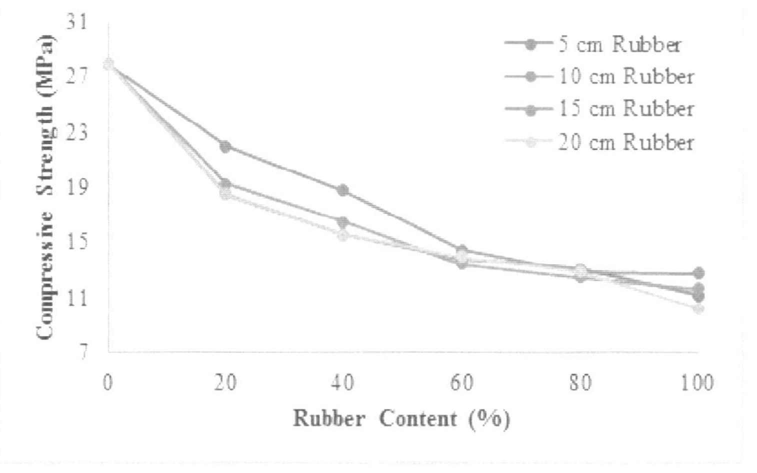

Figure 5. Compressive strength.

3.4. Modulus of Elasticity

Because concrete with waste tire rubber aggregate has a lower compressive strength and there is a relation between the compressive strength and modulus of elasticity, it can be predicted that the concrete will also have a lower modulus of elasticity as shown in Figure 6. The below behaviour can be explained in connection with the lower modulus of elasticity of rubber [22]. The largest decline of modulus of elasticity occurred in concrete with 100% rubber in the form of 5 cm cuboids, i.e. equal to 59.77%. The theoretical modulus of elasticity based on (3) was higher than the modulus of elasticity of the experimental result. The development of (3) resulted in (4) with an accuracy of 84.27% as shown in Figure 7.

$$E = W_c^{1.5} \; x \; 0.0349\sqrt{f_c^{'}} + 1144 \tag{4}$$

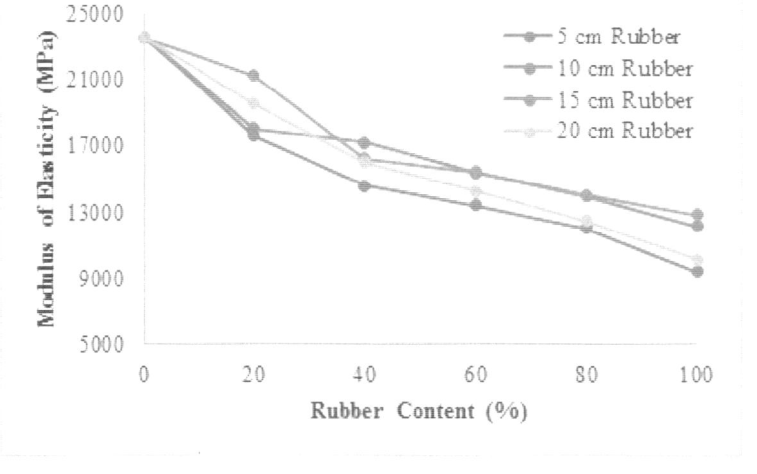

Figure 6. Modulus of elasticity.

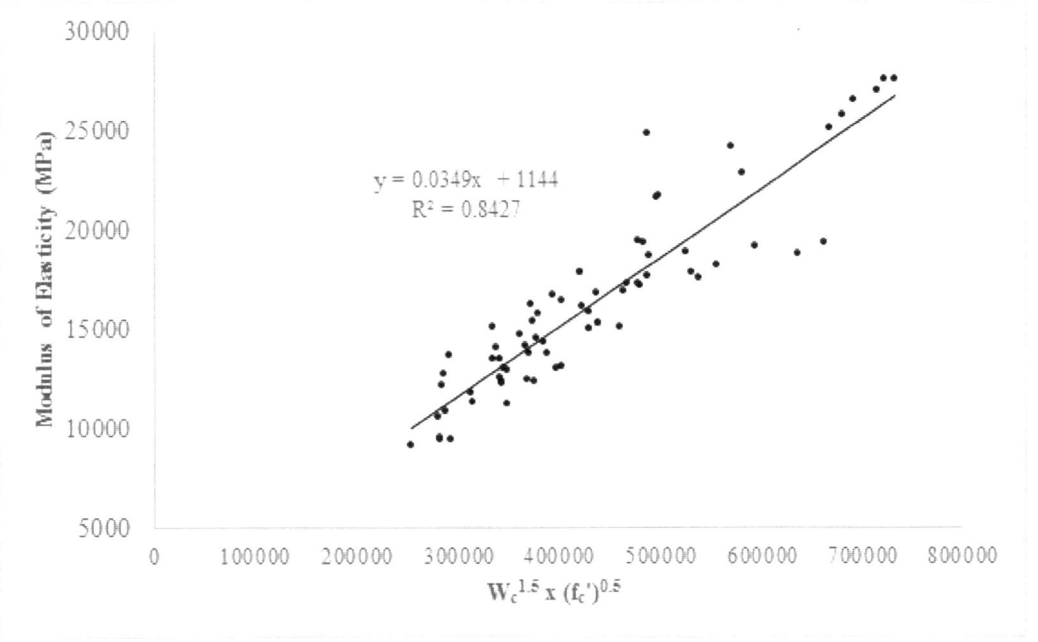

Figure 7. Development of a new theoretical modulus of elasticity equation.

4. Conclusion

Based on the above results, we concluded that:

a. Workability increased with a range of slump values between 125 mm and 151 mm. On the other hand, the density decreased due to the low specific gravity of waste tire rubber aggregate.
b. The compressive strength decreased to 64.34%. If the content of waste rubber tire aggregate is more than 40%, the resulting concrete cannot be categorized as structural concrete.
c. The elasticity modulus decreased to 59.77%. The theoretical equation developed to determine the concrete elasticity modulus with waste rubber tire aggregate had an accuracy of 84.27%.

5. References

[1] Zefanya N A, Noer B A and Hakim M S 2016 *J. Tek. ITS.* **5** A568
[2] Maryoto A, Hermanto N I S, Haryanto Y, Waluyo S and Anisa N A 2015 *Pro. Eng.* **125** 638
[3] Assas M M 2016 *J. Eng. Sci. and Tech.* **11** 899
[4] Shen W, Shan L, Zhang T, Ma H, Cai Z and Shi H 2013 *Const. and Build. Mater.* **38** 667
[5] Pacheco-Torgal F, Ding Y and Jalali S 2012 *Const. and Build. Mater.* **30** 714
[6] Alamo-Nole L A, Perales-Perez O and Roman-Velazquez F R 2011 *J. Hazard. Mater.* **185** 107
[7] Al Bakri A M M, Fadli S A S N, Abu Bakar M D and Leong K W 2007 *Proc. 1st Int. Conf. Sust. Mater. (Penang)*
[8] Bdour A N and Al-Khalayleh Y A 2010 *Jordan J. Civ. Eng.* **4** 55
[9] Mavroulidou M and Figueiredo J 2010 *Glo. Nest J.* **12** 359
[10] Wang H Y, Yung L C and Hua L H 2013 *Const. and Build. Mater.* **41** 665
[11] Issa C A and Salem G 2013 *Const. and Build. Mater.* **42** 48
[12] Nacif G L, Panzera T H, Strecker K, Christoforo A L and Paine K 2013 *Mater. Res.* **16** 25
[13] Siringli G, Abolmaali A and Aswath P B 2015 *The Sci. World J.* **2015** 1
[14] Nemeti K M 2000 *IJRMMS* **37** 133
[15] Emiroglu M, Kelestemur M H and Yildiz S 2007 *Proc. 8th Int. Frac. Conf. (Istambul)*

[16] The National Standardization Agency of Indonesia 2008 *SNI 1972: 2008 Cara uji slump beton*
[17] The National Standardization Agency of Indonesia 2011 *SNI 1974: 2011 Cara uji kuat tekan beton dengan benda uji silinder*
[18] The National Standardization Agency of Indonesia 1996 *SNI 03-4169-1996 Metode pengujian modulus elastisitas statis dan rasio poison dengan kompresor ekstensometer*
[19] The National Standardization Agency of Indonesia 2013 SNI *2814: 2013 Persyaratan beton struktural untuk bangunan gedung*
[20] Isa N F, Zaki M F M, Sofri L A, Ghazaly Z M and Bidin N 2014 *J. Adv. Res. App. Mech.* **4** 12
[21] Siddique R and Naik T R 2004 *Waste Manag.* **24** 563
[22] Anison M 1964 *Mag. Concr. Res.* **47** 73

IOP Conf. Series: Materials Science and Engineering **267** (2017) 012017 doi:10.1088/1757-899X/267/1/012017

Investigation on the fiber based approach to estimate the axial load carrying capacity of the circular concrete filled steel tube (CFST)

B Piscesa[*1,2], **M M Attard**[2], **P Suprobo**[1], **A K Samani**[3]

[1] Civil Engineering Department, Institut Teknologi Sepuluh Nopember, Surabaya 60111, East Java, Indonesia

[2] School of Civil Engineering, University of New South Wales, Sydney NSW 2052, Australia

[3] School of Civil Engineering and Information Technology, Federation University, Ballarat, VIC 3350, Australia

E-mail: piscesa@ce.its.ac.id and bambang.piscesa@live.com

Abstract. External confining devices are often used to enhance the strength and ductility of reinforced concrete columns. Among the available external confining devices, steel tube is one of the most widely used in construction. However, steel tube has some drawbacks such as local buckling which needs to be considered when estimating the axial load carrying capacity of the concrete-filled-steel-tube (CFST) column. To tackle this problem in design, Eurocode 4 provided guidelines to estimate the effective yield strength of the steel tube material. To study the behavior of CFST column, in this paper, a non-linear analysis using a fiber-based approach was conducted. The use of the fiber-based approach allows the engineers to predict not only the axial load carrying capacity but also the complete load-deformation curve of the CFST columns for a known confining pressure. In the proposed fiber-based approach, an inverse analysis is used to estimate the constant confining pressure similar to design-oriented models. This paper also presents comparisons between the fiber-based approach model with the experimental results and the 3D non-linear finite element analysis.

1. Introduction

In the regions with high seismic activities, the design of structures shall comply with rigorous strength and ductility requirements. For instance, the location, the pattern of formation, as well as the behavior of plastic hinges during earthquakes should be well predicted and designed to avoid any brittle failure and sudden collapse of the whole structure. In designing lateral force resisting systems, each structural element should have sufficient strength and ductility to withstand and absorb seismic induced actions in addition to other existing loads. It is important to note that during a major earthquake, the structural elements also experience plastic deformation and damage. The damage in concrete structural elements clearly show a sign of stiffness degradations. Hence, it is important to ensure stiffness degradations and reduction in capacity of concrete elements are gradual and to avoid any sudden capacity loss (due cover spalling or shear failure) of the structural elements.

In the high-rise building, high-strength concrete (HSC) material has become widely popular due to several advantages. By using HSC material, the dimensions of the column can be reduced and thus increase the usable floor area as well as reducing the concrete material usage (1). However, the behavior of HSC columns is different to the behavior of NSC columns and requires further consideration in design. High-strength concrete has more brittle behavior than normal strength concrete (NSC). Hence, the columns built of HSC are less ductile compared to the columns built from NSC. Furthermore, premature cover spalling has been observed in all tests of RC columns made of HSC material. The premature cover spalling results in a sudden drop in strength. Further, in many cases, these columns cannot reach the code predicted strength. Proper detailing of HSC columns will increase their ductility and the strength of the confined concrete core. However, from the experimental investigation (2, 3), it was found that in most cases, the strength recovery due to the confinement of concrete core still does not compensate the cover spalling and the peak strengths are lower than the code predictions.

One of the efficient ways to avoid cover spalling in HSC RC columns and to improve the ductility of these columns is to provide additional external confinement by FRP wrap or steel tube. In this paper, the focus is limited to the mechanical performance of the concrete-filled-steel-tube or CFST column. CFST columns are very convenient construction method. The steel tubes can act as lost formwork while they enhance the strength and ductility of the concrete core as well as circumventing premature cover spalling. In the design of CFST columns, local buckling, large plastic deformation of plates and tube, poor fire resistance and corrosion needs to be well investigated and considered. This study, however, is limited to the mechanical behavior of CFST column under static loading.

The focus of this paper is to investigate the accuracy of a fiber based model in predicting the behavior of CFST columns. Further, this paper also presents the results of three-dimensional nonlinear finite element analysis using 3D-NLFEA for comparison purposes. When predicting the full response of CFST columns, fiber-based model require less computational time than a full three-dimensional (3D) nonlinear finite element analysis. The fast execution time of the fiber-based approach makes it a convenient tool in design. Moreover, the method can be implemented to evaluate the nonlinear behavior of RC frames under static, cyclic or dynamic loads. However, in the fiber-based method, some analytical models are required to simulate the behavior of a three-dimensional structural element into two-dimensional (2D) sectional analysis.

In the proposed fiber based model, Attard and Setunge's (4) empirical constitutive model for the concrete material is used together with the elastic-perfectly plastic model for the steel tube. As highlighted the method is two-dimensional, hence the local buckling of the steel tube and the confinement effect to the concrete core could not be directly modeled. In the analysis, the local buckling of the steel tube is considered by using the effective yield strength based on Eurocode 4 (5). Furthermore, for circular columns, Eurocode 4 allows the inclusion of the confinement effect of the concrete core if the slenderness ratio and the eccentricity ratio are both less than 0.5. To predict the confining pressure of the concrete core, an inverse analysis is carried out by equating the peak axial load capacity of the fiber-based method with the Eurocode 4 prediction. Once the confining pressure is computed, the confining pressure is held constant and is used to predict the behavior of CFST columns. The predictions are then compared with both the experiments and the 3D Non-Linear FEA (3D-NLFEA).

This paper is organized into several sections. The first section was the introduction. The second section contains the formulation for the fiber based method and the Open Source software used during the development and computation. The third section covers a brief description of the constitutive model for the concrete and the steel tube materials. The fourth section discusses the inverse method to compute the confining pressure of CFST column. The fifth section contains the comparison with the experimental results and 3D Non-Linear FEA. The sixth section is the conclusion and possible future works.

2. Fiber based method

In the 2D sectional analysis using a fiber-based method, the section is discretized into smaller elements. In this study, the Constant Strain Triangle (CST) elements are used in which the evaluation of the axial stress and strain is at the center of CST element. The use of higher order elements (such as a Linear Strain Triangle) will increase the accuracy when there is a strain gradient. However, for simulating concentrically loaded CFST columns where all the strain in the elements is equal, there are no differences between the results obtained using the lower or higher order elements. In the analysis, displacement-based control load application is used as in the experimental setup. The fiber-based method developed in this paper can be used not only to evaluate regular/symmetric sections but also in evaluating irregular section. From the fiber-based analysis, the axial force (F_x) and moments in x and y directions (M_{xx} and M_{yy}) can be formulated as:

$$F = \sum_{i}^{nele} \sigma_i A_i \qquad M_{yy} = \sum_{i}^{nele} \sigma_i A_i \left(y_i - \bar{y} \right) \qquad M_{xx} = \sum_{i}^{nele} \sigma_i A_i \left(x_i - \bar{x} \right) \tag{1}$$

where σ_i is the stress in each element; A_i is the area of each element; x_i and y_i are the center coordinate of each element in x and y directions, respectively; $\bar{x}$ and $\bar{y}$ are the centroid of the section in the x and y directions.

The developed code is written in C#.Net (6). The modelling and mesh discretization is prepared using SALOME 7.8.0 (7). The analysis results are then written in a .vtk format file which is accessible in ParaView (8, 9) inside the SALOME 7.8.0. An axial force versus axial strain is dynamically reported during the runtime and is plotted using GNUPlot 5.1.0. Figure 1 shows an example of mesh discretization and axial stress contour of CFST column using ParaView.

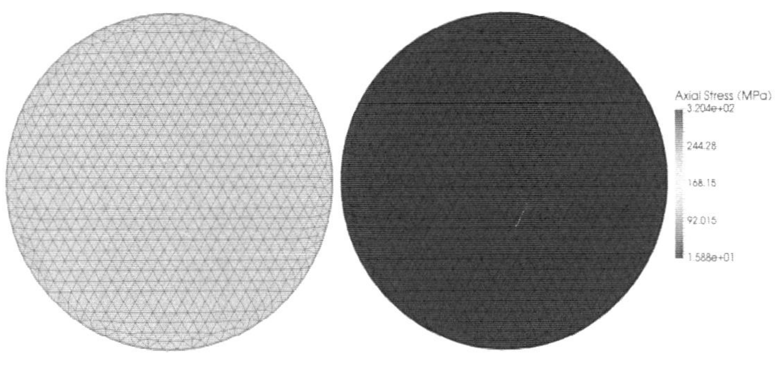

(a) (b)

Figure 1. (a) Mesh discretization of circular CFST column (b) Stresses in the CST elements for circular CFST column.

3. Constitutive models of the materials

The model of Attard and Setunge (4) is adopted for the concrete constitutive model which provides a good prediction of concrete material behavior for a wide range of concrete strength from normal (20 MPa) to a very high concrete strength (130 MPa). The constitutive model is classified as the empirical model where the stress is computed explicitly based on a given axial strain. The formulation for the stress-strain curves is based on the model of Sargin et. al. (10) and is given here as:

$$Y = \frac{AX + BX^2}{1 + CX + DX^2} \quad \text{where} \quad Y = \frac{f}{f_{cc}}, X = \frac{\varepsilon}{\varepsilon_{cc}} \quad \text{and} \quad \forall X \geq 0 \quad 0 \leq Y \leq 1 \tag{2}$$

where f is the axial stress at axial strain ε, f_{cc} is the axial peak stress at axial strain ε equal to ε_{cc}. A, B, C, D are constants that need to found using different boundary condition (see (4) for more detail).

The Attard and Setunge's (4) axial peak stress (f_{cc}) is:

$$\frac{f_{cc}}{f_c^{'}} = \left(\frac{f_r}{f_t}+1\right)^k \quad \text{with} \quad k = 1.25\left[1+0.062\frac{f_r}{f_c^{'}}\right]\left(f_c^{'}\right)^{-0.21} \tag{3}$$

where f_c is uniaxial concrete compressive strength, f_r is the confining pressure and f_t is the uniaxial tensile strength of concrete. Note that the Attard and Setunge's (4) failure surface is extendable to the tension-compression region (11). For more detail and complete formulation, see (4).

The steel tube behavior is simulated by an elastic-perfectly plastic model. The program can also incorporate hardening, or softening modulus in the steel model. The longitudinal steel rebar, if present, is modeled as a discretized solid circle and their behavior can be predicted by similar constitutive models used for the steel tube.

4. Confining Pressure Estimation

In the fiber-based method, normally the confinement is estimated based on the strain compatibility between reinforcement ties and lateral strain of the concrete (12) and simple equilibrium requirement between ties and concrete core similar to the method in (13, 14). However, in CFST columns, the steel tube is under axial compression as well as circumferential stresses at the same time which reduces the steel tube capacity. In this study, confining pressure is back calculated by firstly computing the axial carrying capacity (P_{max}) of the CFST column using Eurocode 4 or directly adopting the data from the test results. In Eurocode 4, the axial carrying capacity of the CFST column is estimated as:

$$P_{max} = P_{steel} + P_{conc} = \eta_s A_s f_y + A_c f_c^{'}\left(1+\eta_c \frac{t}{d}\frac{f_y}{f_c^{'}}\right) \tag{4}$$

where η_s is the strength reduction coefficient for the steel tube, A_s is the cross-sectional area of the steel tube, f_y is the yield strength of the steel tube, A_c is the cross-sectional area of the concrete, f_c is the uniaxial compressive strength of the concrete, η_c is a coefficient enhancement of the concrete strength due to the confinement effect, t is the thickness of the steel tube and d is the outer diameter of the steel tube.

In Eurocode 4, the strength reduction coefficient (η_s) and the concrete enhancement due to the external confining devices (η_c) is computed based on the slenderness ratio (λ) and the ratio of the load eccentricity to the diameter of the specimen (e/d). The expressions for η_s and η_c are:

$$\eta_s = 0.25(3+2\lambda)+(1-0.25(3+2\lambda))(10e/d) \tag{5}$$

$$\eta_c = \left(4.9-18.5\lambda+17\lambda^2\right)(1-10e/d) \tag{6}$$

$$\lambda = \sqrt{\frac{\left(A_s f_y + A_c f_c^{'}\right)L^2}{\pi^2 (EI)_{eff}}} \tag{7}$$

where L is the height of the specimen and $(EI)_{eff}$ is the effective stiffness of the columns. Please note that η_c is only considered if the slenderness ratio is less and equal than 0.5 and also if the ratio of the load eccentricity to the diameter of the specimen is less than 0.5.

Once P_{conc} is computed, the confining pressure is back calculated by simply equating the strength of confined concrete from Attard and Setunge (4) model with the Eurocode 4 confined concrete strength prediction as:

$$\left(1+\eta_c \frac{t}{d}\frac{f_y}{f_c^{'}}\right) = \left(\frac{f_r}{f_t}+1\right)^{k_1} \tag{8}$$

$$k_1 = 1.25\left(1 + 0.062\left(\frac{f_r}{f_c^{'}}\right)\right)\left(f_c^{'}\right)^{-0.21} \tag{9}$$

Obtaining a direct solution for the confining pressure from Eqns. (8) and (9) is rather difficult. Hence, an iterative numerical approach based on Regula-False method is adopted to find the confining pressure at the peak force level. Following the above method to compute the confining pressure, the predicted P_{Max} based on the fiber based method will be the same as the computed P_{max} using Eurocode 4. If P_{Max} is directly extracted from the test results, Equation (8) can be rewritten by replacing the left-hand side of the equation by confined core strength increase in the experiment. The axial force contribution of concrete (P_{Conc}) is then calculated by subtracting the axial force contribution of steel (P_{Steel}) from the total force of the specimen (P_{Max}). Dividing P_{Conc} with the area of concrete times the concrete uniaxial compressive strength will be the core strength enhancement in the experiment. In summary, Equation (8) can be rewritten as:

$$\frac{P_{Max} - P_{Steel}}{A_c f_c^{'}} = \frac{P_{Conc}}{A_c f_c^{'}} = \left(\frac{f_r}{f_t} + 1\right)^{k_1} \tag{10}$$

Confinement will be calculated by iteration from Equation (10) as well.

5. Comparisons and discussion of the fiber-based approach with the finite element analysis and the experimental results

Two experiments of CFST columns from Lai and Ho (15) are modeled to verify the proposed fiber-based approach. The first specimen is identified as CN0-1-114-30 in which the concrete compressive strength is 31.4 MPa with a modulus of elasticity of 19.7 MPa and the uniaxial axial peak strain of 0.0029. The second specimen is identified as CN0-1-114-80 with a concrete compressive strength of 79.9 MPa, a modulus of elasticity 30.7 MPa for concrete material and the uniaxial axial peak strain of 0.0036. Both specimens have the same geometry and material properties of the steel tube. The steel tube has an outer diameter of 111.5 mm, and the thickness of the steel tube is 0.96 mm. The uniaxial yield strength of the steel tube in tension is 370 MPa. The modulus of elasticity and the Poisson's ratio of the steel tube is 206 GPa and 0.29, respectively.

Using Eurocode 4, the axial load carrying capacity of CN0-1-114-30 and CN0-1-114-80 are 418.34 kN and 867.99 kN respectively while the confining pressure for CN0-1-114-30 and CN0-1-114-80 are estimated to be 1.027 MPa and 0.962 MPa respectively. If the confining pressure is computed based on maximum experimental axial load carrying capacity, the confining pressure for CN0-1-114-30 and CN0-1-114-80 are estimated as 2.148 MPa and 1.897 MPa, respectively. Since the confining pressure at peak stress is known, the fiber-based analysis is carried out by assuming a constant confining pressure throughout the loading steps. It is worth mentioning that, in the experimental results (15), after reaching the peak stress and during softening, the confining pressure is still increasing. Hence, it is expected that the load carrying capacity of the column in softening regions predicted from the fiber-based method is lower in comparison with the experimental results.

In addition to the fiber based method predictions, a three-dimensional non-linear finite element analysis of these two specimens are carried out using an in-house FEM program called (16) (3D-NLFEA). The constitutive model used for analysis in 3D-NLFEA is Piscesa's plasticity model (11, 17). The constitutive model for the steel material is Von-Mises criterion or a J_2 plasticity model with zero hardening modulus. In the 3D-NLFEA, the CFST column is modelled using 6,076 hexahedral elements with full integration. Both ends have fixed boundary conditions, and a displacement controlled loading scheme is considered. Figure 2 shows the 3D-NLFEA model of CN0-1-114-30, the hardening parameter, the confining pressure and the axial stress distribution for CFST column at the average axial strain equal to 0.016. The data contour is presented by showing the vertical cross section of the specimen.

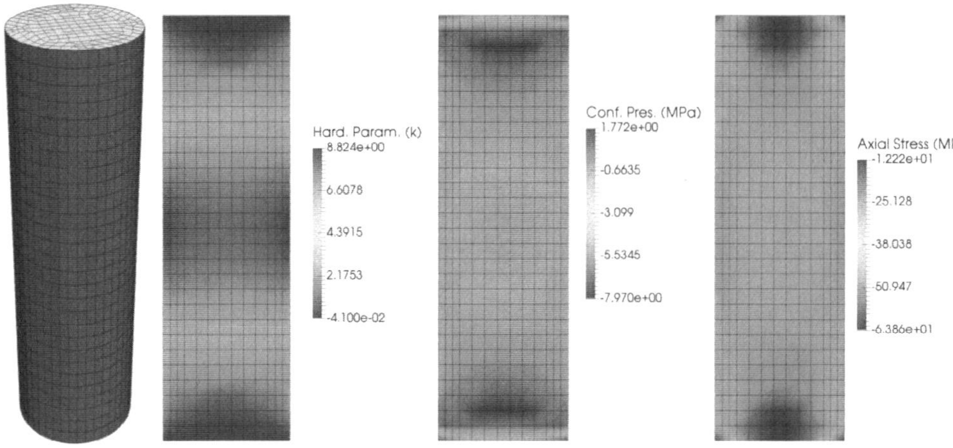

(a) (b) (c) (d)

Figure 2. a) 3D-NLFEA model of CN0-1-114-30 b) Hardening parameter c) Confining pressure (d) Axial stress. The contour is shown at the average axial strain equal to 0.016.

From Figure 2b, the hardening parameter has the highest value at the mid-height which clearly showing the localized region. The distribution of the confining pressure throughout the specimen is not constant with the regions near the platen having higher confining pressure compare to the region at the mid-height of the specimen (see Figure 2c). Figure 2d shows that the axial stress distribution at both end is lower than at the mid-height of the specimen. The lower value of the axial stress distribution at both ends is due to elastic unloading occurring at those regions. These phenomena could not be captured by using a 2D fiber-based model.

Figure 3 and 4 show the comparisons of the axial load versus the axial strain between the fiber-based method, 3D-NLFEA and the experimental results for CN0-1-114-30 and CN0-1-114-80, respectively. Figure 3 shows that the predicted peak axial load carrying capacity using Eurocode 4 (418.34 kN) is lower than the value obtained from the 3D-NLFEA (467.12 kN) and the experimental result (479 kN). However, if the concrete strength enhancement is calculated based on Eqn.(10), the overall predicted response using the fiber-based method is closer to the 3D-NLFEA. It is worth highlighting that both the 3D-NLFEA and fiber based method show lower axial capacity during softening after the peak load. Figure 4 also shows the predicted peak axial load carrying capacity using Eurocode 4 (867.99 kN) is lower than the value obtained from the 3D-NLFEA (997.34 kN) and the experimental result (955 kN). The predicted axial capacity of fiber based model during softening is much lower than 3D-NLFEA and the test results. The lower softening response from the fiber based analysis could be attributed to the constant confining pressure assumption which is more pronounce in high strength concrete specimen. From the discussion above, 3D-NLFEA gives more accurate results and the use of fiber-based model should be improved in the future for CFST column made of high strength concrete.

To study size effect phenomena in the 3D-NLFEA analysis, axial shortening of the column is recorded in two ways. In the first method, the column shortening is a total height shortening between platens thus the end regions near the platens, which are elastically unloading, are also included in the computed average axial strain. While in the second method, column shortening in a distance of two times of the diameter in the middle of the column height is used to compute the average axial strain. Thus, in the second method, the region near the platen is not included when computing the axial strain. The load-strain response by using the first measurement has steeper softening slopes than the second measurement (see Figure 3 and 4) which clearly shows that the 3D-NLFEA can capture the size effect in the analysis. Hence, it is important to note that when comparing the analysis with the experiments, the location of the measurement points is important.

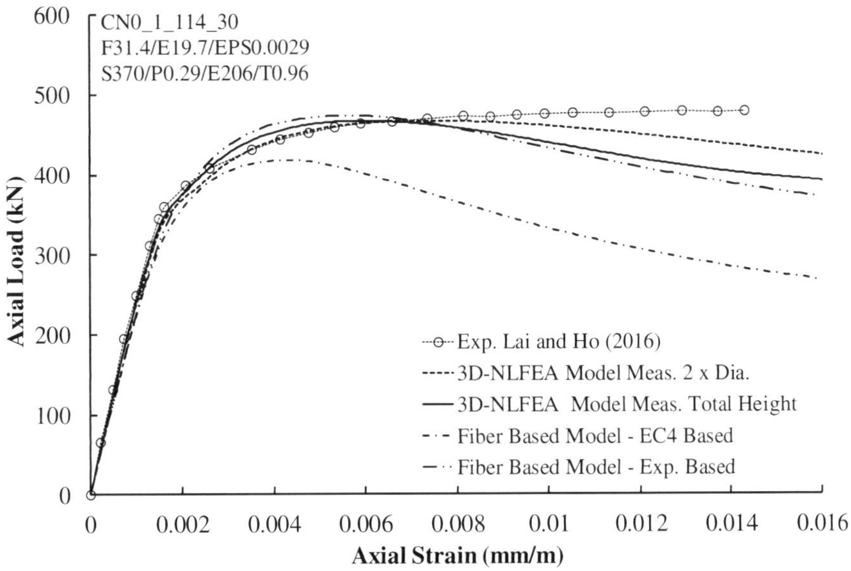

Figure 3. Comparisons of the model with 3DNLFEA and experimental results for CN0_1_114_30 specimen.

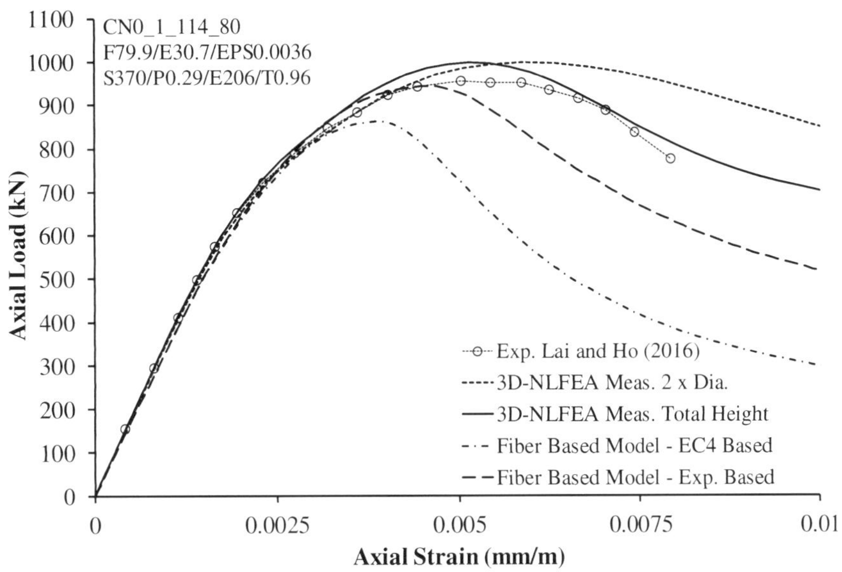

Figure 4. Comparisons of the model with 3DNLFEA and experimental results for CN0_1_114_80 specimen.

6. Conclusion and future work

This paper investigated the performance of the fiber based method in predicting the behavior of concrete filled steel tube (CFST) columns. The use of the fiber-based method is extremely fast and can be extended to a larger scale non-linear analysis such as a non-linear frame analysis. Although the comparisons shown in this paper were limited, the results showed that the use of fiber-based method tended to underestimate the load-deformation response of the CFST columns. There was two main reasons highlighted; the first reason was that the predicted axial load carrying capacity, which was based on Eurocode 4, was lower than the experiments and the second was because of the constant confining pressure assumption throughout the analysis. The first problem could be rectified by using the peak load from the test results, however, experimental data was not always available during design and thus a more robust method to estimate the confinement that should be carried out. On the other hand, the use of full 3D-NLFEA showed an excellent prediction because of the ability of the constitutive model to capture not only the peak response but also the softening response of the selected specimen.

Furthermore, the size effect phenomena in the load-deformation response using 3D-NLFEA is discussed briefly. The future works are focused on improving the formulation to predict the axial load carrying capacity of CFST column and using a more advanced concrete constitutive model which involves the inclusion of the size effect (18) and the lateral strain model (19). The inclusion of the size effect allows the fiber-based method to account for the localized region in the specimen and thus affecting the load-deflection behavior when softened. On the other hand, by using the lateral strain model, the estimate of the confining pressure is more accurate. Further, in the future, a non-circular section and complex configuration of the CFST with steel rebar also worth to be investigated.

7. References

[1] Gjørv O. High strength concrete. Developments in the Formulation and Reinforcement of Concrete. 2008:79-97.
[2] Foster SJ. On behavior of high-strength concrete columns: cover spalling, steel fibers, and ductility. Structural Journal. 2001;98(4):583-9.
[3] Foster SJ, Liu J, Sheikh SA. Cover spalling in HSC columns loaded in concentric compression. Journal of Structural Engineering. 1998;124(12):1431-7.
[4] Attard M, Setunge S. Stress-strain relationship of confined and unconfined concrete. ACI Materials Journal. 1996;93(5).
[5] EN C. 1-1, Eurocode 4: Design of composite steel and concrete structures. Part; 1994.
[6] Microsoft. Visual C# Ver. 6.0. https://www.visualstudio.com/. 2015.
[7] CEA/DEN, EDF_R&D, OPEN_CASCADE. SALOME - The Open Source Integration Platform for Numerical Simulation. http://www.salome-platform.org/. Ver. 7.8.0 ed 2016.
[8] Ahrens J, Geveci B, Law C, Hansen C, Johnson C. 36-ParaView: An End-User Tool for Large-Data Visualization. The Visualization Handbook. 2005:717.
[9] Ayachit U. The paraview guide: a parallel visualization application. 2015.
[10] Sargin M. Stress-strain relationships for concrete and the analysis of structural concrete sections: Solid Mechanics Division, University of Waterloo Waterloo, ON, Canada; 1971.
[11] Piscesa B, Attard M, Samani A, Tangaramvong S. Plasticity Constitutive Model for Stress-Strain Relationship of Confined Concrete. Structural Journal. 2017;114(02):361-71.
[12] Samani AK, Attard MM, Foster SJ. Ductility in concentrically loaded reinforced concrete columns. Australian Journal of Structural Engineering. 2015:1-14.
[13] Sheikh S, Uzumeri S. Analytical model for concrete confinement in tied columns. Journal of the Structural Division. 1982;108(12):2703-22.
[14] Sheikh S, Uzumeri S. Strength and ductility of tied concrete columns. Journal of the Structural Division. 1980;106(5):1079-102.
[15] Lai M, Ho J. A theoretical axial stress-strain model for circular concrete-filled-steel-tube columns. Engineering Structures. 2016;125:124-43.
[16] Piscesa B, Attard MM, Samani AK. Three-dimensional Finite Element Analysis of Circular Reinforced Concrete Column Confined with FRP using Plasticity Model. Procedia Engineering. 2017;171:847-56.
[17] Piscesa B, Attard MM, Samani AK. A lateral strain plasticity model for FRP confined concrete. Composite Structures. 2016;158:160-74.
[18] Samani A, Attard M. A Stress-Strain Model for Uniaxial and Confined Concrete under Compression. Engineering Structures. 2012;41:335-49.
[19] Samani AK, Attard MM. Lateral Strain Model for Concrete under Compression. ACI Structural Journal. 2014;111(1-6).

IOP Conf. Series: Materials Science and Engineering **267** (2017) 012018 doi:10.1088/1757-899X/267/1/012018

Defining Appropriate Temperature for Perfect Erection Time of Steel Arch Bridge Closure to Minimize the Effect of the Thermal Stress. *Case Study: The New Kutai Kartenagara Bridge, Indonesia.*

H Sugihardjo[1,3], Tavio[1], D Prasetya[1], and N I Achmad[2]
[1]Department of Civil Engineering, Institute Teknologi Sepuluh Nopember (ITS), Kampus Sukolilo, Surabaya, Indonesia.
[2]PT. Pembangunan Perumahan (Persero) Tbk, Indonesia
[3]Member of Committee for Bridge and Road Tunner Safety, Indonesia

Corespondence E-mail: hidajat.sugihardjo@gmail.com

Abstract. The impact of temperature change could govern the final stress of structural elements in the long-span steel bridges. The study was conducted to investigate the effect of temperature change during the construction of a new steel arch bridge, namely the Kutai Kartanegara Bridge. The main bridge has a total spanning length of 470 meters. The erection method of the bridge was the cantilever method with temporary towers, mast cranes, and stay cables. The deflections and internal forces of the steel elements were analyzed using the Midas Civil software. The study focuses on the effort to find the perfect erection time for the closure with regards to the temperature. By measuring the temperatures of the steel elements during the construction, it was found that they varied between 19 and 64 degrees Celsius. From the results of the analyses with various temperatures, it can be concluded that the recommended temperature for the closure erection was 44.6 degrees Celsius. This temperature is similar with the air temperature between either 6 AM and 12 noon or 12 noon and 6 PM. During this periods of time, the effect of thermal stress on the final internal forces in the bridge elements was found much lesser than those obtained during any other period of time.

1. Introduction

On November 26th, 2011 the Kutai Kartanegara suspension bridge, called as the Indonesian Golden Gate Bridge, collapsed. The bridge used a closed steel truss system for the stiffening girder with a through deck system. It has a left-hand end span of 100 meters, a main span of 270 meters, and a right-hand end span of 100 meters as shown in Figure 1. It was publicly opened to traffic since 2002. To replace the collapse bridge that connected Samarinda and Tenggarong cities, a new half-through steel arch bridge was constructed and it had been publicly opened to traffic on December 8th, 2015, while it is still using the strengthened existing foundations as shown in Figure 2.

Many investigations on the effects of temperature changes on the bridge structures have been conducted in various methods through the laboratory tests, and the theoretical and experimental investigations. In 1893, it was reported the site experiments conducted on an arch bridge structure at Lyons, France [1]. Parts of the structure that exposed to the direct sunshine had a temperature of 54°C (130°F) and the coldest winter temperature recorded was –26°C (–15°F). An investigation in Austin,

Texas-USA, concluded that the effect of temperature contributes approximately 10 percent of the design due to live load on the structure [2]. An investigation on a composite steel bridge summarized that the environmental influences such as solar radiation, site temperature, and wind speed are crucial variables [3]. A health monitoring was conducted on a Little Mystic Truss Bridge, Massachusetts-USA. It was found that the increase of temperature has caused a camber and the compressive strains in the lower chords, while the decrease of temperature has caused a deflection and the tensile strains in the lower chords [4]. The investigations on the closure erection of a concrete-filled steel tube arch bridge concluded that only by selecting an appropriate erection and temperature of the closure can significantly reduce the thermal stress occurred in the arch bridge elements [5].

Figure 1. The Old Kutai Kartanegara Suspension Bridge in 2007, before Collapsed (Courtesy of Hidajat Sugihardjo).

Figure 2. The new Kutai Kartanegara steel arch bridge in 2015 (courtesy of Hidajat Sugihardjo).

The Forward Assemblage Analysis (FAA) method using a stay-cables or temporary supports and derrick cranes (mast cranes) has been implemented for the erection of many bridges, such as Strömsund Bridge-Sweden, Batman Bridge-Australia, Kniebrücke Bridge-Germany [6], Rainbow Bridge-USA and Lewiston-Queenston Bridge-USA [7] and Gustav Lindenthal New York City's Hell Gate Bridge-USA [8]. Normally, for girder bridges, the roadway of a derrick crane is located on the deck, meanwhile for the arch bridge, it is located on the top of the arch.

On the basis of the previous investigations, this study focused on the prediction of defining the field temperature for erection of the closure. The analysis was performed at the time before the closure was attached by imposing its selfweight and construction load and after closure was mounted due to the live load and superimposed dead load. The paper is the modified and developed the study of the undergraduate thesis conducted by [9]. By obtaining the appropriate temperature, the bridge responses are evaluated against temperature changes of the coldest and hottest temperatures. The effect of thermal stress on the final internal forces in the bridge elements are also evaluated due to the thermal load when the closure is erected during the extreme temperatures.

2. Theoretical Background

A bearing system is one of the structural components that influences the bridge movements. The phenomenon experienced by an arch bridge due to the temperature change are as follows: if the arch is a three-hinged structure (i.e. statically determinate structure), the frame revolves about each hinge, resulting no temperature stress. However, if the arch is a two-hinged structure (i.e. statically indeterminate structure), the frame deflects upward or downward, thus the frame elements slightly bend and further induce the thermal stresses in the frame elements [1]. An increase of temperature causes the compression in the lower elements (intrados) and tension in upper elements (extrados) of the arch crown as illustrated in Figure 3 [1, 10].

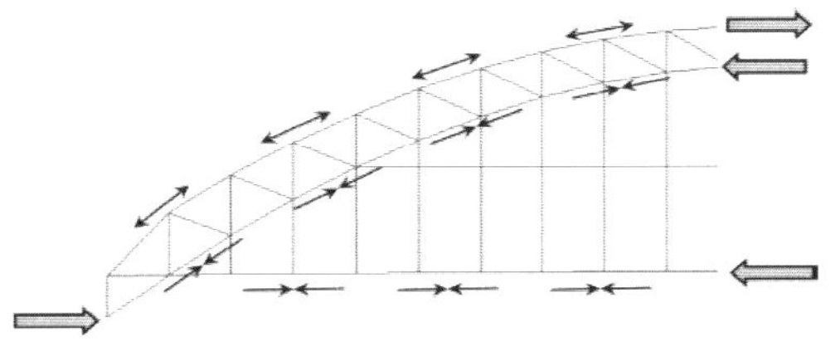

Figure 3. Distributions of thermal forces due to increasing of temperature [1, 10].

From the phenomena of temperature changes on the bridge structures as mentioned above, the axial thermal stress can be calculated by Equation (1). On the basis of this formula, then the bridge structural responses regarding the temperature change during construction is investigated.

$$\sigma_t = E\varepsilon_t = E\alpha(\Delta T)L \quad (1)$$

where:
ε_t = thermal strain; α = coefficient of thermal expansion; ΔT = temperature change; L = original length of element; E = modulus of elasticity of material

3. Bridge Descriptions

The bridge data obtained from the design including the geometry, material properties, cross-sections of the main truss, hangers and bearings are as follows [9, 11]:

- Type: half-through truss arch
- Span: 99.831 + 269.614 + 99.985 meters
- Rise: 57.84 meter; Rise/Span = 1 : 4.7
- No. of lanes of traffic: 2 + sidewalks 2×1.30 meter
- Width: 10.75 meter; Width/Span = 1:25
- Sections of arch truss: box and plate girders
- Materials: SM490YA/YB JIS G 3106
- Hangers: steel wire strands with tensile strength of (1570 ~ 1630) MPa
- Supports: pot bearings of fix-, unidirectional sliding- and multidirectional sliding types
- Loading specification: SNI T-02-2005 [12]

4. Results and Discussion

4.1. Construction Stage

The elevation view of the bridge is shown in Figure 4. The construction stage of the bridge using temporary supports and pylons, stay-cables and mast cranes with a weight of 5 tons is illustrated in Figure 5. The truss elements of the bridge which were erected from the both end spans have been divided into 15 consecutive stages as shown in Figure 6. The erection of the bridge elements in Stages 1 to 6 used the mobile pontoon cranes from both end spans, whereas in Stages 7 to 15 it used the mast cranes. The analysis carried out focused on Stage 11 (see Figure 6(k)), shortly before the closure was connected. In this stage, the bridge could be provided as a structure with fewer redundant than those after the closure was erected [9].

To analyze the temperature impact on the final stress of the bridge structure, the four sides of steel section were measured by four thermocouples during one week in July 2014. Normally, during these months (June to August) in Indonesia, the hottest and coldest temperatures occur during the daytime and the night time or early morning, respectively. The positions of the thermocouples for observations of a steel section element are shown in Figure 7. From the observations, it was found that the temperature change varied between 19° and 64°C [13]. Having the maximum and minimum temperatures of 64° and 19°C during the daytime and the night time or early morning, respectively, thus it can be assumed that the temperature change is approximately 45°C for the analysis. The value was much greater compared with the temperature change according to the Indonesian code, i.e. 25°C [12].

With the measured temperature ranging between 19° and 64°C for the respective minimum and maximum element temperatures, the closure should ideally be erected at a temperature that gives the lowest impact to the expansion and shrinkage of the bridge structural elements, i.e. 19 + (64 – 19)/2 = 41.5°C. However, the ideal temperature still depends on the two end nodes (nodes 29 and 177 at downstream and upstream, respectively) of the left cantilever, shortly before it converges with the right cantilever at the mid-points of the bridge. The superimposed loads at this stage are the gravity load and the thermal load due to the temperature changes during Stage 11 shown in Figure 6(k).

The erection time proposed in this study was begun with the assumptions that the bridge elements were erected at the extreme temperatures in the morning, at noon, and afternoon of 19°, 64°, and 24°C, respectively. These values were set as the initial and final temperatures of any stages in the Midas software [14]. The erection time was assumed to be started in early morning, at noon, and afternoon. For example, if the initial erection of the elements is set in the morning, the temperature change (temperature at noon – temperature in the morning) is assumed as ΔT_1 and then followed by temperature change ΔT_2 (temperature in the afternoon – temperature at noon) as shown in Figure 8.

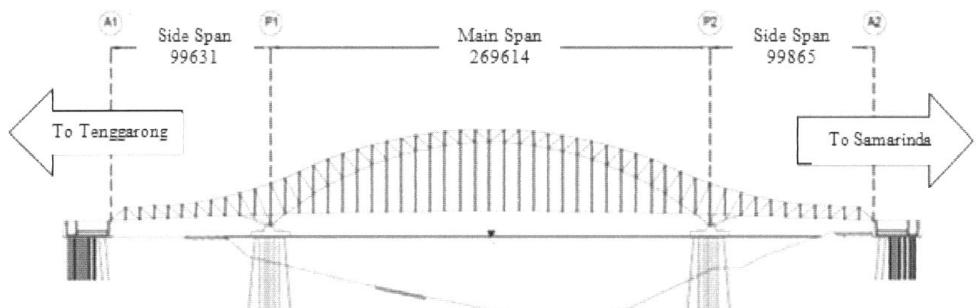

Figure 4. Elevation view of the new bridge.

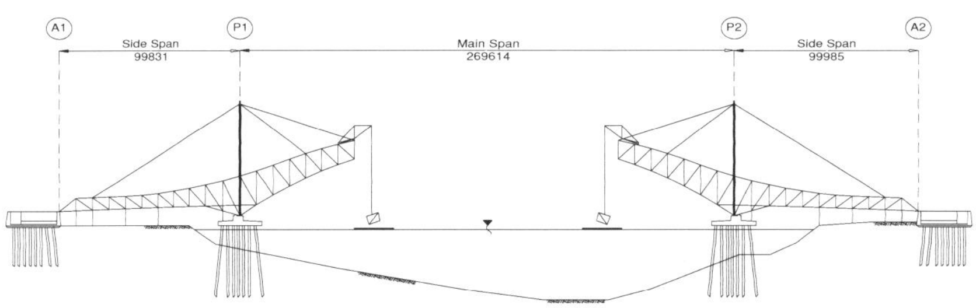

Figure 5. Schematic of erection of the bridge elements.

To illustrate the element responses during the construction of the bridge, an evaluation was conducted on four elements located near the supports, which are BA11, BB3, DG13, and TG11, and four elements at the mid-span, namely BA25, BB17, DG27, and TG27 as shown in Figure 9. The sections and related dimensions of the structural steels used in the bridge structure are listed in Table 1. It can be seen from the table that all the structural steels have the same cross-sectional width of 750 mm to ensure that they can be connected easily. The axial force history of the upper chord BA11 during the construction is given in Figure 10. It can be seen that the forces occured in the elements were still less than their capacities, either in compression or tension.

The evaluations were conducted from Stages 1 to 15 which include the observations of element forces, node deflections, and pretensioning loads of the temporary cables. The imposed loadings were selfweight of the bridge, the mast crane loads, and the load from the temperature changes. As shown in

IOP Conf. Series: Materials Science and Engineering **267** (2017) 012018 doi:10.1088/1757-899X/267/1/012018

Table 2, Stage 7 of the FAA's method sequence in the MIDAS software, for instance, which starts using the mast crane, can be elaborated as follows:

a) Stage 7: Pretensioning and temperature load, ΔT_2, is given, the previous stages are not activated.
b) Stage 7a: The selfweight is activated, followed by temperature load ΔT_1.
c) Stage 7b: Temperature load ΔT_2 is activated simultaneously by deactivating temperature load ΔT_1.
d) Stage 7c: The mast crane is imposed simultaneously by deactivating temperature load ΔT_2.

The deflection history of Node 53 during construction as illustrated in Figure 6(j) is shown in Figure 11. From the figure, it can be seen that the maximum deflections occured when the positions of the mast cranes are all at the cantilever ends. The pretensioning loads applied to Cables Numbers 1 to 6 during Construction Stages 7 to 14 are shown in Figure 12.

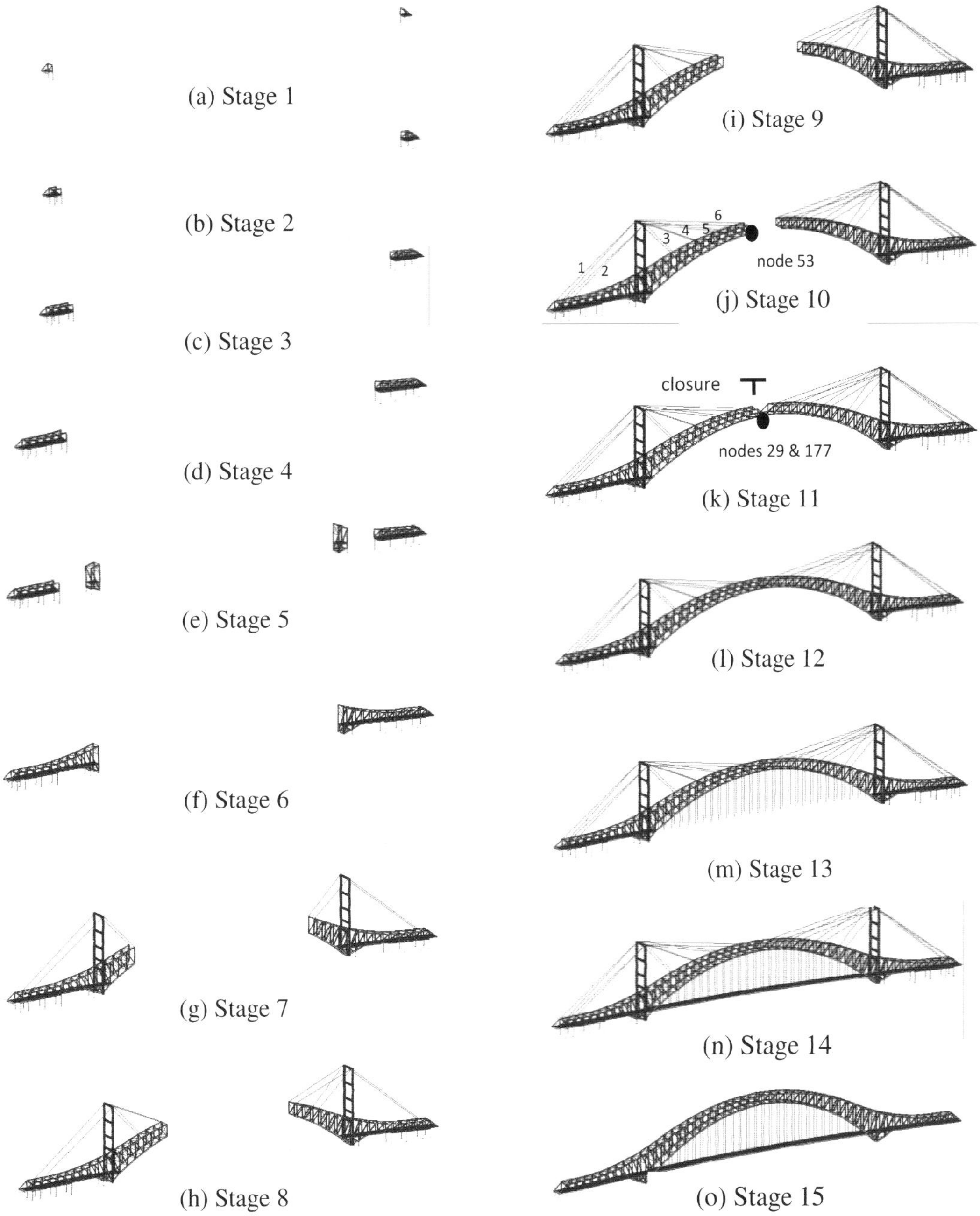

(a) Stage 1
(b) Stage 2
(c) Stage 3
(d) Stage 4
(e) Stage 5
(f) Stage 6
(g) Stage 7
(h) Stage 8
(i) Stage 9
(j) Stage 10
(k) Stage 11
(l) Stage 12
(m) Stage 13
(n) Stage 14
(o) Stage 15

IOP Conf. Series: Materials Science and Engineering **267** (2017) 012018 doi:10.1088/1757-899X/267/1/012018

Figure 6. Modelling of construction stages [9].

The figure depicts the additional (+) and reducing loads (-) that should be applied during the construction stages and they normally do not exceed 50 percent more than the minimum breaking strength of the wire cables [6].

4.2. Closure Erection

The closure was erected at Stage 12. At the end of Stage 11, when the last element was attached, the displacement of Nodes 29 and 177 should be as small as possible or still in the tolerable value. Nodes 29 and 177 are located at the mid-span of the bridge. By obtaining the small displacements of Nodes 29 and 177, it is possible to attach the closure as shown in Figure 6(k). To obtain this ideal condition, it can be done by using the temperature interval $\Delta T = (64 - 19)/4 = 11.25°C$, resulting the elevated temperatures of 19°C, 30.25°C, 41.5°C, 52.75°C, and 64°C for the analytical purpose. In the analytical modelling, the temperature variations were assumed as the final temperatures, while the initial temperatures were 19°C, 64°C, and 24°C, measured in the morning, at noon, and afternoon, respectively.

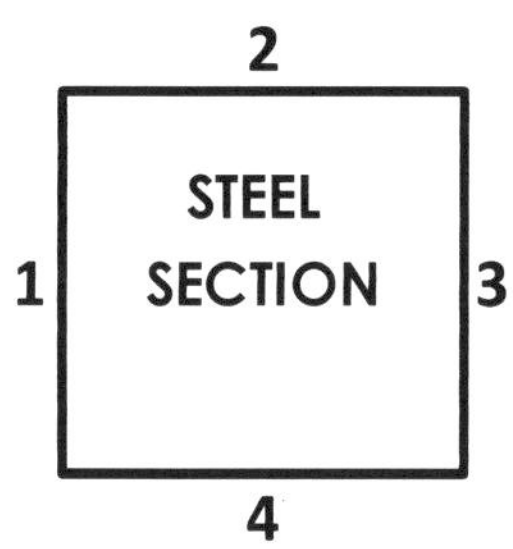

Figure 7. Positioning of four thermocouples on a steel section.

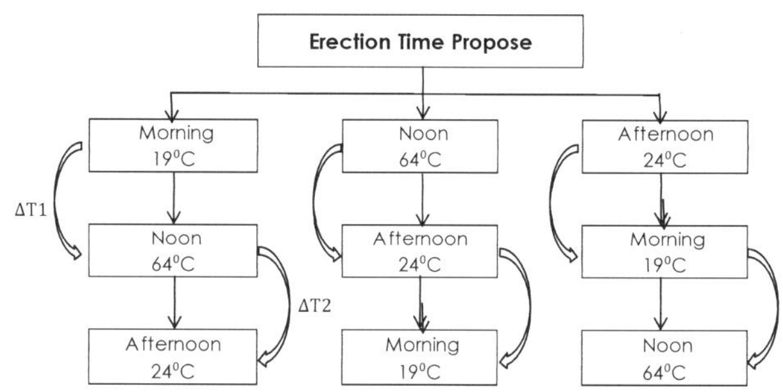

Figure 8. Time variations for erection of bridge elements.

By implementing the analysis using the variations of initial and final temperatures, the average deflections of Nodes 29 and 177 were obtained (Figure 13). From the figure and the assumption that Nodes 29 and 177 had zero deflection in the initial configuration due to its selfweight, resulting the first alternative temperature for closure erection of about 44.6°C. This temperature is similar with the air temperature between 6 AM and 12 noon. Whereas, if the erection of the closure is conducted at an ideal temperature of 41.5°C as described previously, Nodes 29 and 177 (Figure 13) is still deflect downwards of about 16 millimeters from their origins. Furthermore, by evaluating the ratio of the axial force to the axial capacity of steel section, it was found that the erection of closure at temperatures of either 19°C or 64°C yielded the ratio that varied from 3~47 percent greater than those resulted from the erection at the appropriate temperature of 44.6°C, as shown in Table 3.

Table 1. Bridge element sections (not all sections shown).

Element	Location	Positioning	Section		Element	Location	Positioning	Section	
			Type	Dimensions (mm)				Type	Dimensions (mm)
BA11	On support	upper chord	box girder	700x750x11x12	BA25	At midspan	upper chord	box girder	800x750x25x38
BB3		lower chord	box girder	1000x750x25x30	BB17		lower chord	box girder	550x750x10x10
DG13		diagonal	box girder	600x750x18x25	DG27		diagonal	I-girder	450x750x12x20
TG11		vertical	box girder	970x750x20x35	TG26		vertical	I-girder	280x750x10x12

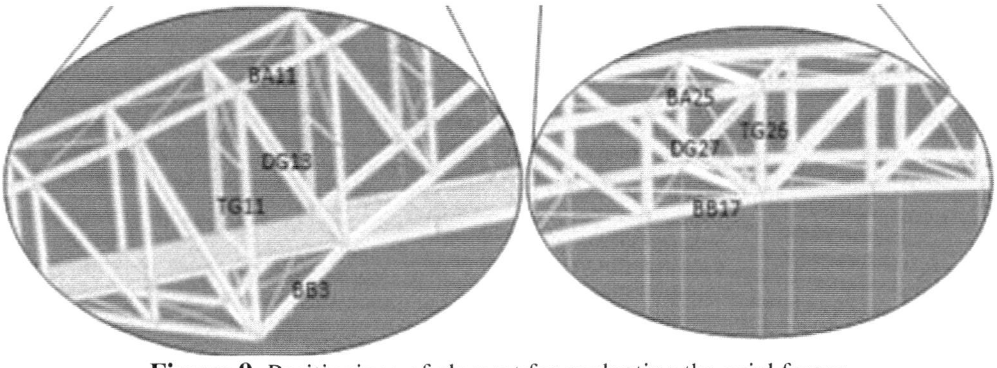

Figure 9. Positionings of element for evaluating the axial forces.

Table 2. Construction stage (not all stages shown).

Stage	Structure Group		Boundary Group		Load Group	
	Activation	Deactivation	Activation	Deactivation	Activation	Deactivation
Stage 7	Stage 7		Temporary Tower		Pretension Load 7	Temperature Load 2
Stage 7a					Temperature Load 1	
Stage 7b					Temperature Load 2	Temperature Load 1
Stage 7c					Mast Crane 7	Temperature Load 2

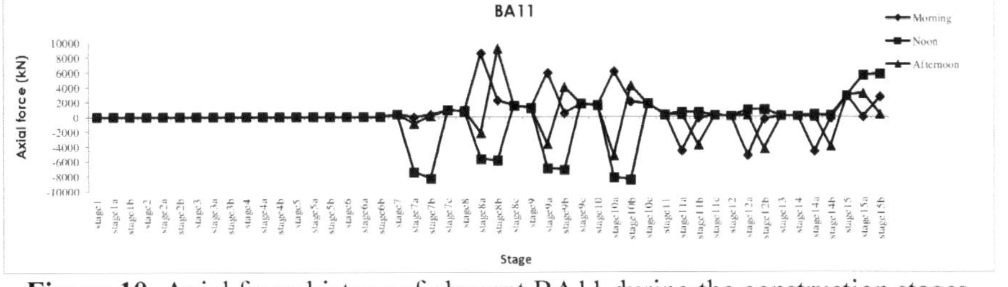

Figure 10. Axial force history of element BA11 during the construction stages.

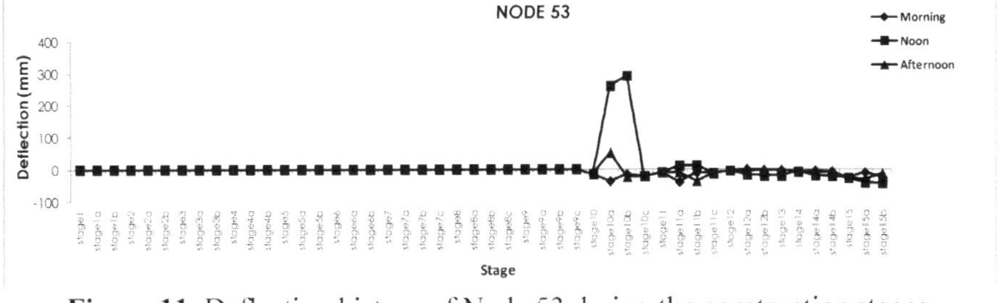

Figure 11. Deflection history of Node 53 during the construction stages.

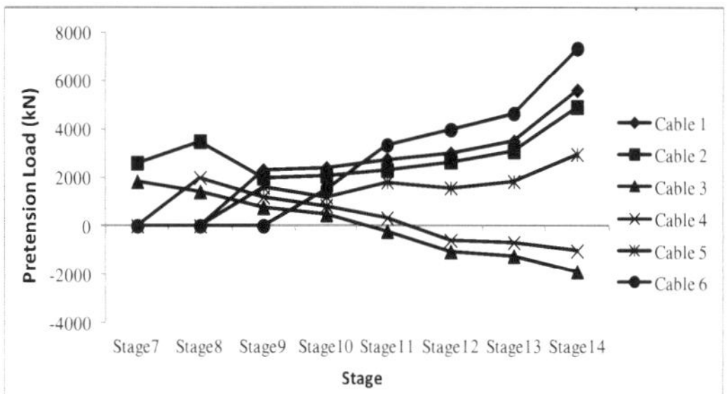

Figure 12. Pretensioning load during the construction stages (+ increase; – reduction).

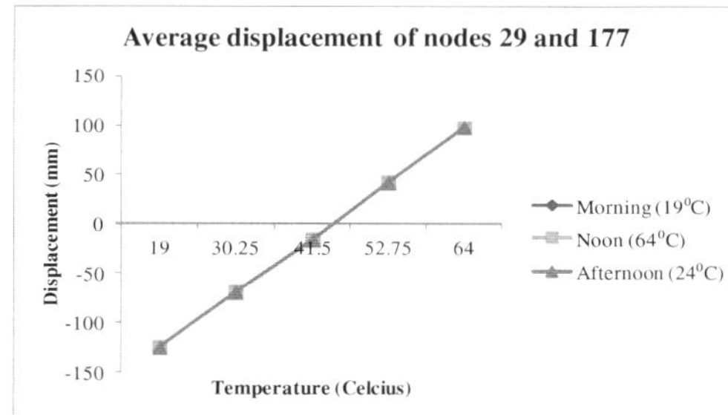

Figure 13. Displacement vs. temperature for defining the appropriate time of closure erection.

5. Conclusions

From the discussion of the current study, the following coclusions can be drawn:

- In general, to define the appropriate temperature for perfect erection time of steel arch bridge closure with the aim to minimize the effect of thermal stress, the actual temperature of bridge elements and the positions of the midspan nodes of the bridge should become the main consideration.
- From the results of the analyses with various temperatures, it can be recommended that appropriate temperature for closure erection is 44.6°C. It means that the closure could be erected during either 6 AM to 12 noon or 12 noon to 6 PM.
- If the closure are erected without considering the appropriate temperature, the axial forces of the bridge elements could increase up to 3 ~ 47 percent higher than those consider the appropriate temperature. The increase of these axial forces should be taken into account accuratelly in the design of a bridge structure.

Table 3. Ratio of thermal axial force to axial section capacity.

Element	Capacity	T-initial=44.6 degree Celcius		T-initial=19 or 64 degree Celcius		Difference
	P (kN)	ΔP (kN)	P/ΔP	ΔP (kN)	P/ΔP	
BA11	9165	2529	28%	9165	75%	47%
BB3	-25306	-5985	24%	-25306	37%	13%
DG13	12173	389	3%	12173	6%	2%
TG11	-14157	-565	4%	-14157	7%	3%
BA25	-26855	-3406	13%	-26855	16%	3%
BB17	7783	1033	13%	-7524	26%	12%
DG27	-4666	-204	4%	-4666	10%	6%
TG26	3506	271	8%	3506	12%	4%

6. References

[1] Hool G A and Kinne W S 1943 Moveable and Long-span Steel Bridges 2nd ed. (McGraw-Hill Inc. New York).

[2] Thepchatri T C, Philip J, and Hudson M 1977 Prediction of Temperature and Stresses in Highway Bridges by a Numerical Procedure Using Daily Weather Reports (Texas, State Department of Highways and Public Transportation).

[3] Yargicoglu A and Philip J 1978 Temperature Induced Stresses in Highway Bridges by Finite Element Analysis and Field Test (Texas, St. Dep. of Highways and Public Transp).

[4] Brenner B R, Sanayai M, Bell E S, Rosenstrauch P L, Pheifer E J and Man W A 2011 The Influence of Temperature Changes on Bridge Structural Behavior.

[5] Yang B, Huang J, Lin C and Wen X 2011 Temperature Effect and Calculation Method of Closure Temperature for Concrete-filled Steel Tube Arch Rib of Dumbbell-shape Section, *The Open Civil Engneering Journal,* **5**, 179-189.

[6] Podolny W, and Scalzi, J B 1976 Construction and Design of Cable-stayed Bridges (John Wiley and Sons, New York).

[7] Durkee J 2000 Steel Bridge Construction, Bridge Engineering Handbook, Ed. (Wai-Fah Chen and Lian Duan, Boca Raton: CRC Press).

[8] Oviatt-Lawrence A 2013 Lindenthal's New York City Hell Gate Bridge-1917, Structure Magazine, Oct. http://www.STRUCTURE.mag.org.

[9] Achmad N I 2015 The Effect of Closure Erection on Steel Arch Bridge Due to Thermal Stress, Case Study, The New Kutai Kartanegara Bridge, Indonesia, Undergraduate Theses, Department of Civil Engineering, ITS Surabaya, Indonesia (in Indonesian).

[10] Yarnold M T, Moon F L, Dubbs N C and Aktan, A E 2012 Evaluation of a Long-span Steel Arch Bridge Using Temperature-based Structural Indentification, London, Bridge Maintenance, Safety Management, Resilience and Sustainability, 2397-2403.

[11] Sugihardjo H, Tavio, Manalu I, and Lesmana Y 2017 Seismic Study of Application of Lead Rubber Bearings In Kutai Kartanegara Steel Arch Bridge, RCEE-ICCER, Surabaya, Ina.

[12] SNI 2005 Indonesian National Standarts T-02-2005, Department of Public Works, Indonesia (in Indonesian).

[13] District Goverment of Kutai Kartanegara 2014 Monitoring of Steel Structure, www.dpukukar.com.

[14] Midas 2006 MIDAS Information Technology Co. Ltd.

IOP Conf. Series: Materials Science and Engineering **267** (2017) 012019 doi:10.1088/1757-899X/267/1/012019

Effect of Casing Imperfection on the Casing Strength in Steam Injection Wells

M I P Hidayat[1], S Irawan[2] and M Z Abdullah[2]

[1] Department of Materials and Metallurgical Engineering, Institut Teknologi Sepuluh
Nopember, Kampus ITS Keputih Sukolilo 60111 Surabaya, Indonesia

[2] Department of Petroleum Engineering, Universiti Teknologi PETRONAS, Bandar Seri
Iskandar, 31750 Tronoh, Perak Darul Ridzuan, Malaysia

E-mail: irfan@mat-eng.its.ac.id,
drsonny_irawan@utp.edu.my,
zaki_abdullah@utp.edu.my

Abstract. In this study, effect of casing imperfection on the casing strength in thermal environment of steam injection wells was investigated. The imperfections were modelled in the form of casing pits which may represent local cavitation, wear or corrosion attacks. Several pits with various depths (*C*) relative to the casing thickness (*t*) and different casing diameter to thickness (*D*/*t*) ratios were considered in this study. For the purpose of investigation, 3D FE models consisting of defective casing-cement-formation system were developed. N80 casing with the casing length of 3.048 m is analyzed in the numerical study. In the analysis, the combined effects of cyclic thermal stress (from 25 °C to 360 °C), collapse pressure (from formation) and defects on casing were taken into account. From the numerical study, it is found that the casing imperfection greatly affected the casing strength in steam injection wells. The presence of pits markedly reduces the casing collapse strength in comparison to that of non-defective casing. For instance, the collapse strength of casing having 2 pits along its circumferential direction with depth C = 0.5*t* reduced about 49% and 43% for D/t = 22.09 and D/t = 17.26, respectively. In addition, the combination of high temperature, imperfection factor and formation pressure induces severe deformation on casing, even at lower formation pressure.

1. Introduction

Casing integrity and its assessment have been always main concerns in the operational of wells in oil and gas industries. This is due to the fact that casing integrity is affecting the oil/gas production rate. In steam injection wells, some important factors related to the casing integrity need to be considered properly so that its assessment can be carried out accurately. They may include: (i) casing thermal expansion and cyclic thermal stress, due to well working temperature during steam injection, production and cooling phases [1, 2], (ii) surrounding formation pressure, due to non-uniform initial in-situ stresses in the rock salt formation, sanding induced cavities or rock flow [3], (iii) internal/burst pressure [4], and (iv) strength degradation of casing material (yield strength/ultimate strength), due to temperature dependent structural and thermal properties [5], such as modulus of elasticity (*E*), density

(ρ), Poisson ratio (μ) and coefficient of thermal expansion (α). All the factors could independently or in combination induce large deformations and stresses to cemented casing, causing strength degradation and integrity losses for the casing [6, 7].

In addition, another factor worth of consideration is imperfection. Imperfection factors may present in the forms of defect, ovality, eccentricity and residual stress. Among others, defect is an important factor that will reduce casing strength as it is thinning out the casing and acting as stress concentrator. Liu et al. n [5] proposed a collapse strength formula to calculate the collapse strength of defective casings in steam injection wells. The formula improved the previous standards and formulations in ISO/TR 10400 [8] by taking into account the effect of temperature along with the presence of pits. It is shown that pit depth and pit circumferential spread are main factors affecting collapse strength of defective casing. In addition, Lin et al. [9] carried out casing integrity evaluation for the application of casing design by considering the effects of wear and corrosion on casing.

In this study, effect of casing imperfection on the casing collapse strength in thermal environment of steam injection wells was investigated. This paper extends the previous study by Hidayat et al. [10] in the sense that defective casing analysis is also related to the thermal cycle experienced by the casing. The imperfections were modelled in the form of casing pits which may represent local cavitation, wear or corrosion attacks. Several pits with various depths (C) relative to the casing thickness (t) and different casing diameter to thickness (D/t) ratios are considered in this study. For the purpose of investigation, 3D FE models consisting of defective casing-cement-formation system were developed. N80 casing with the casing length of 3.048 m is analyzed in the numerical study. In the analysis, the combined effects of cyclic thermal stress (from 25 °C to 360 °C), collapse pressure (from formation) and defects on casing are taken into account. In particular, stress and deformation of defective casing after certain thermal cycle are pointed out, in comparison to that of non-defective casing. It is shown that the combination of high temperature, imperfection factor and formation pressure induces severe deformation on casing, even at lower formation pressure.

2. Thermo-mechanical finite element analysis

Problem of cemented casing in steam injection wells can be represented by coupled thermo-mechanical analysis. Details of the thermo-mechanical analysis can be found in [10], and references therein. Readers are also directed to [11, 12] for further thermal formulation in media or structure.

Here, the coupled thermal-mechanical FE equations are briefly described as follows:

$$^{i+1}[K_1]\{\Delta U\} - {}^{i+1}[K_2]\{\Delta T\} = {}^{i+1}\{R\} - {}^{i}\{R\} \tag{1}$$

where:

$$[K_1] = \int_v [B]_T [D^{ep}]\, [B]\, dV \tag{2}$$

$$[K_1] = \int_v [B]^T [C^{th}]\, [M]\, dV \tag{3}$$

$$\{R\} = \int_s [N]^T \{p\}\, dS + \int [N]^T \{f\} dV \tag{4}$$

$$[D^{ep}] = [D^e] + [D^p] \tag{5}$$

$\{\Delta U\}$ and $\{\Delta T\}$ are the incremental of nodal displacements and temperature, respectively, [B] is the matrix of strain-displacement, $[D^e]$ is the matrix of elastic stiffness, $[D^p]$ is the matrix of plastic stiffness, [C] is the matrix of thermal stiffness, [M] is the temperature shape function, $\{p\}$ is the vector of traction, $\{f\}$ is the vector of body force, i is the current step and $i+1$ is the next step of analysis.

The updated nodal displacements and stresses in the structure are respectively given as:

$$^{i+1}\{U\} = {}^{i}\{U\} + \{\Delta U\} \tag{6}$$

$$^{i+1}\{\sigma\} = {}^{i}\{\sigma\} + \{\Delta\sigma\} \tag{7}$$

$$\{\Delta\sigma\} = [D^{ep}]\,[B]\,\{\Delta U\} + [C]\,[M]\,\{\Delta T\} \tag{8}$$

$$\{R\} = \int_s [N]^T \{p\}\, dS + \int [N]^T \{f\} dV \tag{9}$$

$$[D^{ep}] = [D^{e}] + [D^{p}] \tag{10}$$

3. Collapse analysis of defective casing

3D FE models consisting of casing-cement-formation system were developed in ANSYS, as shown in Figure 1. Element contact with the Coulomb friction model are employed to model interaction between casing and cement as well as cement and formation. Geometrical data and material properties for the system of Grade N80 casing, cement and formation are given in Table 1. Length of the casing is 3.048 m. Imperfections are imposed to the 3D FE model in the form of pits.

Table 1. Geometrical data and material properties (at 25 °C) used in the numerical calculations.

Parameters and properties	Casing	Cement sheath	Rock
Dimension or OD (m)	0.1778	0.2238	1.572
Wall thickness (m)	0.0103	0.023	-
Young's modulus (E, GPa)	209	2.8	100
Poisson ratio (μ)	0.33	0.2	0.31
Density (ρ, kg/m^3)	7850	2400	2650
Thermal expansion (α, 1/°C)	1.69E-6	9E-6	5.4E-6

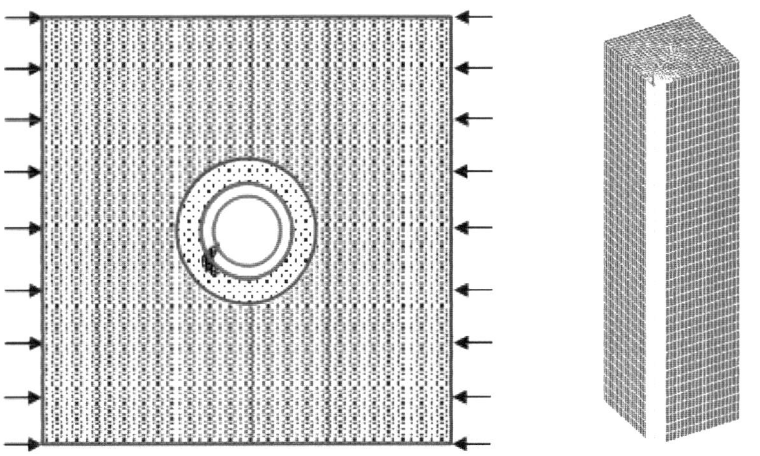

(a) (b)

Figure 1. (a) Schematic of casing-cement-formation system under collapse pressure and (b) three-dimensional FE mesh (Hidayat et al., 2016).

4. Results and Discussion

Several pits with various depths (C) were considered i.e. C = 0.20t, 0.50t and 0.70t, where t is the casing thickness. In addition, the values of D/t = 17.26 and 22.09 were examined. Here, lower D/t value associates with better collapse strength denoting increased casing capability to withstand external or formation pressure. Figure 2 shows defective casing with various pit depths and 2 pits.

In [10], the suitability of the 3D FE model is already shown that the simulation results obtained for non-defective casings are similar with those obtained in [5]. It was obtained previously that collapse strengths for non-defective casings at 25C were 55 MPa and 71 MPa for the values of D/t = 22.09 and 17.26, respectively.

Figure 2. Defective casing with various pit depths of (a) C = 0.20t, (b) C = 0.50t and (c) C = 0.70t.

Table 2 presents the simulation results of defective casing collapse strength with respect to various pit depths and D/t ratios. Note that the results are corresponding with 2 pits shown in Figure 2.

Table 2. Defective casing collapse strength for different D/t ratios (at 25 °C).

D/t = 22.09			D/t = 17.26		
Collapse strength of non-defective casing is 55 MPa	*C*	Collapse strength (MPa)	Collapse strength of non-defective casing is 71 MPa	*C*	Collapse strength (MPa)
	0.2*t*	43.4		0.2*t*	57.6
	0.5*t*	28.2		0.5*t*	40.5
	0.7*t*	19.1		0.7*t*	27.2

It can be seen that the presence of pits affects the casing collapse strength markedly in comparison to that of non-defective casing as given in Table 2. For instance, for C = 0.5t the casing collapse strength reduced about 49% and 43% for D/t = 22.09 and D/t = 17.26, respectively.

Further, the analysis for the defective casing is expanded here i.e. deformation of defective casings is also related to the thermal cycle experienced by the casing. This aspect has not still been considered in the previous studies. For clarity, deformations of non-defective casing (under pressure of 17.2 MPa) and

IOP Conf. Series: Materials Science and Engineering **267** (2017) 012019 doi:10.1088/1757-899X/267/1/012019

some defective casings (C = 0.20t) with various number of pits and different external pressure values are shown in Figure 3.

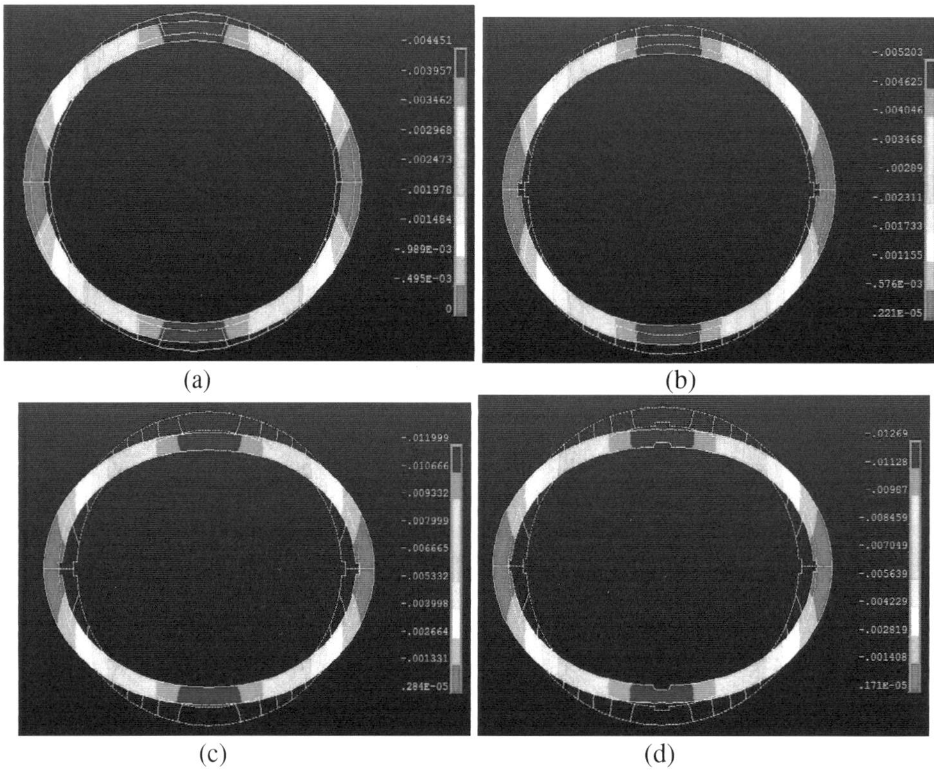

Figure 3. Deformation of (a) non-defective casing (under pressure of 17.2 MPa), and defective casing C = 0.20t) with various number of pits (b) 2 pits under pressure of 17.2 MPa, (c) 2 pits under pressure of 33 MPa, and (d) 4 pits under pressure of 33 MPa.

From the results, it was obtained that defective casing deformation after one thermal cycle was greater than that of non-defective casing after six thermal cycles. For instances: a defective casing with C = 0.20t and 2 pits under 17.2 MPa deformed 5.2 mm after one cycle (Figure 3(b)), while that of non-defective casing under the same pressure was 4.5 mm after six cycles (Figure 3(a)). In addition, defective casings with C = 0.20t and 2 pits and C = 0.20t and 4 pits deformed 12 mm and 13 mm, respectively, after one cycle under 33 MPa (Figures 3(c) and 3(d)), while that of non-defective casing under the same pressure was 10.9 mm after six cycles. Hence, it is clear that the casing imperfection greatly affects the casing strength and casing strength degradation. It is also obvious that the combination of high temperature, imperfection factor and formation pressure induces severe deformation on casing, even at lower formation pressure. It is noted that steady external pressure between 10 and 40 MPa could be experienced by casing due to rock flow caused by creep formation [13].

5. Conclusions

In this study, effect of casing imperfection on the casing strength in thermal environment of steam injection wells has been investigated. The imperfections were modelled in the form of casing pits which may represent local cavitation, wear or corrosion attacks. Several pits with various depths along with different D/t ratios were considered in this study. Deformation of defective casings is also related to the thermal cycle experienced by the casing. It was found that the casing imperfection greatly affected the casing strength in steam injection wells. In addition, the combination of high temperature, imperfection factor and formation pressure induces severe deformation on casing, even at lower formation pressure.

References

[1] Maruyama, K., Tsuru, E., Ogasawara, M., Inoue, Y. and Peters, E.J. (1990), "An experimental study of casing performance under thermal cycling conditions", *SPE Drilling Engineering*, Vol. 5(2), pp. 156–64.

[2] Placido, J.C.R., Ademar P. Jr., Paulo, L.A.B.F., Pasqualino, I.P. and Estefen, S.F., *Stress analysis of casing string submitted to cyclic steam injection*, The Fifth Latin American and Caribbean Petroleum Engineering Conference and Exhibition, 30 August-3 September 1997, Rio de Janeiro, Brazil.

[3] Peng, S., Fu, J. and Zhang, J. (2007), "Borehole casing failure analysis in unconsolidated formations: a case study", *Journal of Petroleum Science and Engineering*, Vol. 59, pp. 226–38.

[4] Dall' Acqua, D., Hodder, M. and Kaiser, T.M.V. (2013), "Burst and collapse responses of production casing in thermal applications", *SPE Drilling and Completion*, March, pp. 93-104.

[5] Liu, S., Zheng, H., Zhu, X. and Tong, H. (2014), "Equations to calculate collapse strength of defective casing for steam injection wells", *Engineering Failure Analysis*, Vol. 42, pp. 240–251.

[6] Li, Z. (2008), "Casing cementing with half warm-up for steam injection wells", *Journal of Petroleum Science and Engineering*, Vol. 61, pp. 94–8.

[7] Lin, Y., Sun, Y., Shi, T., Deng, K., Han, L., Sun, H., Zeng., D and Zhu, H. (2013), "Formula to calculate collapse strength for high collapse casing", *Journal of Pressure Vessel Technology*, Vol. 135(4), pp. 041202-1 - 041202-5.

[8] "Petroleum and Natural Gas Industries-Formula and Calculation for Casing, Tubing, Drill Pipe and Line Pipe Properties," ISO/TR 10400, 2007.

[9] Lin, T., Zhang, Q., Lian, Z., Chang, X., Zhu, K. and Liu, Y. (2016), "Evaluation of casing integrity defects considering wear and corrosion-Application to casing design", *Journal of Natural Gas Science and Engineering*, Vol. 29, pp. 440-452.

[10] Hidayat, M.I.P., Irawan, S. and Abdullah, M.Z. (2016), "Casing strength degradation in thermal environment of steam injection wells", *Journal of Physics: Conference Series*, Vol. 710, pp. 1-10.

[11] Hidayat, M.I.P., Wahjoedi, B.A., Parman, S. and Megat Yusoff, P.S.M. (2014), "Meshless local B- spline-FD method and its application for 2D heat conduction problems with spatially varying thermal conductivity", *Applied Mathematics and Computation*, Vol. 242, pp. 236-54.

[12] Hidayat, M. I. P., Ariwahjoedi, B. and Parman, S. (2015), "A new meshless local B-spline basis functions-FD method for two-dimensional heat conduction problems", *International Journal of Numerical Methods for Heat and Fluid Flow*, Vol. 25, pp. 225-251.

[13] Qian, F. and Gao, D. (2011), "A mechanical model for predicting casing creep load in high temperature wells", *Journal of Natural Gas Science and Engineering*, Vol. 3, pp. 530-535.

Acknowledgment

The present authors gratefully acknowledge grant provided by Universiti Teknologi PETRONAS.

IOP Conf. Series: Materials Science and Engineering **267** (2017) 012020 doi:10.1088/1757-899X/267/1/012020

Digital 3D Microstructure Analysis of Concrete using X-Ray Micro Computed Tomography SkyScan 1173: A Preliminary Study

F D E Latief[1], I H Mohammad[2], A D Rarasati[3]
[1]Physics of Earth and Complex Systems, Faculty of Mathematics and Natural Sciences, Institut Teknologi Bandung, Jalan Ganesha 10 Bandung 40132, Indonesia
[2] Department of Geophysics, Universitas Padjadjaran, Jalan Raya Bandung Sumedang KM 21, Jatinangor, Sumedang 45363 Indonesia
[3]Department of Civil Engineering, Faculty of Engineering, Universitas Indonesia, Kampus UI Depok, Depok 16424, Indonesia

E-mail: fourier@fi.itb.ac.id

Abstract. Digital imaging of a concrete sample using high resolution tomographic imaging by means of X-Ray Micro Computed Tomography (μ-CT) has been conducted to assess the characteristic of the sample's structure. A standard procedure of image acquisition, reconstruction, image processing of the method using a particular scanning device i.e., the Bruker SkyScan 1173 High Energy Micro-CT are elaborated. A qualitative and a quantitative analysis were briefly performed on the sample to deliver some basic ideas of the capability of the system and the bundled software package. Calculation of total VOI volume, object volume, percent of object volume, total VOI surface, object surface, object surface/volume ratio, object surface density, structure thickness, structure separation, total porosity were conducted and analysed. This paper should serve as a brief description of how the device can produce the preferred image quality as well as the ability of the bundled software packages to help in performing qualitative and quantitative analysis.

1. Introduction

Imaging technique for characterization of various type of materials has been advancing in a fast progress along with the development of instrumentation and computer technology. Among the widely used imaging instrument is the X-Ray Micro Tomography Scan (μ-CT). This device has been employed in various field of research such as for digital rock analysis for numerous application such as calculation of porosity, permeability, simulation of fluid flow and various structural description [1-5]. In life science, the method has been applied for various dentistry researches [6-8] as well as bone research. [9-11]. In building materials researches, the μ-CT has been utilized for investigating self-sealing in concrete materials [12], porosity and characterization of building stones microstructure [13], and investigation of water repellent and consolidants in natural building stones [14].

[1] fourier@fi.itb.ac.id
[2] imran@geophys.unpad.ac.id
[3] ayomi@eng.ui.ac.id

This paper is aimed to present an application of μ-CT preliminary analysis on concretes using specific device, i.e. the Bruker SkyScan 1173 X-Ray μ-CT. Procedure of obtaining images through three main stages and the adjustment of corresponding parameters are presented. Qualitative analysis is discussed briefly to describe the capability of the device and the bundled software package.

2. Sample and Method

In this study, a concrete core sample was obtained from an existing building in Jakarta. The sample has a diameter of 1.5 inch and a height of 6.5 cm which was taken by a coring machine from a beam structure (see Figure 1.a.). The sample was scanned using Bruker SkyScan 1173 High Energy Micro Computed Tomography scanning system (Figure 1.b.). Based on the X-Ray attenuation by the sample which is recorded by a flatpanel detector, the device is able to produce microstructural 3D image of the sample. The SkyScan 1173 device is mainly consist of three modules, i.e., the X-Ray source, a rotating sample stage and a CCD flatpanel detector which record and convert the attenuated X-Ray signal into digital image. The tomography concept applied in the system requires the sample to be rotated at a minimum rotation of 180° during the scanning process with the rotation step adjusted according to the chosen image resolution.

(a)

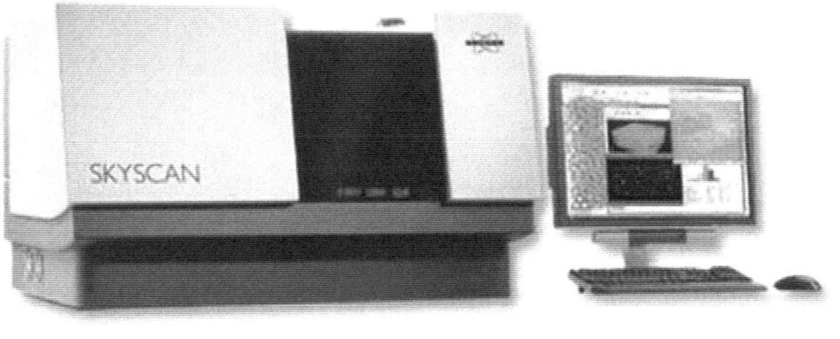

(b)

Figure 1. (a) sample of concrete, (b) the SkyScan 117 Micro-CT scanning device (www.bruker.com).

The detector module of SkyScan 1173 has the ability to produce three different image quality: standard, medium and high which are specified as follows:

- high quality image is produced by using the 1×1 binning system where the process generates a raw projection image with dimension of 2240×2240 pixels and spatial resolution of 5-35 μm,
- medium quality image is produced by increasing the binning size of 2×2 thus generated image with dimension of 1120×1120 pixels and a spatial resolution of 10-70 μm,
- standard quality image is produced by increasing the binning size of 4×4 with a corresponding image dimension of 560×560 pixels and spatial resolution of 20-140 μm.

IOP Conf. Series: Materials Science and Engineering **267** (2017) 012020 doi:10.1088/1757-899X/267/1/012020

The resolution of the image is determined by the position of the sample and the desired field of view (FOV) on the sample. The FOV for a single-scan mode is determined by multiplying image dimension with the desired resolution, thus for example, the highest resolution image will have a FOV of (2240×2240)×5 μm = (11.2×11.2) mm. For a large sample (up to a maximum diameter of 14 cm and height of 18 cm), SkyScan 1173 system provide extended capability of non-single-scan modes, i.e., the wide scan, a (vertically) oversize scan, and the combination of both. Wide scan is achieved by moving the CCD to the left and right position thus producing two half-sides projection images with dimension of 4480×2240. Oversize scan is done by moving the sample holder vertically. Another possible mode is the spiral scan where the scanning process is done by rotating the sample holder and at the same time moving the sample holder vertically thus producing a spiral movement.

Producing a good 3D image of the structure means generating images with a good contrast ratio of the components and the high signal to noise ratio. By adjusting the scanning parameters, which are the energy of the X-Ray source (in kV), the current of the source (in μA), the exposure time of the detector (in ms), the type of X-Ray filter, the rotation step, and the amount of frame averaging accordingly, the good contrast ration can be achieved. For a high density sample such as concrete, a high energy X-ray is required as well as a high current and exposure time to reduce the noise. Brass-type filter is also required to reduce the noise produced from low energy X-Rays. In this study, the sample was scanned using three different sets of parameters. The adjusted scanning parameters for each scanning process are listed in Table 1.

Table 1. The adjusted scanning parameters.

Parameter	**Scan 1**	**Scan 2**	**Scan 3**
Energy [kV]	100	130	130
Current [μA]	40	61	61
Exposure time [ms]	125	225	1500
Filter type	brass 0.25 mm	brass 0.25 mm	brass 0.25 mm
Rotation step	0.3°	0.1°	0.2°
Camera binning	4×4	4×4	1×1
Dimension of projection image	560×560	560×560	2240×2240
Raw image resolution [μm/pixel]	89.78	119.71	9.98
Frame averaging	10	10	10
Wide scan	Yes	No	No

For Scan 1, the sample was positioned to be quite close to the source. For Scan 2, the sample was scanned in a position where the total height of the samples is within the field of view (FOV) of the CCD. For Scan 3, the sample is positioned as close as possible to the source thus producing highest resolution, however in this scanning mode, the produced image is the centre (partial) volumetric region of the sample.

From the scanning process, raw projection images were produced. These images are basically similar to the ones obtained from X-Ray radiograph which display the attenuated X-Rays that are captured by the detector. The images portray the nature of the sample's structure vaguely, because they display the projected structure in 2D space. A complete 3D structure of the sample can be further obtained by performing a reconstruction of the projection images (sometimes also referred as sinograms) which is basically a numerical inversion based on the Feldkamp backpropagation algorithm. The reconstruction is done by the NRecon software (Bruker Micro-CT) which uses the GPUReconServer that utilizes the GPU processors to accelerate the core reconstruction process. During this process, several parameters can be adjusted to enhance the quality of the generated images. The adjustment of these parameters level can be done by first performing a reconstruction of one 2D visual of the trans-axial slice as a preview. A fine tuning of the parameters can be done by NRecon to ensure that the artefact removal can be done accurately. The parameters are as follows:

IOP Conf. Series: Materials Science and Engineering **267** (2017) 012020 doi:10.1088/1757-899X/267/1/012020

1. Post-alignment/misalignment compensation, which is basically a correction of raw image position that might be caused by object's movement during the scanning process.
2. Ring artefact reduction. A process of removing circular noise/pattern which is basically appears due to rotating sample. The level of this artefact might vary according to sample and holder state. Increasing the number of random movement usually decrease the level of the artefact quite significantly.
3. Beam hardening correction. Beam hardening artefact appears in the reconstructed image mainly caused by the density or the attenuation coefficient of the sample. Most high density objects that are scanned using SkyScan 1173 shows the pattern where the centre part of the sample usually has significantly lower grayscale value compared to the ones in the outer part of the image.
4. If the object is larger than the field of view, an option named 'Object Bigger Than FOV' should be selected, and the contrast ratio of the generated image would be enhanced.

Other parameter that can be adjusted during the reconstruction process is the reconstructed image pixel size. This option provides the possibility of producing reconstructed images with lower spatial resolution compared to the one that was adjusted for the scanning process. This can be done by means of undersampling method which is basically a method of choosing the voxel size of the reconstructed image and then pixels of the raw projection images are backprojected into that chosen area during the reconstruction process [15]. A smoothing can also be done if necessary, and the suitable amount of smoothing can be adjusted by observing the preview image. The smoothing reduces noise, however, the use of the smoothing kernels (Box (asym.), Box (sym.), and Gaussian) it might also promote blurring for highly detailed structures. A certain degree of static rotation can also be selected if we want to produce a reconstructed image that has different angle with the actual angular position of the scanned sample. Another important thing to consider is that the reconstruction process is done using real number (floating-point) while the produced images consist of integers where the range is determined by the output file format. The default output format is 8-bit BMP, which means that the output voxels have the value between 0 and 255. The dynamic range of data of the image can be tuned by adjusting the range from the histogram of the previewed slice.

The output from the reconstruction process is a series of greyscale images, which is a stack of 2D slice in the z direction (x-y plane) or most of the time referred as trans-axial images. These 8-bit images contain data of relative density which represented by the voxel value of the images. Qualitative analysis through visual observation can be done in 2D, orthoslice view, as well as in 3D volumetric rendered images. Quantitative characterization is commonly done on binary images, which mean that the grayscale image needs to be converted through thresholding process. Various thresholding scheme are available in CTAn software and they can be generally categorized into global histogram-based thresholding and local-adaptive threshold. Depending on the nature of the samples, both of them can be chosen appropriately. Using the binary images, various characterization of the sample can be done. Table 2 lists the common characteristics that can be measured using CTAn software.

3. Results and Discussions

Scanning process produces a series of raw projection images in 16-bit TIFF file format. The preview of the images can be seen in Figure 2 (a,b,c). The interior microstructure of the sample can't be observed only from these images. By reconstructing the projection images, we obtained the grayscale images as shown in Figure 2 (d,e,f). The 2D slices of Figure 2.d-f were chosen for the approximately the same vertical position. We could see a similar structure appeared in these images. Scan 1 and Scan 2 covered the same area but they had difference in the resolution and the rotation step during the scan.

The result from Scan 1 and Scan 2 showed a compatible comparison which involves the effect of X-Ray energy source, rotation step and the resolution. Scan 1 was done using energy of 100 kV while Scan 2 using 130 kV. Scan 1 was rotated through a rotation step of 0.3° while Scan 1 was scanned using rotation step of 0.1°, and the voxel size of Scan 1 is ¾ that of Scan 2.

Table 2. Some typically calculated morphometric parameters.

Parameter name, symbol and unit	Brief description
Total VOI volume, TV, mm^3	Total volume of the volume-of-interest (VOI), measured using marching cubes volume model approach.
Object volume, Obj.V, mm^3	Total volume of the solid within the VOI, measured using marching cubes volume model approach.
Percent object volume, Obj.V/TV, %	Ratio of the solid volume to the total VOI.
Total Porosity, Po(tot), %	Ratio of the pore volume to the total VOI volume, expressed in percent.
Total VOI surface, TS, mm^2	Surface area of the total 3D VOI, measured by marching cubes method. Very useful for irregular VOI.
Object surface, Obj.S, mm^2	Surface area of the 3D solid objects inside the VOI, using the marching cubes method.
Object surface/volume ratio, Obj.S/Obj.V, mm^{-1}	Specific surface area: ratio of the solid wall surface the total measured VOI.
Object surface density, Obj.S/TV, mm^{-1}	The ratio of surface area to total VOI volume.
Structure Thickness, St.Th, mm	Thickness of the solid part of the sample, measured based on method defined in [16], [17] and [18].
Structure Separation, St.Sp, mm	Thickness of the void space which is measured by the same method to the measurement of the structure thickness which applied to the void voxels.

Generally, higher resolution provides better result in terms of quality, however from this comparison we could observe that the rotation step also gave significant effect to the reconstructed image. By doing the scan using higher rotation step, it is shown that certain kind of noise is introduced in the final reconstructed image. This is shown by unsmooth image of Scan 1. Comparing Scan 1-2 to Scan 3 (Figure 2f.), it is obvious that lower binning of the camera produces better image in terms of the depicted details of the composing objects as well as the sharpness of the object boundaries. These pattern is also observable in the 3D visual of the samples.

Volumetric visual of the three sets of digital data are shown in Figure 3. We can see from Figure 3 that each scanning mode produced different quality. Scan 1 provided a slightly better detail compared to Scan 2, however Scan 1 introduce noise that reduces the sharpness of the image. In addition to that, lower energy of the X-Ray source used in Scan 1 caused the luminosity of image to be lower (the image appear darker) compared to image obtained from Scan 2. This pattern appeared in both the 2D and 3D visual (Figure 2 and Figure 3).

Afterwards, quantitative analysis can be done by picking out the proper VOI and performing appropriate thresholding (binarisation) method. For the samples, VOI selection was done using regular cylindrical sample which clipped a small amount of outer area of the sample, especially for Scan 1 and Scan 2. Sequentially, automatic thresholding using global Otsu method was applied on the digital samples. The 3D analysis was later conducted using CTAn software and several characteristics are measured and listed in Table 3. Several discrepancies of the parameters were observed. A more elaborated analysis of the measured characteristic in the context of different resolution is required to decide which result to use according to a specific intention. Such analysis was already performed in rock sample by Latief et al. [19].

As an additional analysis of the sample listed in Table 3, some simple and direct measurement can also be done on the grayscale image, such as measurement of length and angle of orientation as shown in Figure 4. CTAn provides several ways of conducting such simple measurements on the grayscale image.

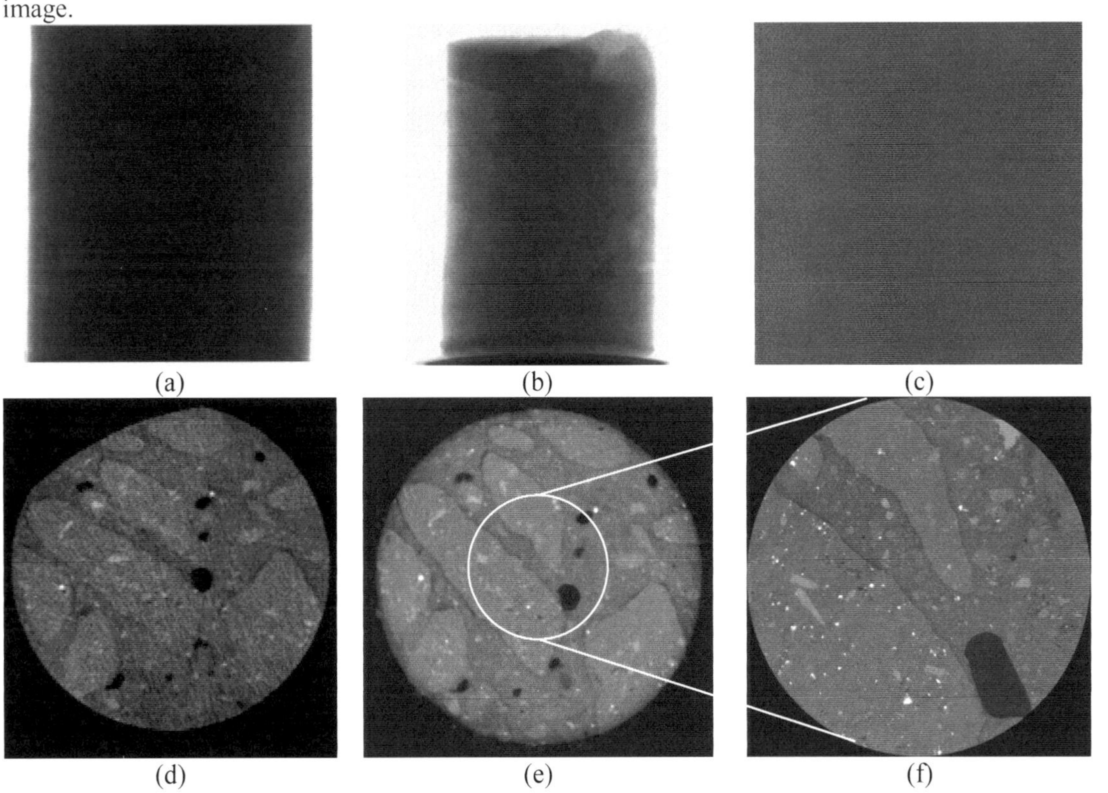

(a) (b) (c)

(d) (e) (f)

Figure 2. Projection image of the three scanning modes (a) Scan 1, (b) Scan 2, and (c) Scan 3.

(a) (b) (c)

Figure 3. Volumetric 3D visual of the three scanning modes (a) Scan 1, (b) Scan 2, and (c) Scan 3.

Other facts that needs to be put into a consideration is that a certain scanning mode is highly related to certain needs in investigating the structure. In our case, the highest resolution (Scan 3) was obtained by placing the sample in a very close position thus only display a partial area of the sample. In contrast to Scan 3, Scan 2 provide a description of the whole sample. Scan 2 has lower resolution compared to Scan 3, which means, some details would not be able to be analysed only by using data from Scan 2. Considering that, the data that was generated from Scan 3 must also be carefully interpreted. In Figure 2.f a thin gap is visible across the diameter of the sample. We must be careful not to hastily interpret this as a fracture inside the sample. If we observe carefully the image from Scan 1 and Scan 2, this gap is actually a border of a large aggregate which could not be visualized on the whole due to its size which is larger than the designated FOV.

Table 3. Calculated morphometric parameters for the corresponding scan modes.

Parameter name, symbol and unit	Scan 1	Scan 2	Scan 3
Total VOI volume, TV, μm^3	5.23×10^{13}	6.13×10^{13}	6.80×10^{12}
Object volume, Obj.V, μm^3	5.16×10^{13}	6.04×10^{13}	5.07×10^{12}
Percent object volume, Obj.V/TV, %	98.33	93.72	74.57
Total VOI surface, TS, μm^2	8.08×10^{9}	9.27×10^{9}	2.06×10^{9}
Object surface, Obj.S, μm^2	1.48×10^{10}	2.79×10^{10}	1.83×10^{10}
Object surface/volume ratio, Obj.S/Obj.V, μm^{-1}	2.88×10^{-4}	4.62×10^{-4}	3.60×10^{-2}
Object surface Density, Obj.S/TV, μm^{-1}	2.83×10^{-4}	4.33×10^{-4}	2.69×10^{-2}
Structure Separation, St.Sp, μm	1.22×10^{3}	8.60×10^{2}	8.13
Total Porosity, Po(tot), %	1.67	6.28	25.43

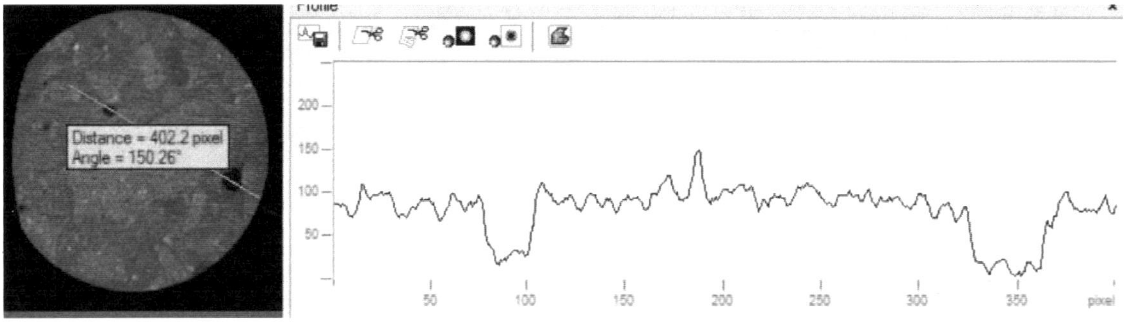

Figure 4. Measuring the distance between two points (length of an object) inside the sample as well as the angle of orientation. Graph on the right is the grey-level value of the pixels along the yellow line.

From Table 3 we can also see the pattern of the measured (calculated) parameters which are highly dependent to the resolution (or to be exact, the quality) of the produced image dataset. In other words, thorough analysis should be taken carefully in order to produce accurate measurement and interpretation of the sample. Other various possible scanning modes or schemes are also important to be studied in order to obtain a more detailed insight of what could be achieved from each of the scanning schemes.

4. Summary

Procedure of conducting digital analysis on a concrete sample using Bruker SkyScan 1173 has been presented. Three different sets of scanning parameters were used to provide a depiction of how the produced images was affected by the parameters as well as to deliver a brief portray of what sets of parameters should be adjusted for a particular purpose. Brief qualitative analysis through visual inspection and quantitative analysis via some calculated parameters were presented. We can summarize that higher resolution does not always guarantee to generate images which can display good detail of the sample. This paper should serve as a brief description of how the device can produce the preferred image quality as well as the ability of the bundled software packages to help in performing qualitative and quantitative analysis.

5. Acknowledgments

This research is partially funded by the research program "Riset Desentralisasi DIKTI-ITB (PUPT) 2017".

6. References

[1] Bornert M 2010 X-ray micro CT for studying strain localization in clay rocks under triaxial compression *Adv. X-ray Tomogr. Geomater* **118** 35

[2] Dong H, Fjeldstad S, Alberts L, Roth S, Bakke S and Øren P-E 2008 Pore network modelling on carbonate: a comparative study of different micro-CT Network extraction methods. In: *International Symposium of the Society of Core Analysts Abu Dhabi, UAE,*

[3] Sok R M, Knackstedt M A, Varslot T, Ghous A, Latham S and Sheppard A P 2010 Pore scale characterization of carbonates at multiple scales: Integration of Micro-CT, BSEM, and FIBSEM *Petrophysics* **51**

[4] Mostaghimi P, Blunt M J and Bijeljic B 2013 Computations of absolute permeability on micro-CT images *Mathematical Geosciences* **45** 103-25

[5] Latief F D E, Fauzi U and Feranie S 2012 Digital isolation technique for reconstruction and visualization of cracks in micro-CT images of geothermal reservoir rock *Microscopy and Analysis* 13-7

[6] Jung M, Lommel D and Klimek J 2005 The imaging of root canal obturation using micro-CT *International endodontic journal* **38** 617-26

[7] Magne P 2007 Efficient 3D finite element analysis of dental restorative procedures using micro-CT data *Dental Materials* **23** 539-48

[8] Peters O, Peters C, Schonenberger K and Barbakow F 2003 ProTaper rotary root canal preparation: effects of canal anatomy on final shape analysed by micro CT *International endodontic journal* **36**

[9] Vandenberghe B, Luchsinger S, Hostens J, Dhoore E, Jacobs R and Consortium S P 2014 The influence of exposure parameters on jawbone model accuracy using cone beam CT and multislice CT *Dentomaxillofacial Radiology*

[10] Latief F D E, Dewi D E O, Shari M A B M, Djamal M, Vierdayanti K, Alamsyah I M, Mukti R R, Indrasari W and Islahuddin M 2014 Three-dimensional visualization and characterization of bone structure using reconstructed in-vitro μCT images: A pilot study for bone microarchitecture analysis. In: *AIP Conference Proceedings*: AIP) pp 418-21

[11] Ibrahim N, Parsa A, Hassan B, Stelt P, Aartman I H and Wismeijer D 2014 Accuracy of trabecular bone microstructural measurement at planned dental implant sites using cone-beam CT datasets *Clinical oral implants research* **25** 941-5

[12] Fukuda D, Nara Y, Kobayashi Y, Maruyama M, Koketsu M, Hayashi D, Ogawa H and Kaneko K 2012 Investigation of self-sealing in high-strength and ultra-low-permeability concrete in water using micro-focus X-ray CT *Cement and Concrete Research* **42** 1494-500

[13] Cnudde V, Cwirzen A, Masschaele B and Jacobs P 2009 Porosity and microstructure characterization of building stones and concretes *Engineering geology* **103** 76-83

[14] Cnudde V, Cnudde J, Dupuis C and Jacobs P 2004 X-ray micro-CT used for the localization of water repellents and consolidants inside natural building stones *Materials characterization* **53** 259-71

[15] Feldkamp L, Davis L and Kress J 1984 Practical cone-beam algorithm *JOSA A* **1** 612-9

[16] Hildebrand T and Rüegsegger P 1997 A new method for the model-independent assessment of thickness in three-dimensional images *Journal of microscopy* **185** 67-75

[17] Ulrich D, Van Rietbergen B, Laib A and Ruegsegger P 1999 The ability of three-dimensional structural indices to reflect mechanical aspects of trabecular bone *Bone* **25** 55-60

[18] Rémy É and Thiel É 2002 Medial axis for chamfer distances: computing look-up tables and neighbourhoods in 2D or 3D *Pattern Recognition Letters* **23** 649-61

[19] Latief F, Fauzi U, Irayani Z and Dougherty G 2017 The effect of X-ray micro computed t omography image resolution on flow properties of porous rocks *Journal of microscopy* **266** 69-88

Odua Weston Jambi Hotel's Structural Building Design with Prestressed Concrete Slab System Approach

R Bayuaji [1], M S Darmawan[1], M A Rofiq[1], S E Santoso[1] and E Hardiyanto[1]
[1]Department of Civil Infrastructure Engineering Institut Teknologi Sepuluh Nopember Indonesia

E-mail: bayuaji@ce.its.ac.id

Abstract. Odua Weston Jambi Hotel is an eight-floor hotel and located in a prone to earth-quake area. This building used conventional concrete to its structural beam and column. This research's purpose was to maximize the second-floor's function by modifing its architectural design. *Special Moment Resisting Frame System (SMRFS) approach* was used in the structural design, referred to SNI 03-2847-2013 dan SNI 1726-2012 and to compensate the needs of a spacious hall without any column in the centre of the hall, so therefore, prestressed concrete plate is used to solve this problem.

1. Introduction

Odua Weston Jambi Hotel is an eight floor building with conventional and structural concrete building with the height of 35.3 meter tall excluded the basement. Based on a review, this building used reinforced concrete applied to its beam and column. In order to maximize its space function, a modification was planned by designing a non-column hall at the second floor, which is used for meeting room. To compensate that design, prestressed concrete slab is applied because it can resist a bigger deflection.

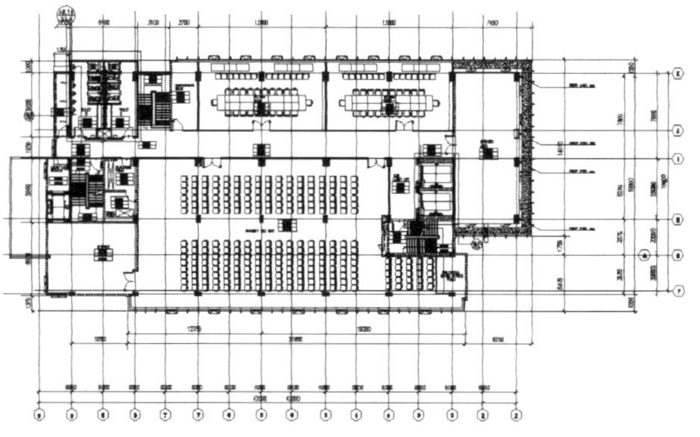

Figure 1. Odua Weston Jambi Hotel's second floor.

Figure 1 shows column segments at Odua Weston Jambi Hotel's Second Floor. To maximize its floor space function, H5, H6, H7 column would be eliminated. This structural building modification are being planned to use conventional and structural concrete and prestressed concrete plate with the area of 24 meter × 12 meter landscape constructed at third floor and 5,2 meter x 24 meter plate constructed from fourth floor to eight floor.

2. Methodology

The installation of prestressed concrete plate are as follows:

2.1 Section Design

Post-tensioned prestressed concrete was choosen in this design because, if prestressed concrete construction is too wide and the structural component is long and heavy, it must be best casted locally or casted part-by-part, and it should be prestressed with postension system, otherwise pre-tensioned concrete is unecconomical (Lin and Burns, 1996). For prestressed concrete design, groating is choosed to be created because of its steel-concrete mixture and to increase prestressed loss. This design used partial prestressed component which allow tensile stress towards its working load, and in the pre-tensioned areas, additional reinforcement is used to be added to non prestressed reinforcement (Lin and Burns, 1996).

2.2 Wire Layout Arrangement

Wire type area stipulation limitation should meet design criteria and permitted requirements. Wire type and quantity stipulation determines wire setting and its limited and permitted requirement.

2.3 Prestressed Loss

Prestressed loss occur because of continous transfer according to time function. This calculation is used to determine prestressed effective force. Prestressed loss effects can lead to concrete elasticity abbreviation, friction, wooble effect, holder armature, creep, shrinkage and steel relaxation.

2.4 Stress Control

Controlling stress in the beam at jacking and service step is a critical step in the design. Control function is to measure wether dimension of the plate can receive given amount of stress and wether the amount of received stress is fit the stress design.

2.5 Deflection Control

Occured deflections are calculated and controled so it would not exceed the stipulated constraint. Deflection is calculated according to loading models which was affected by its own load and external load.

3. Two-Way Direction Tendon Prestressed Plate Design

Noted that prestressed plate size with L_x= 12 m and L_y= 24 m with concrete tension is equal to 30 Mpa and concrete elasticity, E_c= 27691 MPa. Plate with width of 250 mm with 4 strands tendon type strand-7 Ply and strand type A.S.T.M A 416/80 grade 270 kpsi 1.7 mm is designed. Tendon with 4 × 183,7 kN which is equal 729 KN forces is used with several specifications; three tendons per meter, number of tendon in short side is 12 meter × 3 tendons per meter = 36 pieces and for the higher side is 24 meter × 3 tendons per meter = 72 pieces.

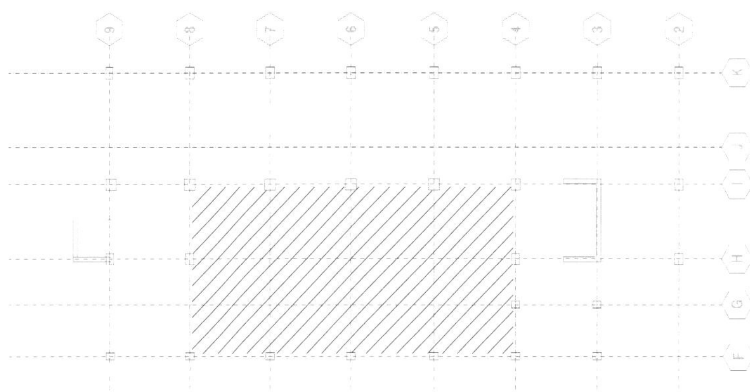

Figure 2. Odua Weston Jambi Hotel's Third Floor.

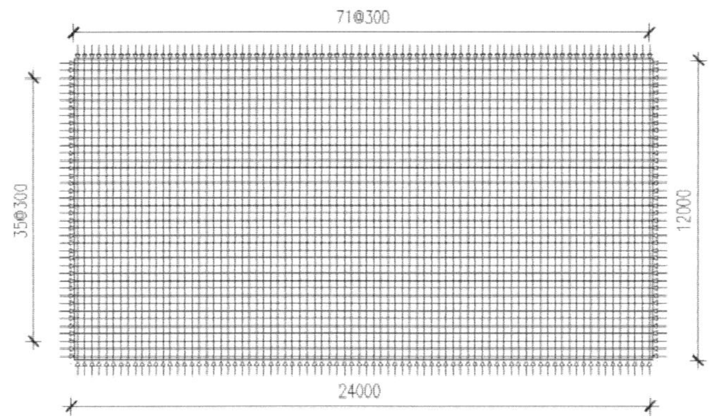

Figure 3. Prestressed concrete slab tendons layout 24 meter x 12 meter.

4. One-Way Direction Tendon Prestressed Plate Design

Noted that prestressed plate size is L_x= 12 m and L_y= 24 m with concrete tension = 30 Mpa and concrete elasticity, E_c= 27691 MPa. Plate with the width of 250 mm with 4 strand(s) tendon(s) type strand-7 Ply and strand type A.S.T.M A 416/80 grade 270 kpsi 12.7 mm is designed. Tendon with 4 × 183,7 kN = 729 KN forces is used with several specification; three tendons per meter, number of tendon in short side is 5.2 meter × 3 tendons per meter = 16 pieces.

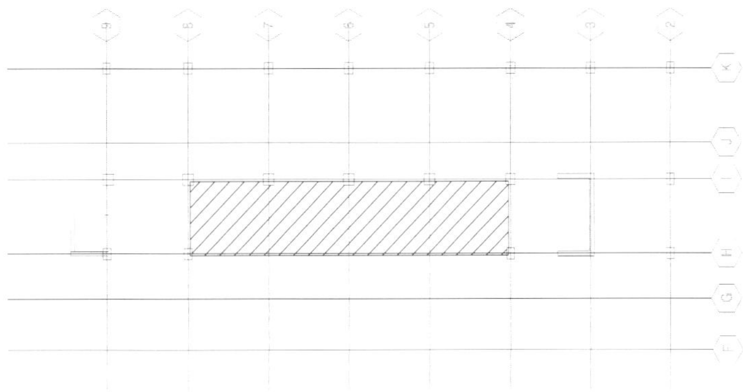

Figure 4. Odua Weston Jambi Hotel's Fourth-Eight Floor.

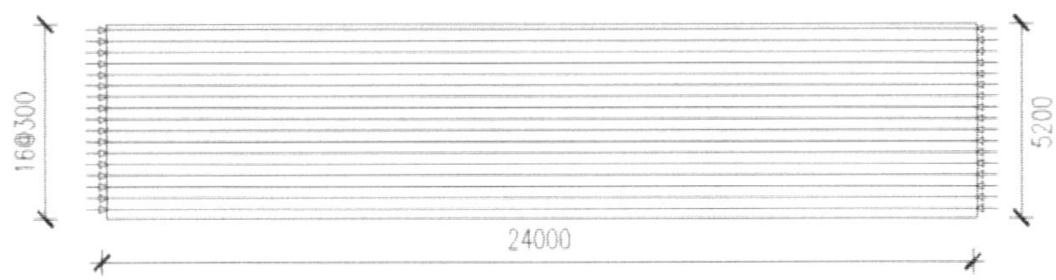

Figure 5. Prestressed concrete slab tendons layout 24 meter x 5,2 meter.

5. Tendon Anchor Design

According to the calculation result, three dashes is required to be assembled within 0,2 H – 1 H range, or with assumption that 1 H = 250 mm then 0,8 H = 200 mm, so dash space is 200 mm/3 which is equal with 66,667 mm. So that dash (Ø) which is 10 mm and dash range which is 60 mm can be used (Freyssinet Spesification).

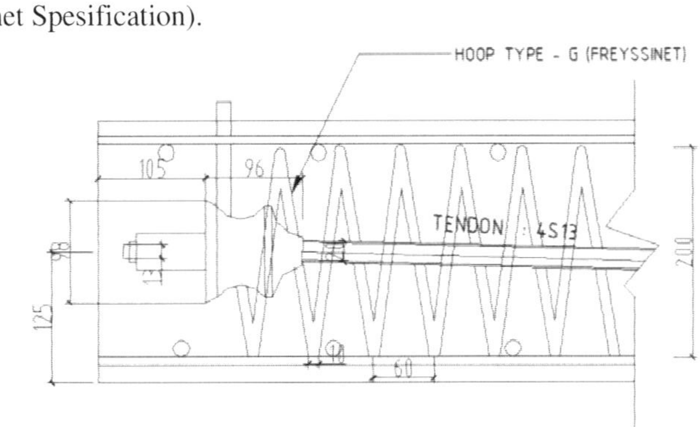

Figure 6. Detail of active anchor.

6. Prestressed Plate Details

Prestressed plate is designed to unmonolithycal concentrate towards its beam by installing a pegs at one side and rubber sheets at both side. According to calculation result, D13 peg is used with the distance of 1.200 mm. Figure 6 shows 12 m landscape of prestressed plate's section. Raised floor is used to overcome floor elevation difference. Parabolical prestressed plate tendon design is shown in Figure 7.

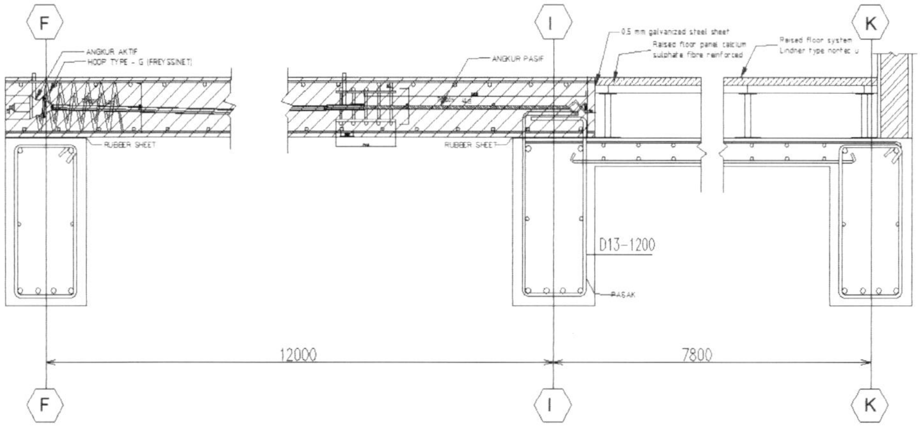

Figure 7. Prestressed concrete slab piece.

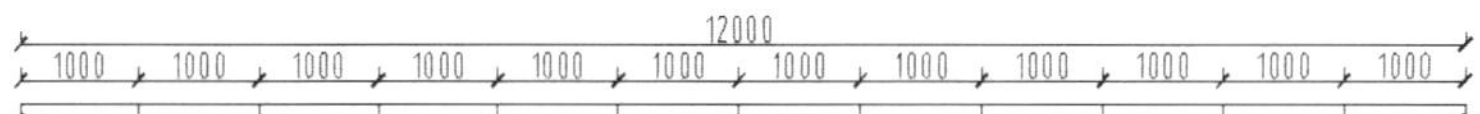

CABLE 12	4S ASTMA416	X	0	1000	2000	3000	4000	5000	6000	7000	8000	9000	10000	11000	12000
		Y	125	112.778	102.778	95	89.444	86.111	85	86.111	89.444	95	102.778	112.778	125
		Z	1000	1000	1000	1000	1000	1000	1000	1000	1000	1000	1000	1000	1000

Figure 8. Tendon coordinate.

7. Prestressed Plate Beam Support Reinforcement 70/35

Torque value that was obtained from auxiliary program is used to calculate the required number of the reinforcement. After that, beam reinforcement should be designed according to SNI 2847-2013. This picture below shows field reinforcement support requirement and its friction from the calculation results.

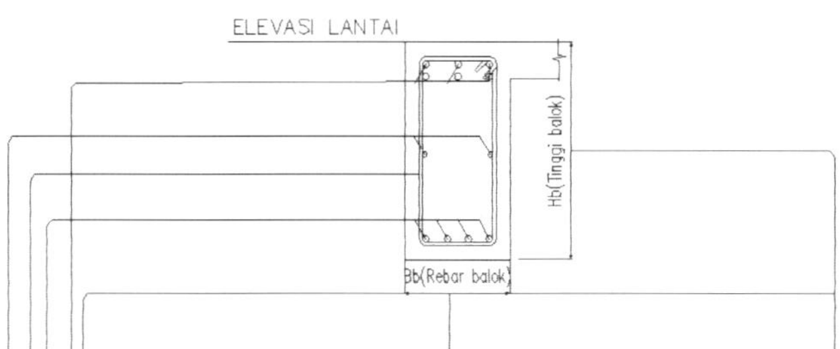

TIPE / LANTAI	POSISI				
		T:	AS 6I–5I		
		P:	TUMP. DALAM	LAPANGAN	TUMP. DALAM
	DIMENSI	D:	350 X 700		
	TULANGAN ATAS	TA:	6 D19	2 D19	6 D19
	TULANGAN BAWAH	TB:	4 D19	4 D19	4 D19
	SENGKANG	SK:	D10–100	D10–200	D10–100
	TUL. PINGGANG	TP:	2 D13		

Figure 9. Beam reinforcement.

IOP Conf. Series: Materials Science and Engineering **267** (2017) 012021 doi:10.1088/1757-899X/267/1/012021

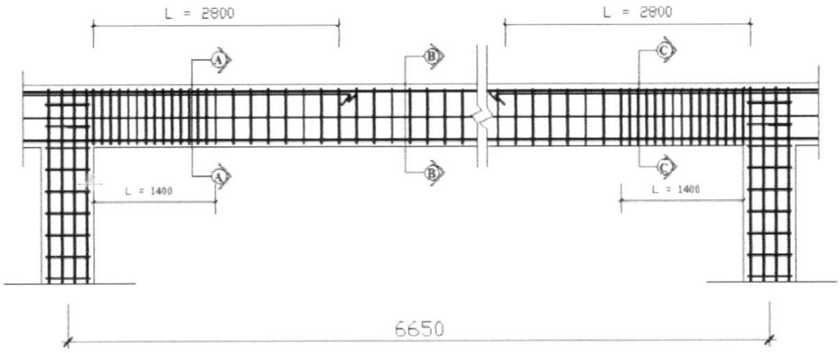

Figure 10. Cross section of beam.

8. Column Reinforcement

Column dimension is 550 mm x 880 mm. Rebar area that was obtained from auxiliary program is used to calculate required number of reinforcement. After that, column reinforcement should be designed according to SNI 2847-2013. This Figure 11 and Figure 12 below show required number of reinforcement according to the calculation result.

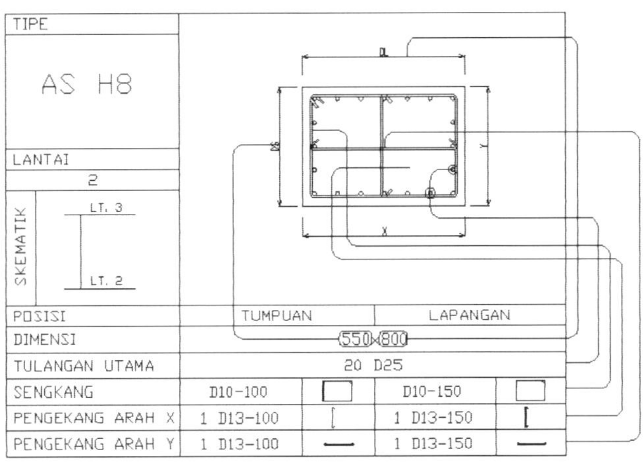

Figure 11. Column reinforcement.

IOP Conf. Series: Materials Science and Engineering **267** (2017) 012021 doi:10.1088/1757-899X/267/1/012021

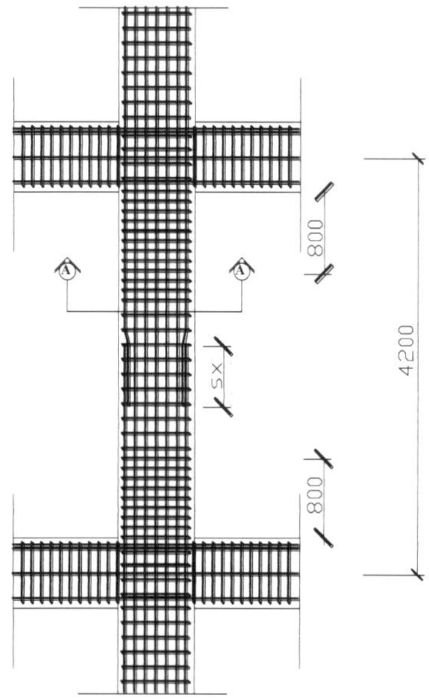

Figure 12. Cross section of column.

9. Story Drift Control

Elasticity of movement value and δ_{xe} from structural analysis should be determined to obtain interstage deviation value. Then, δ_{xe} value is multiplied by magnifying factor C_d/I_e. After that, interstage deviation value can be obtained by calculating the difference between magnified elasticity movement towards one of level up-value with magnified elasticity movement on one level-down value. Then, this deviation value is controlled with deviation limit of 0,02 h_{sx}. Table 1 shows deviation values that is resulted from seismic forces x.

Table 1. Deviation values.

Lantai	Elevasi (m)	Tinggi antar tingkat (m)	δe (mm)	δxe (mm)	δx (mm)	δa (mm)	Ket
	1	2	3 (Output SAP)	4 (Δtiap Lantai)	5 = Cd x 4 / I	6 = 0,02 x 1	
lt. LG-G	-2.60	4.20	0	0	0	0	OK
lt. G-1	1.60	4.80	0.3	0.3	1.65	84	OK
lt. 1-2	6.40	3.60	2.3	2	11	180	OK
lt. 2-3	10.00	4.20	4.2	2.2	12.1	252	OK
lt. 3-4	14.20	3.40	6.6	4.4	24.2	336	OK
lt. 4-5	17.60	3.40	8.7	4.3	23.65	404	OK
lt. 5-6	21.00	3.40	10.8	6.5	35.75	472	OK
lt. 6-7	24.40	3.40	12.9	6.4	35.2	540	OK
lt. 7-8	27.80	3.40	14.9	8.5	46.75	608	OK
lt. 8-9	31.20	3.40	16.9	8.4	46.2	676	OK

10. Two-Way Prestressed Plate Deflection Control

10.1. Condition 1

Condition after 25% stressed with load combination of 1 DL + 0,25 Prestress. Deflection occured downwards in amount of 11 mm (SAP2000 output < 100 mm).

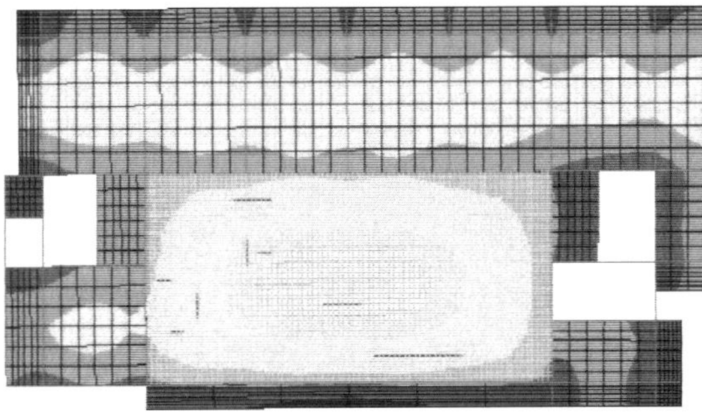

Figure 13. First condition of slab deflection.

10.2. Condition 2 (Service)

Condition after 100% stressed with load combination of 1,4 DL + 1,6 LL + 1 Prestress. Deflection occured downwards in amount of 6 mm (SAP2000 output < 100 mm).

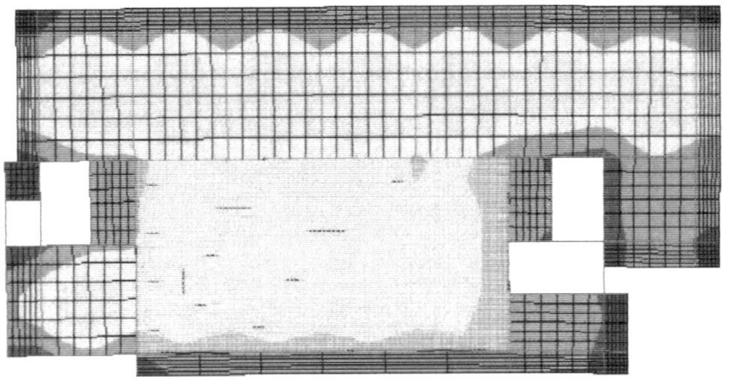

Figure 14. Second condition of slab deflection.

11. Tendon Prestressed Plate Stressing Excecution

According to calculation result, 12 m x 24 m plate with 250 mm width's casting duration is 6.5 hours. Stressing is done after the concrete aged 14 days (80% strength) with pre-tension of 0,7 f_{pu}. Stressing is done step by step for every 45 MPa corresponding to its power pack stressing strength. Stressing used E.O.H.P power pack equipment, MK-I type, K-100 jack type with stressing strength 45 MPa. Strand extension should be watched during tendon stressing process to know the prestressed loss that was caused by anchor slip and tendon friction.

$$\Delta l = \frac{P}{E} \times l_x \qquad (1)$$

$$P_{loss} = \frac{\Delta l \times E}{l_x} \qquad (2)$$

Where:

Δ l = Strand elongation
P = Prestress force
E = Strand elasticity

Table 2.Strand elongation.

Pressure applied	Elongation	Elongation Observed	Pressure observed	Loss
Mpa	mm	Mm	Mpa	%
45	2.8272251	2.5	39.79166667	11.574
90	5.6544503	5.1	81.175	9.8056
135	8.4816754			
270	16.963351			
405	25.445026			

According to calculation result, 12 m tendon strand extension is 60,89 mm. Tendon stressing duration for 72 tendons with each length of 12 m are 13,2 hours meanwhile grouting duration are 15,4 hours per 72 tendons.

12. Conclusion

Refer to this research purpose, then the conclusions are as follows:

1. According to prestressed plate calculation, effective dimension to overcome architectural challenge is 250 mm with 12 m x 24 m landscape.
2. Odua Weston Jambi's Second floor's colum elimination is possible due to prestressed plate utilization.
3. According to control examination; double system control, mass participation value control, last spectrum respons value control, and drift control, designed structure met the reuqirement.
4. According to stressing excecution method, stressing duration time is 13,92 hours/ 72 tendons with tendon's length 12 m (4 strand 7 wire tendon).
5. According to grouting excecution method, grouting duration time is 15,4 hours/ 72 tendons with tendon's length 12 m (4 strand 7 wire tendon).

13. References

[1] Ministry of Public Works, 2013, *Indonesian National Standard:Structural Concrete Requirement for Building* (SNI 2847-2013*)*, Building Problem Investigation Directorate, Jendral Cipta Karya Directorate, Public Works and Electricity Departement, Bandung.

[2] Ministry of Public Works, 2013, *Indonesian National Standard:Minimum Load for Building Design and Another Structure* (SNI 1727-2013*)*, Building Problem Investigation Directorate, Jendral Cipta Karya Directorate, Public Works and Electricity Departement, Bandung.

[3] Ministry of Public Works, 2012, *Indonesian National Standard:Earthquake Resistance Design Procedur for Building and Another* (SNI 1726-2012*)*, Building Problem Investigation Directorate, Jendral Cipta Karya Directorate, Public Works and Electricity Departement, Bandung.

[4] Ministry of Public Works, 2012, *Indonesian National Standard:Precast and Prestressed Concrete Design for Building* (SNI 7833-2012),Building Problem Investigation Directorate, Jendral Cipta Karya Directorate, Public Works and Electricity Departement, Bandung.

[5] Ned, T.Y.Lin and Burns, H, 1996, *Prestressed Concrete Structure Design,* Yogyakarta: Erlangga.

[6] Nawy, Edward G, 2001, *Steel Construction Profil Table,* Jakarta: Erlangga.

[7] Iswandi Imran and Fajar Hendrik, 2009, *Earthquake Resistance Reinforcement Concrete Structure According to SNI 03-2847-2002,* Yogyakarta : Erlangga.

[8] Budiadi, Andri, 2008, *Prestressed Concrete Practical Design According to SNI 03-2847-2002,* Yogyakarta: ANDI.

[9] Raju, N Khrisna, 1988, *Prestressed Concrete Second Edition,* Yogyakarta: Erlangga.

IOP Conf. Series: Materials Science and Engineering **267** (2017) 012022 doi:10.1088/1757-899X/267/1/012022

Seismic performance evaluation of a pile-supported pier in Aceh, Indonesia

Y N Hanifah[1,2], A Budipriyanto[1,2] and I P Rahardjo[1]
[1]The Department of Civil Infrastructure Engineering, Institut Teknologi Sepuluh Nopember, Kampus ITS Surabaya 60116, Indonesia.

[2]E-mail: yudnina13@mhs.ce.its.ac.id, agungbp@ce.its.ac.id

Abstract. The seismic performance of shore structures located in regions having high seismicity is still attracted the interest of researchers. Earthquakes could bring about serious damages on the structures which might lead to failure. Thus the structures' performance under the ground motions need to assessed. This paper describes results of study on performance of a pile supported pier which was constructed in Aceh, Indonesia under seismic motions. In this study the performance evaluation of the pier was conducted using displacement based method of ASCE 61-14 code. 3D numerical models of the pier structure were developed utilizing a structural analysis software package. To account for soil-structure interaction under simulated seismic loads, the nonlinear p-y spring model recommended by American Petroleum Institute was applied. Nonlinear static pushover analysis was performed to obtain structure's capacity. Respons spectrum analysis and nonlinear static demand analysis were carried out to obtain the structure's demand capacity. Results of this study showed that under the 475 and 950-year return periods of earthquakes, the pile-supported pier could be categorized as Minimal Damage. Nevertheless, when it was under 2475-year earthquake return period its performance could be categorized as Control and Repairable Damage; These met the requirements stated in ASCE 61-14.

1. Introduction

Seismic performance of shore structures situated in high seismicity regions is still attracted the interest of researchers. Earthquake events could cause important shore structures, such as pier and wharf structures, suffer serious damage that might lead to failure. Therefore, the behavior of the structures under the ground shakings need to be studied and their performances can be assessed. Methods for assessing quay walls and pile supported wharves under seismic loads were explained in [1]. The assessment was performed according to analysis methods recommended by Technical Standard for the results were presented in [2]. Eight specimen of connections widely used in practice were tested in the laboratory to investigate their performances under simulated seismic loads. They concluded that these types of connections could resist large cyclic loads and non-elastic deformations. However, there were substantial stiffness and resistance reductions in them. In addition, due to axial load on pile, moment capacity of the connections increased but it caused the resistance deteriorated more rapidly.

Mondal and Rei [3] reported damages happened on pile-supported harbor structures in Andaman Islands due to the 2006 Sumatra earthquake. They noted that mayor damages of the structure were

[2] To whom any correspondence should be addressed.

because of pounding between two segments of the structure, short pile effects on piles deployed on slope near shore, improper design and poor maintenance which led to corrosion of steel reinforcement and concrete cover spalling, and poor performance of some the structures under the earthquake. Tests on piles on rock fill under lateral loads were performed in [4] to study the relations of the piles' resistance and deformation when the soil particle size was large and might be comparable to the pile diameter. This would be typical open type wharf structure constructed in rock fill dike. The author proposed a curve of soil resistance and deformation, viz. p-y curve where the interlocking between rock particles was taken into consideration. Brunet and de la Llera [5] reported successful application of seismic isolation in pile-supported pier at Coronel port in Chile. The pier was in operational condition after an earthquake while other ports close to the pier were damaged significantly. Donahue et al. [6] conducted numerical analysis of a pile supported wharf using recorded strong motion data. The data were obtained from measurements of acceleration responses of the wharf during the M7.0 Loma Prieta earthquake. 3D numerical models of the wharf were utilized to investigate its responses during the earthquake. The model responses under the earthquake loads were compared with the measured responses. They pointed out that one of factors influencing the structure's model accuracy was the model representing pile-soil interaction. Doran et al. [7] presented evaluation of seismic performance of two pile supported wharves using Pushover analyses according to Turkish code. The results of the study concluded that these two wharves were met minimum requirement of the code.

This study was aimed at evaluating seismic performance of a pile-supported pier located in the high seismically regions in the province of Aceh, Indonesia, see the location in figure 1. The pier was designed for serving loading and unloading bulk coal of 10,000 DWT ship and Portland cement of ship ranging from 20,000 to 40,000 DWT. The structure was modelled numerically using a computer software package; effects of soil liquefaction and lateral spreading were not taken into consideration in this study. Nonlinear Pushover analysis was conducted for assessing the structure under earthquake loads using procedures recommended by the ASCE code for wharf and pier design.

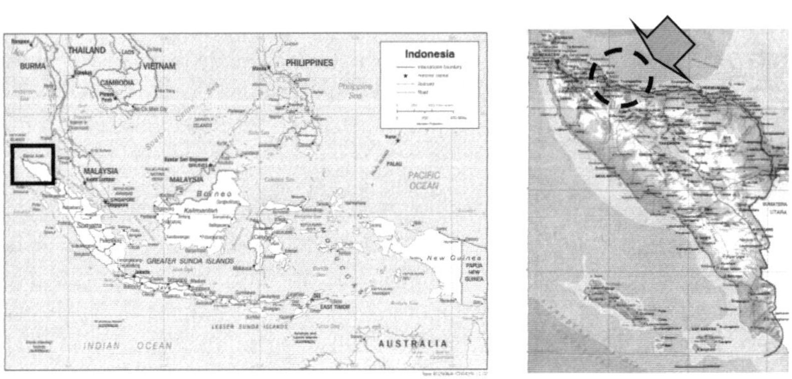

Figure 1. The pier location (http://www.maps-of-the-world.net).

2. Method

As stated earlier the pier evaluated for this study is located in high seismicity regions. It was designed for berthing of bulk carriers, especially for coal and cement. This important structure has to be in operational condition after being subjected to designed earthquakes or at least damages occurred in the structure can be repaired shortly. In the following sections steps conducted in this study are described.

2.1. Data collecting

The piled-supported pier has dimension of 102m in length and 25m in width. It was constructed on steel pipe piles with the diameters of 1.0m and 0.8m. The steel pile had yield strength of 310 MPa. The pier deck was concrete having compressive strength of 29.05 MPa and rebar yield strength of 400 MPa. Figures 2, 3, and 4 show the plan view, longitudinal and transverse cross sections of the pier.

IOP Conf. Series: Materials Science and Engineering **267** (2017) 012022 doi:10.1088/1757-899X/267/1/012022

The connection of pile and pile-cap was composite material of steel pipe and reinforced concrete; It spanned from the pile cap to about -1.0 m below the low seawater level. Reinforcement details of the connection are presented in figures 5 and 6. Available geotechnical data on site was standard penetration test (SPT); the soil profile below the sea bed in term of SPT values is presented in table 1.

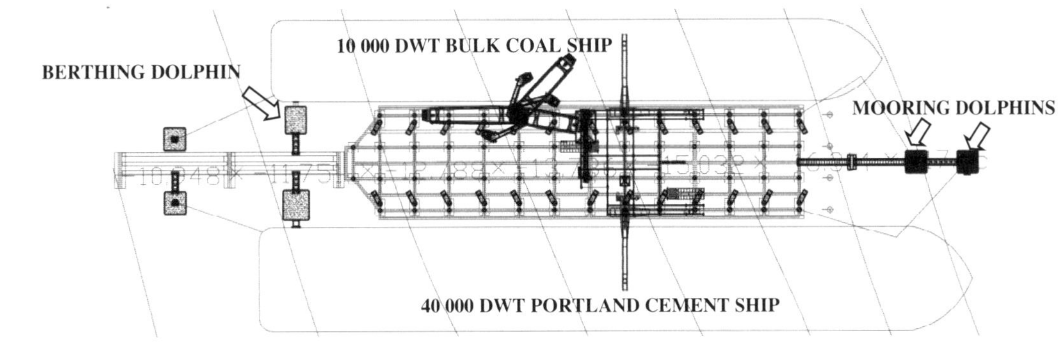

Figure 2. Plan view of the pier.

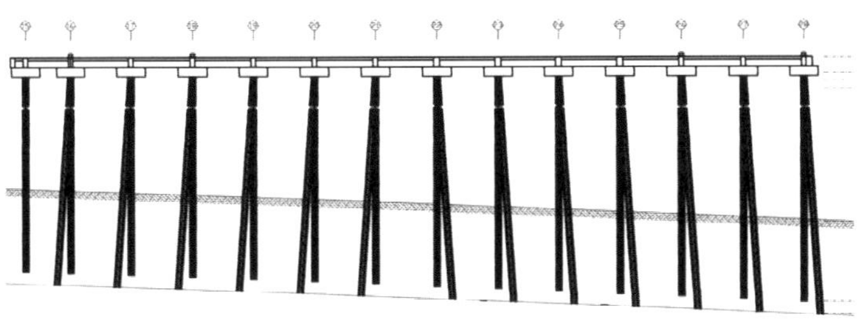

Figure 3. Longitudinal cross section of the pier structure.

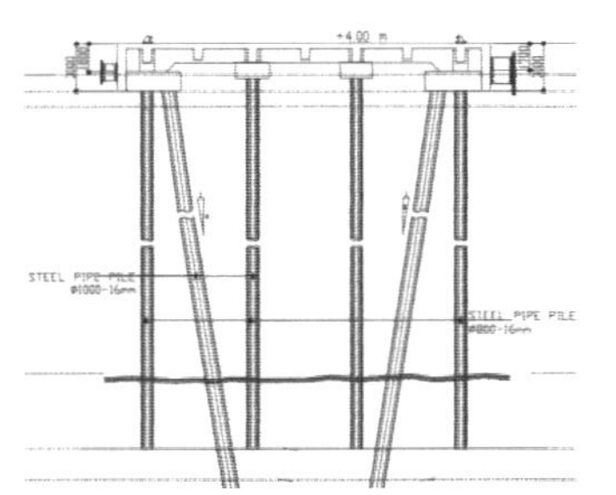

Figure 4. Transverse cross section of the pier structure.

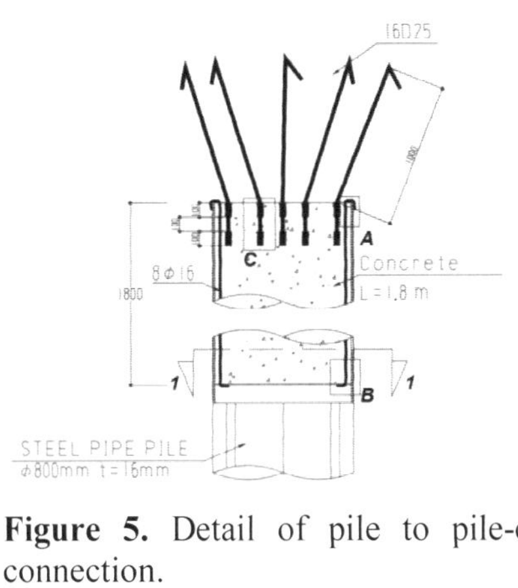

Figure 5. Detail of pile to pile-cap connection.

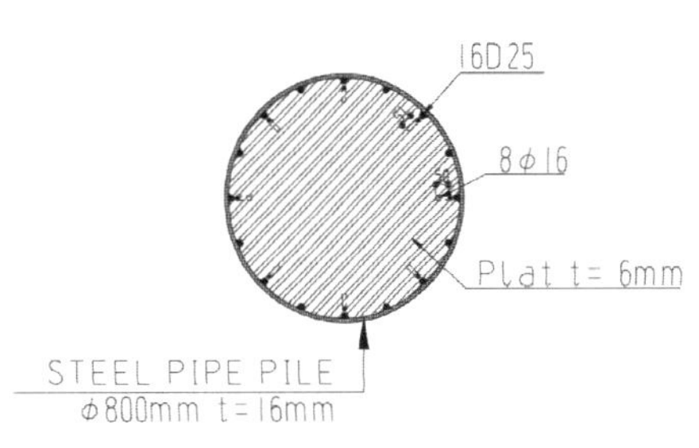

Figure 6. Cross cection of pile to pile-cap connection.

Table 1. Typical standard penetration test (SPT) values at the pier location.

Depth below the seabed (m)	2	4	6	8	10	≥12
N SPT Value	17	18	26	35	45	60

IOP Conf. Series: Materials Science and Engineering **267** (2017) 012022 doi:10.1088/1757-899X/267/1/012022

2.2. Structural modelling

Three-dimensional models of the pier were developed using SAP2000 structural analysis software [8]. Soil-structure interaction was modelled using nonlinear p-y spring method recommended by American Petroleum Institute [9]. Figures 7 and 8 show curves of soil resistance as function of deflection for the piles of the pier at different depths below the seabed. Lower-bound and upper bound spring values were computed to take into account uncertainties of the soil characteristics and soil variability [10]. In this study response spectra were obtained using two different codes, i.e., RSNI 2833-201X [11], the Indonesia's bridge seismic design code, and ASCE 7-05 [12], the code for minimum load recommended for building and other structure. These spectra for 475, 950, and 2475-year earthquake return periods are presented in figures 9 and 10.

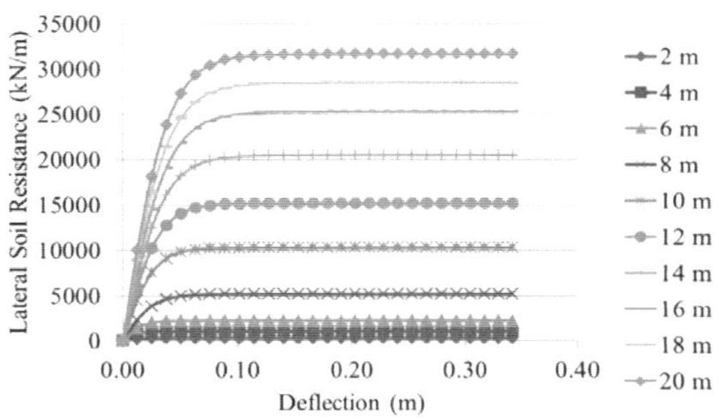

Figure 7. p-y curves of 0.8m-diameter pile at different soil depths below the seabed.

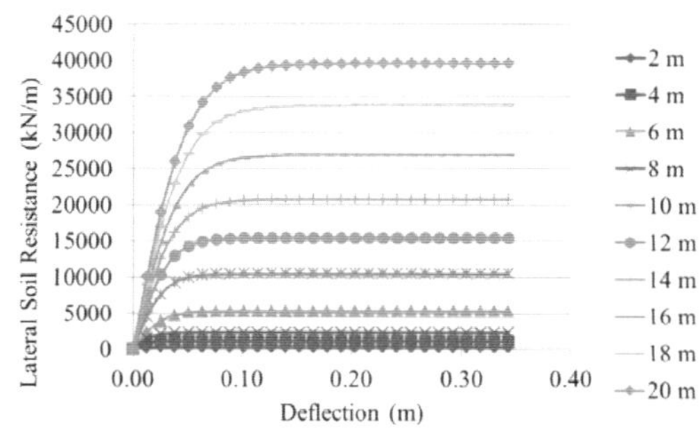

Figure 8. p-y curves of 1m-diameter pile at different soil depths below the seabed.

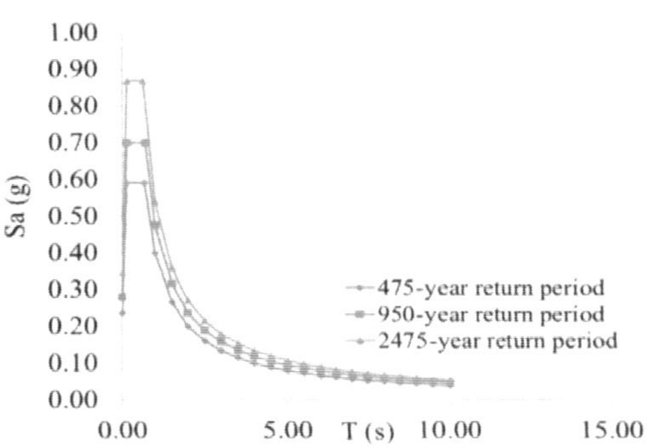

Figure 9. Response spectra for 475, 950, 2475–year return periods using RSNI 2833-201X.

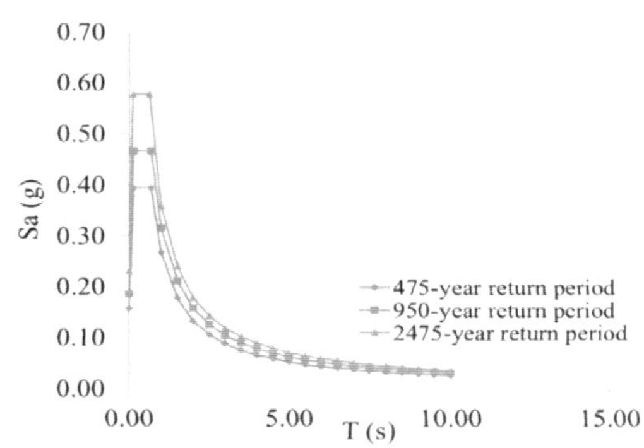

Figure 10. Response spectra for 475, 950, 2475–year return periods using ASCE 7-05.

2.3. *Structural capacity analysis*

Nonlinear Static Pushover analysis were performed to obtain structural capacity under the seismic loads. The analysis was conducted following steps described below,

2.3.1. Expected material strength. Expected material properties were determined to assess Structural capacity and Demands under the earthquake loads. The expected strength for each material was defined according to ASCE 61-14 [13].

2.3.2. Plastic hinge and strain limit. In many cases plastic hinges were formed at (a) connection of pile to pile-cap, and (b) pile beneath the seabed where the large bending moment occurs. To define the plastic hinge properties, in this study cross section analyses were conducted to obtain the moment curvatures of plastic hinge at the connection of pile to pile-cap and the one below the seabed.

In the connection of pile to pile-cap, the steel pile was considered to be embedded; it implied that it was a full-moment connection. ASCE 61-14 [13] permits Method B for moment-curvature analysis for this type of connection. The allowable strain limits for different performance levels of pier structure are presented in table 2 [13].

Table 2. Strain limits for different performance levels [13].

Performance level	Strain limit	
	Top of pile	In ground
Minimal Damage	$\varepsilon_s \leq 0.010$	$\varepsilon_s \leq 0.010$
Controlled and Repairable Damage	$\varepsilon_s \leq 0.035$	$\varepsilon_s \leq 0.025$
Life Safety Protection	$\varepsilon_s \leq 0.050$	$\varepsilon_s \leq 0.035$

2.3.3. Loads applied on the structure. During the earthquake, the loads applied on the pier were the structure self-weight, weight of equipment that was attached on the structure constantly, and 10% of the live load. The design loads for the pier were computed according to POLB [10] and AASTHO LRFD [14]. The seismic mass of the pier considered in this study were therefore made up of mass of these loads.

2.3.4. Lateral load distribution pattern. To model lateral load pattern during earthquake events, load patterns used in this study were uniform acceleration and modal patterns. In uniform acceleration pattern, the force pattern was assigned proportional to the mass distribution in the corresponding direction. Whereas in modal pattern, modal analysis results were to be assigned to the corresponding mode load case; In this study, the first two modes of the structure were considered in the analysis. Figures 11(a) and (b) depict the deformed shape of longitudinal and transverse modes.

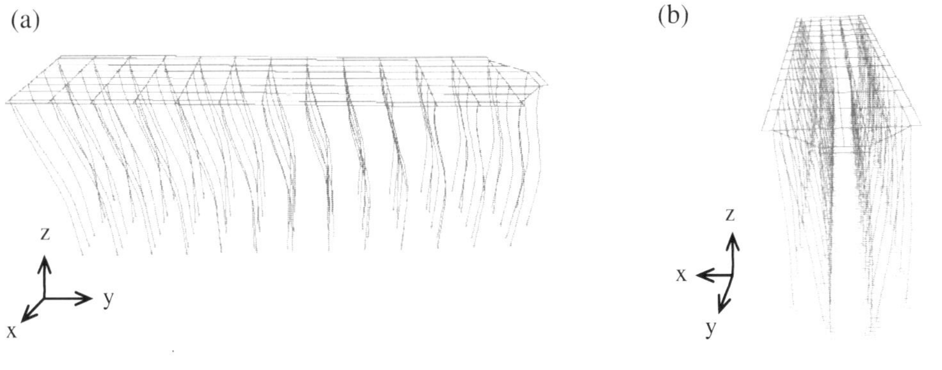

Figure 11. Deformed shape of (a) longitudinal mode (in the-y axis direction), and (b) transverse mode (in the-x axis direction).

2.4. Demand capacity analysis

Response spectrum analysis and nonlinear static demand analysis were used to obtain demand capacity. These analyses were performed by applying the procedure described in ASCE 61-14 [13].

IOP Conf. Series: Materials Science and Engineering **267** (2017) 012022 doi:10.1088/1757-899X/267/1/012022

3. Results and Discussion

Results of this study are reported and discussed in the sections below.

3.1. Capacity curve

Capacity curve represents the function of base shear and displacement of the pier deck. In this study capacity curves of the pier were obtained by applying Nonlinear Static Pushover analysis under load applied in the transverse and longitudinal directions. Figures 12(a) and (b) present capacity curves of the pier structure obtained using upper bound and lower bound spring models for two lateral load distribution patterns, i.e., uniform acceleration and modal load distribution patterns. It can be seen from these figures that for the same deformation values, the base shear forces obtained by the uniform acceleration pattern were higher than those computed by the modal pattern. In addition, the load applied on the tranverse direction resulted in larger displacements on the structure for the same base shear. Plastic hinges developed under loads applied during the Nonlinear Static Pushover analysis could also be monitored. Figures 13(a) and (b) show the first yield plastic hinge occurred in the pier structure for loads applied in the transverse (the–x axis) and the longitudinal (the–y axis) directions using lower bound spring model. The results for upper bound spring model was similar therefore it is not presented in this paper.

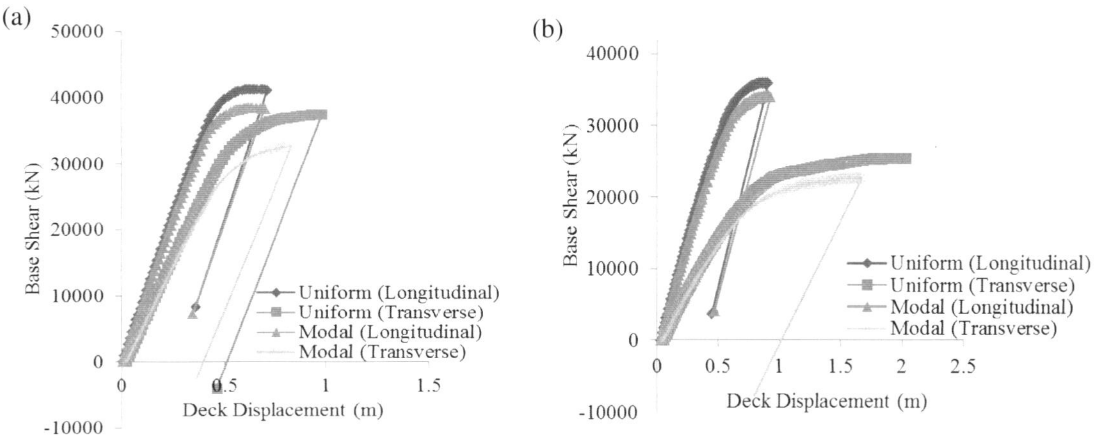

Figure 12. Capacity curves of pier obtained using (a) upper bound, and (b) lower bound spring models in the transverse and longitudinal directions by applying uniform acceleration load pattern (uniform) and modal load pattern (modal).

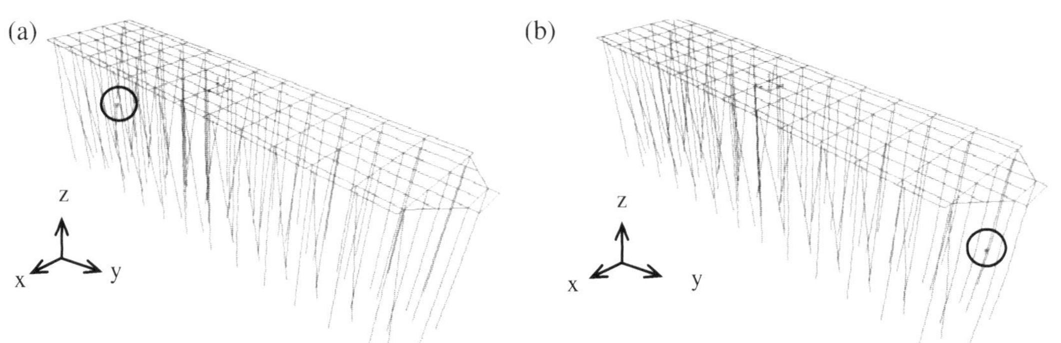

Figure 13. First yield-hinge under the load applied in the (a) transverse (the –x axis) , and (b) longitudinal (the –y axis) directions of the pier using lower bound spring model.

3.2. Seismic demand

To obtain the seismic demand, first the respons spectrum analysis was conducted to investigate wether the structure was still in its elastic state under the ground motions or not. Then when the structure behaved beyond its elastic limit, the nonlinear static demand analysis was performed. The nonlinear static demand analysis was conducted by using capacity spectrum approach available in SAP2000; numbers of iterations might be needed to satisfy the conditions required in ASCE 61-14 [13].

3.3. Performance levels of the structure

The pier was subjected to 475, 950, and 2475-year earthquake return periods calculated using procedures recommended by RSNI 2833-201X [11] and ASCE 7-05 [12]. The minimum performance level of the pier required the ASCE 61-14 [13] was Controlled and Repairable Damage under 475-year return period of ground motion and Life Protection under 950 and 2475-year return period of ground motions. When the response spectra of the ground motions having 475 and 950 year return periods were computed using procedures described in [11] and [12] it was observed that performance of the pier structure could be categorized as Minimal Damage. Pier's performance level under 2475-year earthquake return period was also Minimal Damage when the spectrum was calculated using procedure in [12]. Minimal Damage can be desribed as the damage that can be identified with initial cracking and spalling of the pile and/or pile cap [13]. Tables 3 and 4 present the performance levels of the pier under 2475-year return period of the ground motion in the transverse and longitudinal directions using two load distribution patterns when the spectrum was obtained using procedure stipulated in [11]. The pier's performance levels under the 2475-year earthquake return period are also

Table 3. Performance levels of the pier in the transverse direction under the 2475-year return period seismic ground motion for different lateral load distribution patterns.

Lateral load pattern	Displacement capacity limit (m)				Seismic Demand (m)	Strain at yield (m/m)	Performance Level
	MD [e]	CRD [f]	LSP [g]	Collapse			
Uniform-UB [a]	0.4441	0.5545	0.6079	0.6612	0.4090	0.0081	MD
Uniform-LB [b]	0.7061	1.0081	1.1126	1.2200	0.6450	0.0051	MD
Modal-UB [c]	0.3986	0.4936	0.5419	0.5953	0.3960	0.0042	MD
Modal-LB [d]	0.5830	0.8740	0.9741	1.0711	0.6750	0.0171	CRD

[a] Uniform acceleration load pattern with upper bound spring model
[b] Uniform acceleration load pattern with lower bound spring model
[c] Modal load pattern with upper bound spring model
[d] Modal load pattern with lower bound spring model
[e] Minimal Damage
[f] Controlled and Repairable Damage
[g] Life Safety Protection

Table 4. Performance levels of the pier in the longitudinal direction under the 2475-year return period seismic ground motion for different lateral load distribution patterns.

Lateral load pattern	Displacement capacity limit (m)				Seismic Demand (m)	Strain at yield (m/m)	Performance level
	MD [e]	CRD [f]	LSP [g]	Collapse			
Uniform-UB [a]	0.2986	0.4119	0.4719	0.5392	0.2970	0.0079	MD
Uniform-LB [b]	0.3734	0.5357	0.6135	0.6930	0.3950	0.0139	CRD
Modal-UB [c]	0.2982	0.4112	0.4712	0.5382	0.3050	0.0125	CRD
Modal-LB [d]	0.3727	0.5329	0.6077	0.6930	0.4020	0.0156	CRD

[a] Uniform acceleration load pattern with upper bound spring model
[b] Uniform acceleration load pattern with lower bound spring model
[c] Modal load pattern with upper bound spring model
[d] Modal load pattern with lower bound spring model
[e] Minimal Damage
[f] Controlled and Repairable Damage
[g] Life Safety Protection

IOP Conf. Series: Materials Science and Engineering **267** (2017) 012022 doi:10.1088/1757-899X/267/1/012022

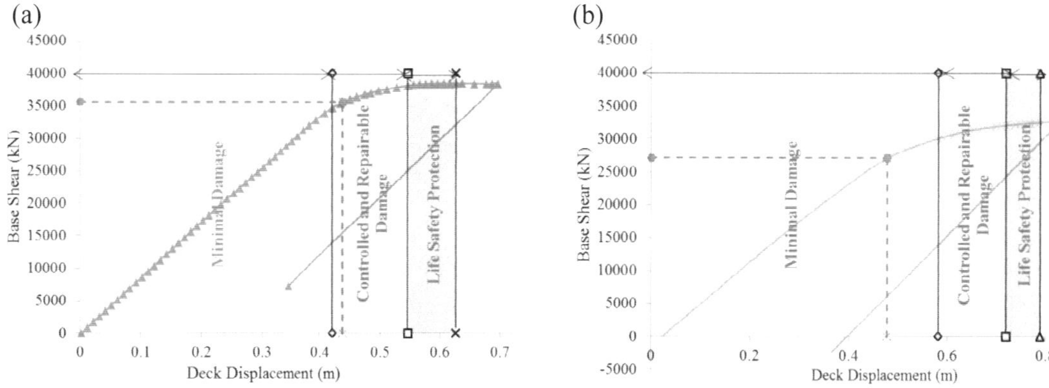

Figure 14. Performace levels of the pile-supported pier under the 2475-year earthquake return period obtained using upper bound spring model for modal load pattern applied in the (a) longitudinal (the-y axis), and (b) transverse (the-x axis) directions.

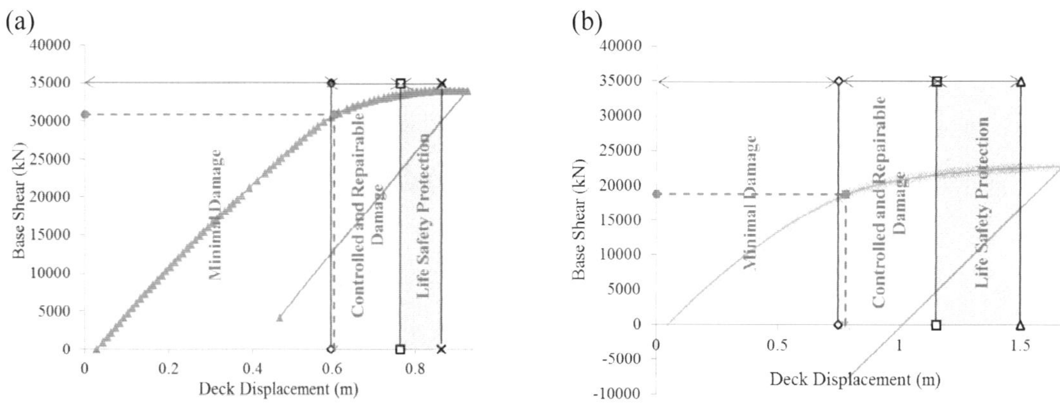

Figure 15. Performace levels of the pile-supported pier under the 2475-year earthquake return period obtained using lower bound spring model for modal load pattern applied in (a) the longitudinal (the-y axis), and (b) the transverse (the-x axis) directions.

presented in figures 14 and 15 for clarity; Modal load pattern was applied for obtaning the performance levels presented in these figures. The performance level of the pier structure under the earthquake was Controlled and Repairable Damage. This type of damage could be indicated by exposure of reinforcement bar of the deck due to considerable spalling occurred in the deck particularly at the region adjacent the pile. Thus the performance levels of the pile-supported pier met the requirement stated in the code [13].

4. Conclusion

In this study, a pile-supported pier structure located in Aceh - a high seismicity region in Indonesia – was evaluated for its seismic performance. 3D models of the pier were developed using SAP2000 structural analysis software. To account for soil-pile interaction, nonlinear spring model of API [9] was applied. The structural capacity was obtained using nonlinear static Pushover analysis and demand capacity was computed using respons spectrum analysis and nonlinear static demand analysis described in [13]. The pile-supported pier was subjected to 475, 950, and 2475-year return periods of earthquakes which spectra were calculated using two different methods, viz., RSNI 2833-201X [11] and ASCE 7-05 [12].

IOP Conf. Series: Materials Science and Engineering **267** (2017) 012022 doi:10.1088/1757-899X/267/1/012022

Using response spectrum analysis along with nonlinear static demand analysis, it was observed that under the 475 and 950-year earthquake return periods the structure's performance could be categorized as Minimal Damage. However, under 2475-year earthquake return period which spectrum computed using procedure in [11], the pier's performance could be categorized as Controlled and Repairable Damage. Therefore the seismic performance level of the pile-supported pier structure met the requirement stated in the ASCE 61-14.

References

[1] Nozu A, Ichii K and Sugano T 2004 *J. of Japan Assoc. for Earthquake Engineering* **4** 3
[2] Roeder C W, Graff R, Soderstrom J, and Yoo, J H 2004 *Proc. 13th World Conf. on Earthquake Engineering*, Vancouver, B.C., Canada, Paper No. 2570
[3] Mondal G and Rai D C 2008 *Engineering Structures* **30** 174–182
[4] Kawamata, Y 2009 PhD Dissertation Oregon State University, Corvallis, the USA
[5] Brunet S and de la Llera J C 2012 *Proc.15th World Conf. on Earthquake Engineering*, Lisbon, Portugal, Paper No. 3420
[6] Donahue M J, Dickenson S E, Miller T H and Yim S C 2005 *Earthquake Spectra* **21** 3 617–634
[7] Doran B, Shen J and Akbas B, 2015 *Earthquake Spectra* **31** 2 1177-94.
[8] Computers and Structures Inc. 2000 SAP2000: Static and Dynamic Analysis of Finite Element Structures Nonlinear (Berkeley, CA).
[9] American Petroleum Institute 2002 *Recommended practice for planning, designing and constructing fixed offshore platforms: Working stress design, in API Recommended Practice 2A-WSD (RP 2A-WSD)*, 21st Edition (Washington, D.C: API Publishing Service) 242 pp
[10] Port of Long Beach 2012 *Port of Long Beach Wharf Design Criteria* Version 3.0 (California: Port of Long Beach) 39pp.
[11] Badan Standarisasi Nasional 2013 *RSNI 2833-201X Design of bridge under earthquake loads* Jakarta: Badan Standarisasi Nasional 73pp (in Indonesia Language).
[12] American Society of Civil Engineers 2005 *Minimum Design Load for Building and Other Structure* (Reston, Virginia: American Society of Civil Engineers) 419 pp.
[13] American Society of Civil Engineers 2014 *Seismic Design of Wharves and Piers* (Reston, Virginia: American Society of Civil Engineers) 90 pp.
[14] American Association of State Highway and Transportation Official 2012 *AASTHO LRFD Bridge Design Specification* (Washington, D.C: American Association of State Highway and Transportation Officials) 1661pp.

IOP Conf. Series: Materials Science and Engineering **267** (2017) 012023 doi:10.1088/1757-899X/267/1/012023

Schedule Risk Analysis Of Southern Mainway Construction In Jember Regency

K Susilo[1], I P A Wiguna[1] and T J W Adi[1]

[1]Department of Civil Engineering, Faculty of Civil Engineering and Planning, Institut Teknologi Sepuluh Nopember, ITS Campus Sukolilo, Surabaya, 60111, Indonesia

Email: logawadian@gmail.com

Abstract. In Jember Regency, it has been built Southern Cross Road (JLS) as part of regional project. On the implementation of previous construction, there were still some events which gave negative impact to the project. The purpose of this research is to analyze risk and its effect on schedule at the construction phase of JLS at Jember Regency. Risk identification process is carried out by site survey, literature studies and supporting data. The use of Probability and Impact Matrix were aimed to obtain the level of risk. Based on the analysis, it was obtained six highest risk that could affecting schedule, such as difficult access locations, heavy rains, increases of material price, broken road pavement work, change order, and work accident. Risk responses were proposed by applying agreement to guarantee stock and price of materials, prioritized drainage, and constructing bridge to solve difficult access. An intense coordination in the site, routine checks of quality, manufacturing of retailing walls were also needed to reduce possibility of distruption to pavement work. To avoid work accident, it is needed to socialize about harsh terrain condition, mutual allertness among supervisor, worker and the others, and also all personals must comply with savety rules.

1. Introduction

Project risk is an uncertain event or condition which may have an impact on the objectives of the project that cover scope, schedule, cost and quality [1]. It is an unexpected event or condition beyond what has been planned, or becoming an opportunity of both unwanted and profitable events and affect the objectives of the project. According to [2], risk in the project is the measurement of the probabilities and consequences and unachieved goal of a project. At the other reference by [3], if we relate the risk to the concept of opportunity, it can be said that risk is probability of occurance of unexpected condition with all possible consequences which may result in delay or project failures.

In Jember Regency, there has been built JLS which belong to the part of the RRDP (Regional Roads Development Project). The problems faced are such as southern contour conditions, limited funds, land acquisition, the need for bridges and issues of annual contracts. This research as shown in figure 1 was conducted in Jember Regency with the total road constructional plan of 83,5 km. Until this moment, most of the constraction is still in a land-road form and aggregate, which almost have been released and only a few kilometers were finished in flexible pavement. Further work is planned for this year up to 2019.

In the continued development of the project, there is any possibility of various problems encountered, including the risks which is occurred in the construction phase. In addition, with the existing condition of JLS project near the beach and hill, the risk which threatens on the implementation of road construction can be landslide, water channel disruption, land degradation, abrasion and others. Although many research has solved the risk analysis at the road construction project, the different site and external factor of every project could be different. According to [4], there are three external factors and five internal factors that could affect the road construction phase. The external factors are politics and economic factor, nature, and culture / society. The internal factors are owner / consultan factor, contractor, labour, subcontractor, and material.

On the other studies by [5] about risks of the road works, there were found 23 possible risks. Those are as follows: the different site conditions from assumptions, the limits on the working hours, the unworkable design, the quality of work which is not appropriate to the standards, the structural damage and so on. From these studies, it can be seen that different site conditions with assumptions, design changes, force majeure, bad weather, and workplace accidents are also the relevant risks toward this research. While the restriction on working hours, unsuitable materials supply, labor disputes, structural damage, substandard quality and labor strikes do not occur in this study. It is rare and considered as a risk which can be avoided.

From the literature review above, there are several similarities to this research, namely political policy changes, inflation, weather factor, land condition factor, force major, and other technical factor. The difference between this research to previous study is the risk analysis using research variable based on General Affair of Road Department of East Java to ensure that this research is more relevant. Today, the same research about construction phase of JLS is still quite few while the next construction also has a lot of risk. Based on the conditions above, the research is needed to determine what risks can be occurred in the continuation of the project and how far it can affect the schedule. So, it can be planned an appropriate response.

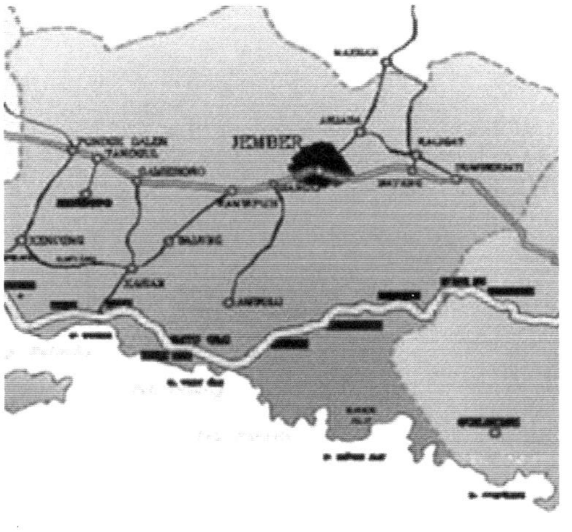

Figure 1. JLS in Jember Regency.

2. Methods

In this study, first step of risk analysis is risk identification process that carried out by site survey, literature studies and supporting data. The field survey was done in order to observe directly the condition of research object, such as topography in JLS environment, while literature review is done to support relevant risk identification process and relevant analysis method. Regarding the supporting data, the completed map from current research location condition is needed so that the risk mapping can be illustrated.

Furthermore, at the main survey in this research used some questionnaire towards some experts to get the value of probability (P) and impact (I). The questionnaire in this study is aimed for those who already expert and have experience in this field for years, they are project manager, General Affair of East Java Road Department and supervisor consultan. It is important to ensure that their experience could give valid answers. The value of probability and impact is aranged from 1-5, and can be seen at table 1 and table 2.

Table 1. Probability Ratings [6].

Ratings	Descriptor	Description
A	Almost certain	Expected to occur in most circumtances
B	Likely	Will probably occur in most circumtances
C	Possible	Might occur at some time
D	Unlikely	Could occur at some time
E	Rare	May occur only in exceptional circumtances

Table 2. Consequences Ratings [6].

Ratings	Descriptor	Description
1	Insignificant	No injuries, low financial loss
2	Minor	First aid threatment, on site release immediately contained, medium financial loss
3	Moderate	Medical treatment required, on site release contained with outside assitance, high financial loss
4	Major	Extensive injuries, loss of production capacity, off site release with detrimental effects, high financial loss
5	Catastrophic	Death, toxic release off-site with detrimental effect, huge financial loss

Table 3. Probability and Impact Matrix [6].

		Consequences				
		1	2	3	4	5
Likelihood	E	major	major	catastrophic	catastrophic	catastrophic
	D	moderate	major	major	catastrophic	catastrophic
	C	low	moderate	major	catastrophic	catastrophic
	B	low	low	moderate	major	catastrophic
	A	low	low	moderate	major	major

The next step is to search for mean value of probability and impact of every risk from respondents by using mean value method as shown in table 4, to represent some values into single value or number.

Table 4. Mean Value Method [7].

Mean	1≤x<1,5	1,5≤x<2,5	2,5≤x<3,5	3,5≤x<4,5	4,5≤x<5
Value	1	2	3	4	5

From the tables above, the next step is to multiply probability (P) and impact (I) of every risk which result risk (R).

$$R=P*I \tag{1}$$

From the calculation of risk (R), it was obtained the six highest risk that could affect schedule of the next construction. The next step is to plan the risk response, to reduce / avoid the highest risk. The risk response is also required as a handling if the highest risks occur.
According [8], there are four action about risk response:

1. Risk rotation, is the action to take and hold risks, that could be done if the risk doesn't make any loose.
2. Risk reduction or risk mitigation, which is an action to reduce probability of risk and distrupted impact of a risk.
3. Risk transfer, is an action to transfer risk to other party, either through contract or hedging.
4. Risk avoidance, is and action to reject identified risk which mean to reject the offered project.

The explanation of the response was obtained through discussions with same respondents / expert to follow up the risks in which the impact is moderate and high level.

3. Results and Discussions

In this research, the data was successfully obtained from 10 respondents. Respondents with experience in the road construction in this study was divided into 4 categories: <5 years, 5-10 years, 10-15 years and >15 years. In this study, the respondents are the project manager, Head of Working Unit, Decision Maker Officer, Site Engineer, Head of Road Planning, and from Supervision Consultant. The respondents are educated at bachelor and master degree level of civil engineering and they have a long experience enough regarding road construction. This questionnaire is also aimed to respondents who have long experience in the field to ensure their experience and position in order to provide accurate information about JLS condition in Jember area.

According to risk identification process by literature study and site survey, they were 18 risk that was relevan to this research. The respondents also provided additional information that there were some events (incidents which were negative in the initial construction phase) that could be added as a risk to the future. This is because there are topographic equations and other factors which are predicted to be occurred again in the phase of advanced trace construction.

The second step is the assessment of probability (P) and its impact (I) by the respondents with scale 1-5. The next step is multiplication of mean value between the probability and the impact of these variables, so that it can be known some risks with moderate and high levels which require the future risk response planning. It will produce what risks which are most likely to be occurred and what risks which have most impact to the schedule of the project. Different scores by the respondents will be averaged by using the mean value method.

Table 5. Relevan Risk at Construction Phase of JLS in Jember Regency.

No	Risk	P	I	R
1	Flood	1	2	2
2	Landslide	2	2	4
3	Abration	2	2	4
4	Storm	1	1	1
5	Heavy rain	3	2	6
6	Wildfires	2	1	2
7	Inflation	2	1	2
8	Increase of material price	3	2	6
9	Difficult access to the project	3	3	9
10	Politic policy changes	1	2	2
11	Less coordination	2	1	2
12	Cost estimate failure	1	1	1
13	Cost overrun	1	2	2
No	Risk	P	I	R
14	Bankrupt of contractor	1	1	1
15	Disturbance to temporary office work	2	2	4
16	Disturbance of land clearing	2	2	4
17	Disturbance of material mobilization	2	2	4
18	Disturbance of fill and cut	2	2	4

19	Disturbance of drainage work	2	2	4
20	Distruption of pavement work	3	2	6
21	Disturbance to minor work	1	2	2
22	Change order	3	2	6
23	Unsuitable work method	2	2	4
24	Error / broken equipment	2	2	4
25	Difficult topography in the project	2	2	4
26	Land condition below stadart	2	2	4
27	Productivity below standart	1	1	1
28	Work accident	3	2	6
29	Rework	2	2	4
30	Broken / damaged bridge	1	3	3
31	Protests from society	2	2	4
32	Crime	2	2	4
33	Theft of material	2	2	4

Furthermore, this data below shows that it was obtained six highest risk to schedule and shown in table 6.

Table 6. Variable with Highest Risk to Schedule.

No	Variable	P	I	R	Level
1	difficult access locations	3	3	9	High
2	heavy rains	3	2	6	Moderate
3	increases of the material price	3	2	6	Moderate
4	broken road pavement work	3	2	6	Moderate
5	change order	3	2	6	Moderate
6	work accident	3	2	6	Moderate

From the table 6, those risks need to be followed up with the planning of the risk response:

1. JLS Jember condition, especially near the border of Banyuwangi is in the form of hills which have high potential of slopes. The response planning to reduce the difficulty of access to project sites can be done by improving access or finding other road routes. Another alternative way is to build a connecting road such as a bridge, so that the process of material and equipment mobilization and routine inspection can be done without any significant obstacles. The use of costs to realize the response is also the responsibility of the contractor and the allocation of the risk is acceptable.
2. The occurrence of heavy rain is an acceptable event which its response can also be done with the completion of the drainage system. Particularly on the 2015 budget road packages, the road works were done in the rainy season, so the impacts were sufficient to affect the quality and timing of the work. For example, when the rain is going on, the compaction work cannot be done because the soil becomes softer, so that the work can be done at the overtime hours if it is needed. Looking at this situation, in the next road construction work schedule is expected to be prepared not during the rainy season because of the occurrence of the damage road, the landslides in the hills and the potential of rework which can be occurred again.
3. The increase of the material prices is a condition which can be said to be reasonable because it is influenced by various factors and it can be predicted. In responding this situation, the government as the owner and project implementer can create an agreement with the supplier with the aim to ensure more efficient price and the availability of guaranteed road construction materials. In this case, the source of the high quality sand material is located in Lumajang Regency, especially around the foot of Semeru Mountain. The area supplies the sand which will be distributed to various parts of East Java, so if there is any occurrence or rise in material prices, buyers continue

IOP Conf. Series: Materials Science and Engineering **267** (2017) 012023 doi:10.1088/1757-899X/267/1/012023

to prioritize quarry in that area. However, in the case of a very significant case such as sand mining conflicts, local and provincial governments should also take into account the search for other sand quarry, such as in Mayang and Puger, Jember District. The allocation of risks for this risk is accepted by volume adjustment or diverted at lower material prices of similar quality through the tests.

4. There are several factors that cause the distruption on road pavement process, namely work quality or material that does not meet the requirement, frequent rain fall, landslides, and abration. Things that can be done to solve the problems include the creation of retaining wall to prevent landslide and geotextile, improvement of road alignment and finishing channel edge (drainage). In case of landslide, the soil material covering the road can be cleaned with bulldozer, then relocated to vacant land with agreement. So that, the allocation is risk reduction.
5. The agreed contract in this project is the unit price, on which the adjustment work can still be done based on Contract Change Order. Like what happened in several some road trace points, soil conditions do not meet the technical requirement due to the presence of sand and mud, so that excavation and landfill are required to do. Another way that can be done is the use of geomembrane so that the muddy soil does not slide down. Thus, Decision Maker Officer and the contractor can make changes to selected landfill volumes based on field conditions. The allocation of risk is acceptable.
6. Work accidents that may occur in road construction project are such as struck landslide material in the form of soil or rock, exposed to dust / dirt because of wrong usage on heavy equipment, exposed to liquid asphalt, injury, illness, even death. Implementation of work savety system should always be implemented in order to avoid material or physical damage of all parties in the field. Operators, workers, foreman and other personnel in the field must always maintain a safe distance from heavy equipment and compactors, and should use other equipment according to its function. Use of personal protective equipment is also required in every activity undertaken by all personnel in the field. The allocation of risk is avoided.

4. Discussion

Previously, it was explained that there were some incidents which caused losses in JLS Lumajang-Jember project in 2007-2015 period, of which those variables could be added back to the future risk identification. This was due to the similarity of topographic conditions between the completed road package and the next road package. This step distinguishes this research from other similar studies, in which past event can be added as risk for the future analysis.

From the table 6, the increases of the material prices are a possible condition which can happen for many times, where the sand material is a factor which needs to be focused. Quarry sand at the foot of Semeru Mountain in Lumajang with a high quality is always prioritized, so if there is a problem such as a mine conflict can affect the supply of the material. An intense coordination between the Provincial Government of East Java and Lumajang Regency is needed to find a solution to the problem so that an early anticipation can minimize the losses for the government and the contractor. The socialization to the local community near the sand quarry location of Semeru Mountain should be prioritized, so that all parties can understand that the sand in that area is needed as the main material for the development of the area itself and other areas in East Java. The supervision and application of legal rules on sand mining activities needs to be tightened in order to avoid similar conflicts. Furthermore for other factors such as the possibility of heavy rain, increases of the material price, broken road pavement work, change order, and work accidents, when those are compared with advanced construction field, they also have the potential to be occurred. With the topography in which located in Jember Regency, the form of hills with less stable soils and clays at some point reaching 12%.

Other similar research is also conducted by [9] which identifies and analyzes the risk of road projects in Lingkar Nagrek V Bandung. The results of the analysis indicate that the dominant risks to the project are landslide on excavation, landslide on slope casting, landslide on landfill work, collapse due to shifting of the soil on geomambran work and the delay of the landfill work. The risks are caused by rain and human error due to lack of reinforcement of construction and unclear images. Another cause is the shifting of soil which can lead to collapse in the construction of *bronjong* and less material in the field causes the risk of the delay in the work of embankment. The results of this study indicated the similarity of difficult terrain characteristics in the hills which potentially cause landslides and work accidents. Mitigation which is done with the use of personal protective equipment must also be applied as part of the OSH standard rules in construction projects. Landslide in JLS development in Jember Regency is not categorized as a dominant risk. This is because of the intensity of landslides which are considered small by the PU personnel and contractors in the field and rare cases of human error. The construction drawings are also considered clear, so the method and manufacture of landslide retaining wall can be well implemented. The risk of an earthquake can also be eliminated because Lumajang and Jember areas are not considered as the large earthquakes areas. Besides, the delay on the pile work is happened because of the distant ground quarry. In addition, this study there is no a dominant risk where less land material in the field is rare. The search for land quarry in Jember Regency area is quite easy. So, there is no delay in the pile work.

On another research by [10] which analyzed the risk of road construction in Jayapura, the study suggests that the main risks include delayed material delivery, deficit in material supply, technology implementation problem, and an increase in material cost. These risks are caused by difficult terrain condition, shortage of material and equipment provision as well as inexperienced worker. Similar to the risk analyzed in this research, material expense increase, the calculation has resulted in a high probability and impact of project schedule delay. The other similar risk factor is topography condition where the site consist of hilly terrain that makes land clearence for road difficult to access. On the other hand, there is a significant difference in the risk factor of technology and human resources, in this study both factors are not considered to be major cause.

Based on the explanation above, it can be concluded that the problems which need to be considered in the future is how to do the appropriate handling measures if the identified dominant risk is really happened. The involvement of all personnel in the field in carrying out their respective tasks needs to be monitored so that the work and rework accidents can be avoided. While the increases of the material price, heavy rain, difficult access and change orders are a situation and conditions which must be accepted by all related parties, where the problems are often exist and difficult to be avoided.

5. Conclusion

1. From the analytical process, is was obtained 6 highest risk, they were difficult access locations, heavy rains, increases of the material price, broken road pavement work, change order, and work accident
2. Risk responses were proposed by applying agreement to guarantee stock and price of materials, prioritized drainage, and constructing bridge to solve difficult area. An intense coordination in the site and routine checks of quality were needed to avoid rework. Supervision of quality of materials, manufacturing of retailing walls were also needed to reduce possibility of distruption to pavement work. To avoid work accident, it need to socialize about harsh terrain condition, mutual allertness among supervisor, worker and the others, and also all personals must comply with savety rules.

6. References

[1] Poject Management Institute 2013 *A guide to the project management body of knowledge (pmbok@guide)-fourth edition* (Pennsylvania: Project Management Institute, Inc)
[2] Kezner H 2001 *Project management 7th edition* (New York: John Wiley & Sons, Inc)
[3] Gray C F and Larson E W 2000 *Project Management, 1st edition* (Boston: Irwin McGraw-Hill College)

[4] Wang M T and Chou HY 2003 Risk allocation and risk handling of handling of highway projects in taiwan *J. Management in Engineering.* **19** No 2 60-68
[5] Yuwana P D P and Wiguna I P A 2013 *Analisa resiko proyek infrastruktur jalan dengan sistem performance based contract studi kasus peningkatan jalan demak-trengguli Proc. Seminar Nasional* (Surabaya: MMT-Institut Teknologi Sepuluh Nopember)
[6] Australian Standard 2004 *AS/NZS 4360: 2004 Risk Management* (Strathfield: Standards Association of Australia)
[7] Sukaarta I W 2012 Analisis resiko proyek pembangunan dermaga study kasus dermaga pehe di kecamatan siau barat kabupaten kepulauan sitaro *J. Ilmiah MEDIA ENGINEERING.* **2** No 4 257-266
[8] Flanagan E and Norman G 1993 *Risk Management and Constraction* (London: Blackwell Science)
[9] Octavia R D 2012 *Identifikasi dan Analisa Risiko Konstruksi dengan Metode Failure Mode and Effect Analysis (FMEA) dan Fault Tree Analysis (FTA) pada Proyek Pembangunan Jalan Lingkar Nagrek V Bandung Under Graduate* (Surabaya: Institut Teknologi Sepuluh Nopember)
[10] Mabui D S S 2006 *Analisis risiko pelaksanaan peningkatan jalan di jayapura Thesis* (Surabaya: Institut Teknologi Sepuluh Nopember)

IOP Conf. Series: Materials Science and Engineering **267** (2017) 012024 doi:10.1088/1757-899X/267/1/012024

Collaborative Decision Model on Stockpile Material of a Traditional Market Infrastructure using Value-Based HBU

C Utomo[1], Y Rahmawati[2], D L Pararta[1] and A Ariesta[1]
[1] Department of Civil Engineering, Institut Teknologi Sepuluh Nopember, Indonesia
[2] Department of Civil and Environmental Engineering, Universiti Teknologi PETRONAS, Malaysia

E-mail: christiono@ce.its.ac.id; yani.rahmawati@utp.edu.my

Abstract. Readiness of infrastructure establishment is needed in the early phase of real estate development. To meet the needs of retail property in the form of traditional markets, the Government prepares to build a new 1300 units. Traditional market development requires infrastructure development. One of it is the preparation of sand material embankment as much as ± 200,000 m^3. With a distance of 30 km, sand material can be delivered to the project site by dump trucks that can only be operated by 2 trip per day. The material is managed by using stockpile method. Decision of stockpile location requires multi person and multi criteria in a collaborative environment. The highest and the best use (HBU) criteria was used to construct a value-based decision hierarchy. Decision makers from five stakeholders analyzed the best of three locations by giving their own preference of development cost and HBU function. Analytical Hierarchy Process (AHP) based on satisfying options and cooperative game was applied for agreement options and coalition formation on collaborative decision. The result indicates that not all solutions become a possible location for the stockpile material. It shows the 'best fit' options process for all decision makers.

1. Introduction

Mechanical land removal is used by companies to implement business strategies that are preceded by the readiness of infrastructure in the development of traditional market, one of that is the preparation of material of sandstone of ± 200.000 m^3. The sandstone can be transported by fleet of dump truck with capacity of 5-7 m^3 from the origin to project location. With a distance of 30 km, the fleet of dump trucks can only be operated by two trips (count unit for delivering material using truck) per day. With the assumption that 25 units of dump truck fleets are needed to transport 200,000 m3 sandstones to the project location, it can be calculated that the material delivery takes 571 days. Related with this condition, it will be necessary to select and determine the location of stockpile. The choice of stockpile location depends on the characteristics of each field condition. The implementation of construction project is appeared to be always encountering with time and cost constraints. These constraints are able to cause delay in the project, so that the time required in completing the project will be able to exceed the employment contract. In this research, there are three alternatives of places as the location of the stockpile, these alternatives have different characteristics. Generally, location selection is determined by several factors such as local input, local demand, transferred input, and external demand [1]. Industrial site selection is based on the principle of cost minimization. Weber [2] stated that the location of each industry depends on the total cost of transportation and labour where the sum of both must be low. The minimum cost of transportation and labour is identical to the maximum profit level. In the concept of the highest and the best use (HBU), the factors are grouped in two criteria which are cost of development and function. This is in accordance with the definition of location that can be interpreted as the science of geographical allocation of scarce resources, and the relationship or influence on the location of various businesses or other activities. Based on the previous problem formulation, this research purposes to choose the best stockpile location that will be used in the project development of traditional market in East Java.

2. Conceptual

2.1. Collaborative Decision

Collaborative design is essential in the design process of project development. Decision for best option for sustainable design should apply this approach [3, 4, and 5]. Collaborative design is an effort to integrate various participants to obtain the best design by concerning the best shared-goal that is suitable with all requirements from the participants [6]. Integration of tangible and intangible aspects is important for collaborative design. There are two main aspects that are needed to be integrated, which are tangible and intangible aspects [7]. During the design process, each expertise is encouraged to share their knowledge. The knowledge is included the intangible aspect, where it is also classified into tacit and explicit knowledge [8], meanwhile the tangible aspect consists of design drawings, and the results of design developments. The two aspects are needed to be integrated in order to achieve successful collaboration. The involvement of various participants makes it difficult to be realized, in which it is not only based on their different backgrounds, but also their different availabilities in conducting the design meetings. Previous studies developed systems and tools to support the design collaboration, especially the virtual design [9, 10, and 11]. Based from review conducted by [7], it is found that decision making and negotiation are two vital activities in supporting the collaboration as well as the integration, especially in finding the best design option for the sustainability. The importance of it also can be found in the previous studies, in which the advance researches or applications of collaboration in decision making and negotiation can also be found [12, 13, and 14].

2.2. Value-based

Value-based design decision is a structural and analytical process that seeks to achieve value by identifying all necessary functions at the lowest cost, while maintaining with the required levels of quality and performance [15]. The decision has been widely adopted in many countries over several decades as a very effective tool to meet the increasing demands for value. Kirk, et al. [16] described value based on approach as new approach and methodology that involves the use of a multidisciplinary team. In the natural characteristic of construction, it means that a tool for decision team is necessary. Cooperation is the nature in team work [17]. Decision analysis techniques can be applied to determine the relative value of the alternative solutions for performing function. Many studies in value-based decision apply multi criteria decision making [12] such as in assessment of exterior building wall, in material design of concrete and in a modification of value engineering in petrochemical industry.

2.3. The Highest and the Best Used

The definition of the highest and the best use is a concept of appraisal that can be applied to land or buildings that is usually interpreted as land use [18]. It will maximize the wealth of the owner through the use of the most profitable land. The highest and the best use can also be defined as a possible legal and logical use of land, which allows physically, reasonably, and financially permitting to provide the highest value. To determine the highest and the best use in a plot of vacant land, some testing steps must be performed, the testing process involves four criteria with each requirement that must meet:

legally permissible, physically possible, financially feasible, and maximally productive [19]. Selection process based on these criteria is often done in sequence. The first and second criteria test must be done before the next criteria. This is because it is possible that the use of the financial aspect may be allowed but from legal aspect and physical aspect is impossible. By using this method, the value of the land can be known if the value of the building is also known with certainty because the building is relatively new. The value of the building multiplied by the capitalization rate of the building (the interest rate plus the rate of return on the building) obtained the net income of the building. Net income of property minus net income of buildings acquired net income from land. It is capitalized at the interest rate on the value of the land

2.4. Stockpile

Stockpile management serves as a buffer between delivery and process as a strategic stock against the short-term or long-term disturbances. Stockpile also serves as a process of homogenizing and/or mixing sandstone to prepare the required qualities. Homogenization aims to prepare the product of one type of material where fluctuations in sandstone quality and size distribution are equated. The process of homogenization is done through mixing techniques. According to [20], stockpile management is very important part in the process of handling sandstone either from the side of mining until it is received and used by the user. Generally, the mining industry of sandstone and sandstone users utilize stockpile in its industrial facilities. This becomes part of securing the supply of both quantity and quality of use of sandstone. Stockpiles may account for half of the total land base. In the dense urban areas, stockpile placement requires careful consideration. Handling and storage is a large cost element, therefore, there should be a plan for each project that shows how and where the materials and products will be handled and stored. A storage plan will help to ensure the quality control of the product. When stock is placed between extraction, processing and loading facilities, it will minimize hauling distance, saving time, fuel, and machine maintenance.

3. Method

3.1. Function Analysis of the Highest and the Best Use

Function analysis is a technique that oriented to the required functions on each item or system reviewed to produce the desired product value. The analysis is used to describe the main function of the product, it also describes the classification of the primary functions and the secondary functions to obtain a comparison between cost and worth required. For some functions it may be decided that a set of generic process is needed to perform the function, each of that will increase the associated set of possible specific processes. Based on the process of function analysis system technique (FAST) [21], there are four HBU functions that are legally, physically, financially, and productively maximum. A number of processes identified as being probable alternatives for performing the function

3.2. Life Cycle Cost of Development

Life cycle cost (LCC) is a technique to evaluate the economic value of stockpile location by calculating all relevant costs during the investment period through adjustment to time value of money. The LCC application method involves a combination of managerial, financial and technical expertise in all phases of life-cycle. The main purpose of this LCC is to evaluate alternative solutions to specific design problems. LCC in this paper refers to the development process [22] of a property building. The process follows the development stages in the form of inception, design, formal negotiation, construction, and property management. LCC as development cost consist of capital expenditure, energy cost, operation and maintenance (O&M) cost, and replacement cost. In the collaborative decision, perspective of LCC is different among stakeholders.

IOP Conf. Series: Materials Science and Engineering **267** (2017) 012024 doi:10.1088/1757-899X/267/1/012024

3.3. Decision for Material Location
This section deals with the definition of the criteria for evaluating a location of stockpile. In this paper, the criteria are taken from the basic theory of function and cost. The functions will make the technical solutions worth considered and proceeded to become attributes of the decision. The determinants of the stockpile location shown are the result of literature studies, previous researches, and discussions with expert taken from practitioners. To obtain a good representation of a problem, it has to be structured into different components called activities. Figure 1 shows three levels of decision hierarchy. The goal of the problem (G ="Select the best value of stockpile location material of a traditional market infrastructure") is addressed by three alternatives (A = a1; a2; a3). Two evaluation criteria are used to select the best stockpile location. Then, implementation of analytical hierarchy process (AHP) [23] can be started with compilation of the hierarchy model.

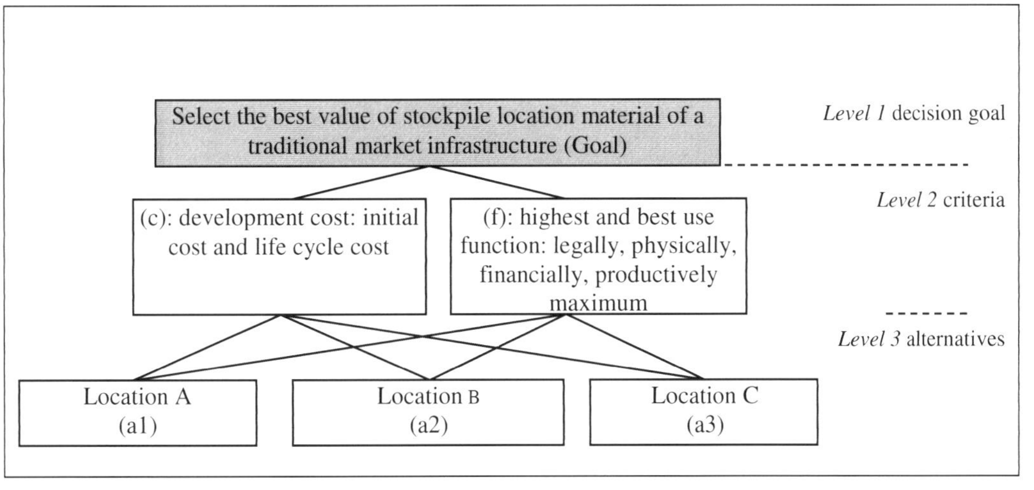

Figure 1. Decision hierarchy for stockpile location.

4. Result and Discussion

4.1. Value-based Decision
The technical solution options were categorized into 'Cost' that was identified by development cost; and 'Function' by the highest and the best use functions. Table 1 shows the selected ability (Ps) and rejected ability (Pr) of a satisficing option [24] that represents function and cost of technical solution for stockpile location respectively. The value of each alternative is calculated based on the value equation that is function/cost. The greater the number the greater the value. The best alternative is alternatives with the highest value. The table presents the difference of best option among stakeholder. The choice of government is different from the choice of tenant. This can be understood because of differences of interest. There are two judgments involved in this decision - the first is criteria judgment for each decision makers and the second, technical solution judgment for each criterion. The process of weighting factor is presented in Table 1.

4.2. Agreement Options and Coalition
The framework of collaborative decision is developed based on agreement option and coalition. It is applied for selection of stockpile location on traditional market. Five stakeholders (Figure 2) were involved to give their own preference. Once every stakeholder is aware of the options, they analyze to determine what they will get, gain or loss if each alternative is selected. This agreement options process provided additional functionality to negotiate a joint representation of the problem. Stakeholder of multi-criteria decision making problems usually evaluates the alternative solution from

IOP Conf. Series: Materials Science and Engineering **267** (2017) 012024 doi:10.1088/1757-899X/267/1/012024

different perspective, making it possible to have a dominant solution among the alternatives. Only promising solution were available to stakeholders for detailed collaboration. Collaboration has coalition formation algorithms. This research employs multiple coalition formation algorithms. The stakeholders are the parties in the stock pile location selection. There are three stages to determine the collaborative decision. The first stage came from individual decision. The results from this first stage are used to determine the agreement options in the last two stages. The stages are:

1. Determining and scoring of every alternative for every stakeholder. Stakeholders had different best option as an alternative solution.
2. Determining the optimal solution (payoff optimum) for each stakeholder in a coalition is based on cooperative multi-person games [25]. A linear programming formula is used to determine the Pareto optimal for each stakeholder in each coalition.
3. Stage three analyzes the best fit options for every coalition and grand coalition. The result is presented on Table 3. The result shows that all solutions became a possible location for the stockpile material. B location (a2) is the best choice of stockpile location for the group. Observed on a3, the result is interesting. Even though this solution has no first priority by any decision maker, this solution is chosen by many coalitions as the best fit option. The best means that the option is the most selected in all individual and also all coalition.

Table 1. Value of alternatives by each stakeholder.

SH1: Equipment manager			
Alternatives of material location	Ps (Function)	Pr (Cost)	Value = Function/Cost
(a1) Location A	0.612698	0.309324	1.980769
(a2) Location B	0.117989	0.450973	0.261633
(a3) Location C	0.269312	0.239703	1.123524
SH2: Operational manager			
Alternatives of material location	Ps (Function)	Pr (Cost)	Value = Function/Cost
(a1) Location A	0.137288	0.093241	1.472393
(a2) Location B	0.239488	0.494470	0.484332
(a3) Location C	0.623225	0.412289	1.511622
SH3: Finance manager			
Alternatives of material location	Ps (Function)	Pr (Cost)	Value = Function/Cost
(a1) Location A	0.093382	0.093241	1.001514
(a2) Location B	0.685294	0.49447	1.385916
(a3) Location C	0.221324	0.412289	0.536817
SH4: Project manager			
Alternatives of material location	Ps (Function)	Pr (Cost)	Value = Function/Cost
(a1) Location A	0.127949	0.093241	1.372233
(a2) Location B	0.784870	0.494470	1.587296
(a3) Location C	0.087181	0.412289	0.211456
SH5: Project director			
Alternatives of material location	Ps (Function)	Pr (Cost)	Value = Function/Cost
(a1) Location A	0.583152	0.109964	5.303130

(a2) Location B	0.109964	0.583152	0.188568
(a3) Location C	0.206884	0.486232	0.425484

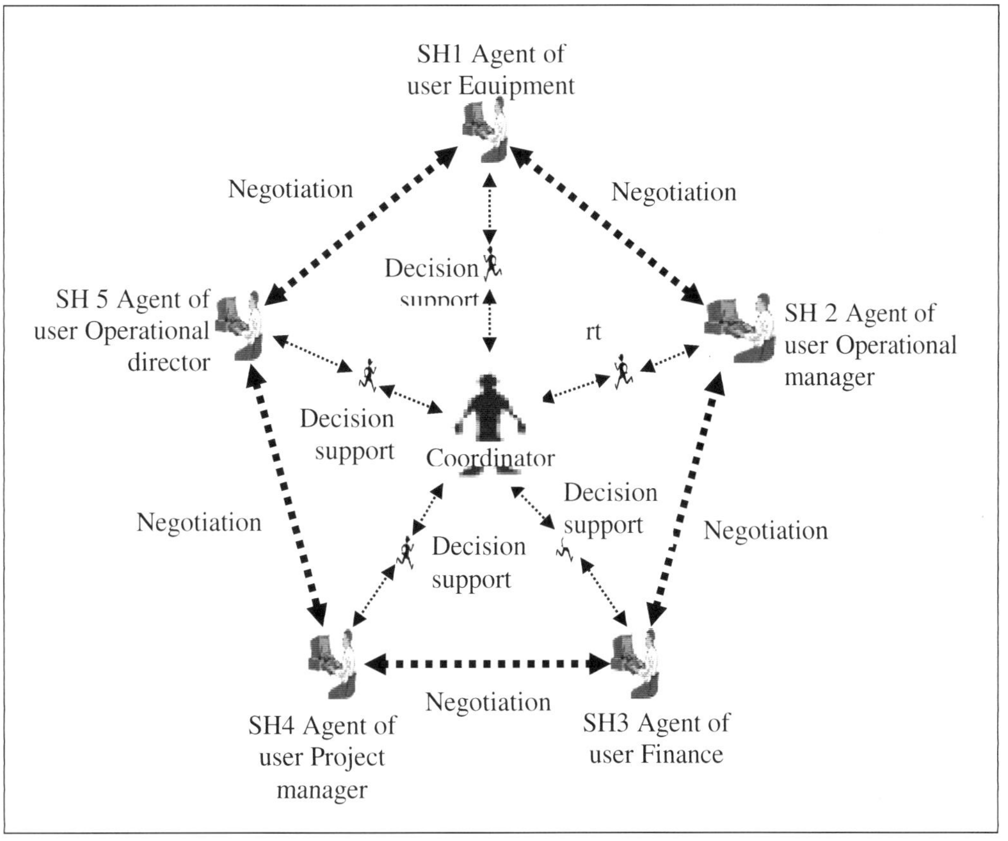

Figure 2. System architecture collaboration.

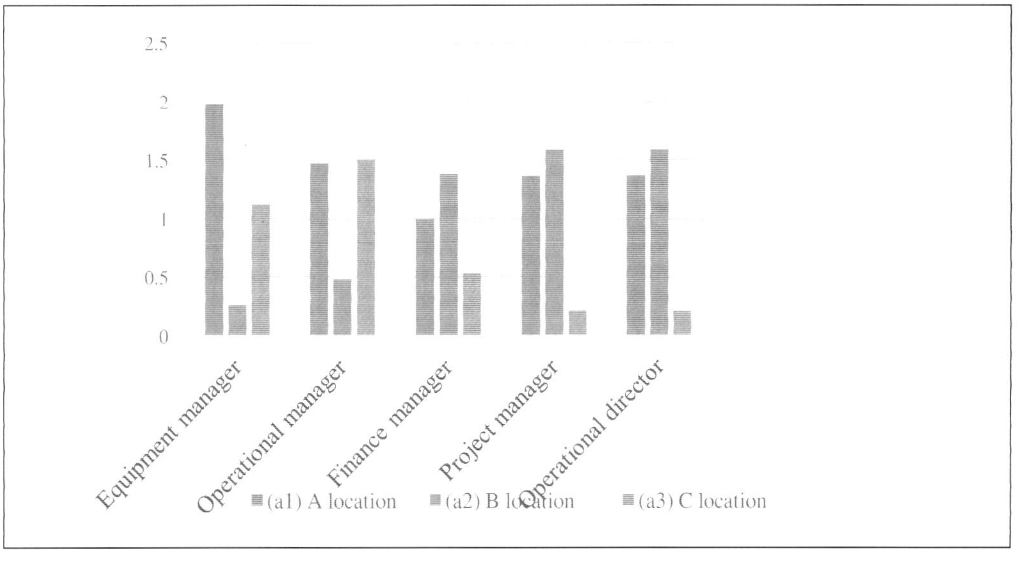

Figure 3. Value of alternatives for each stakeholder.

Table 2. Ranking of stockpile location solution for each coalition.

	Coalition	Alternatives of stockpile location		
		a1	a2	a3
1	0	0	0	0
2	SH 1 (Equipment manager)	1st	2nd	3rd
3	SH 2 (Operational manager)	2nd	1st	3rd
4	SH 3 (Finance manager)	2nd	1st	3rd
5	SH 4 (Project manager)	3rd	1st	2nd
6	SH 5 (Operational director)	1st	3rd	2nd
7	Coalition SH1, SH2, SH3, SH4	3rd	2rd	1st
8	Coalition SH1, SH2, SH3, SH5	3rd	1st	2nd
9	Coalition SH1, SH2, SH4, SH5	1st	2nd	3rd
10	Coalition SH1, SH3, SH4, SH5	3rd	1st	2nd
11	Coalition SH2, SH3, SH4, SH5	3rd	2rd	1st
12	Grand Coalition SH1, SH2, SH3, SH4, SH5	3rd	1st	2nd
RESULT		2nd	1st	3rd

5. Conclusion

The paper deals with a technique during selection of a stockpile location alternative for infrastructure of traditional market. The process was conducted by identifying the agreement options among stakeholders. Value of the best fit option was formed by the comparison between the highest and the best used function and its development cost. Advanced research is needed, primarily in the study of automation on collaborative decision. Future research in the application of this methodology in many field of decision will build a wide range of knowledge to solve the theoretical and practical gap in automated design and automated decision.

6. Acknowledgement

The authors appreciate the recognition and awards in forms of research grant and fellowship from "Penelitian Unggulan Perguruan Tinggi (PUPT) 2017 based on contract number 563/PKS/ITS/2017"

7. References

[1] Hoover E M and Giarratani M 2007 *An Introduction to Regional Economics*
[2] Weber A 1909 *Theory of the Location of Industries* (Chicago: The University of Chicago Press)
[3] Rahmawati Y, Utomo C, Anwar N, Nurcahyo C.B, and Negoro N P 2014 Theoretical framework of collaborative design issues *Jurnal Teknologi* **70 (7)** 47-53
[4] Lu S C Y, Elmaraghy W, Schuh G., and Wilhelm R 2007 A scientific foundation of collaborative engineering *Annals of the CIRP* **56 (2)** 605-634
[5] Detienne F 2006 Collaborative design: managing task interdependencies and multiple perspective *Journal of Interacting with Computer* **18** 1-20
[6] Utomo C, Rahmawati Y, Suhartono, and Negoro N P 2015 A concept toward decision support for collaborative urban heritage selection *Journal of Sustainable Development* **8(8)**
[7] Rahmawati Y, Utomo C, Anwar N, Negoro N P and Nurcahyo C B 2014 a framework of knowledge management for successful group decision in design process *IEEE Conference on Open Systems* Subang Jaya 26-28 Oct. 2014
[8] Rahmawati Y and Utomo C 2014 The influence of knowledge management to successful collaborative design *IEEE Conference on Information Technology, Computer and Electrical Engineering, Semarang*, 7-8 Nov. 2014

[9] Gross M D, Yi-L D E, McCall R J Citrin V W, Hamill P, Warmack A, and Kuczun, Collaboration and coordination in architectural design: approaches to computer mediated teamwork *Automation in Construction* **7** 465-473
[10] Kvan T 2000 Collaborative design: what is it? *Automation in Construction* 409-415
[11] Utomo C and Rahmawati Y 2014 Knowledge and protocol on collaborative design selection *IEEE Conference on Information Technology, Computer and Electrical Engineering*, Semarang, 7-8 Nov. 2014
[12] Utomo C, Zin R M, Zakaria R and Rahmawati Y 2014 A conceptual model of agreement options for value-based group decision on value management *Jurnal Teknologi* **70(70)** 47-53
[13] Utomo C, Krisnawaty N and Rahmawati Y Automated negotiation for non-producing management decision *Int'l Conference on Advanced Mechatronics, Intelligent Manufacture, and Industrial Automation*, Surabaya 15-17 Oct. 2015
[14] Rahmawati Y and Utomo C 2015 Media support for elicitation in virtual multidisciplinary design *Int'l Conference on Advanced Mechatronics, Intelligent Manufacture, and Industrial Automation*, Surabaya 15-17 Oct 2015
[15] Kaufman J and Woodhead R 2006 *Stimulating Innovation in Products and Services with Function Analysis and Mapping* (Hoboken NJ: Wiley-Interscience)
[16] Kirk S J, Turk R G and Hobbs R.W 2007 *Value Based Team Design Decision Making* (USA: The American Institute of Architects)
[17] Sanchez M, Prats F, Agell N and Ormazabal G 2005 Multiple-criteria evaluation for value management in civil engineering *Journal of Management in Engineering* **21(3)** 131-137
[18] Munizzo M A, Musial L V, *General Market Analysis and Highest and Best Use* (Mason, OH USA: Cengage Learning, 2010) p 10
[19] *The Appraisal of Real Estate* 2013 (the Appraisal Institute)
[20] American Concrete Pavement Association 2004 Clay Ball Prevention and Repair: Stockpile Management is Key (Illionis, USA)
[21] Bytheway C W 2007 *FAST Creativity and Innovation: Rapidly Improving Processes, Product Development and Solving Complex Problems* (Florida: J.Ross Publishing)
[22] Miles M E, Laurence M N and Adrienne S 2007 *Real Estate Development: Principles and Process.* (Urban Land Institute)
[23] Saaty T L 2004 Decision making – the analytical hierarchy process and network process (AHP/ANP) *J. of System Science and System Engineering* **13(1)** 1-34
[24] Stirling W C 2003 *Satisfying Games and Decision Making with Applications to Engineering and Computer Science* (Cambridge, UK: Cambridge University Press)
[25] Kelly A 2003 *Decision Making Using Game Theory: an Introduction for Managers* (Cambridge, UK: Cambridge University Press)

IOP Conf. Series: Materials Science and Engineering **267** (2017) 012025 doi:10.1088/1757-899X/267/1/012025

Building Structure Housing: Case Study of Community Housing in Kendari City

M Z Umar[1], A Faslih[1], M Arsyad[1], A S Sjamsu[1], I Kadir[1]

[1] Department of Architecture, Vocational Education Program, Halu Oleo University Jl. H.E.A Mokodompit, Andounouhou-Kendari 93232

Email: muzakum.uho@gmail.com

Abstract. Housing development has been pioneered through a simple home construction program to reduce the production cost. Simple housing program was developed in Kendari City. The purpose of this study is to show the principles of reducing the cost production for the type 36 homes, in Kendari City. The selected architectural objects are the lower, middle and the upper structure of type 36 house. The data collection was done by observation and in-depth discussion with construction workers. The analysis technique used in this research was a descriptive narrative analysis technique in the form of tabulation data. This study concluded that there are several principles of price reduction in the structure of public housing buildings. Quick principles exist in constructing techniques such as using cigarette packs as a foundation pad, mortar usage for rapid wall standing, and the spacing of mortars could be done manually by using two fingers on a human hand. Economic principles could be used for material matters, such as eliminating the use of gravel for concrete, the use of sand material to contain the soil, the foundation does not use sand and empty stone, and the shape of the ring beam was made using triangle reinforcement.

Keywords: Housing structure

1. Introduction

Nowadays, the housing ownership has become the national problem in Indonesia, especially in urban areas [1]. Kendari city community for instance, has difficulty to own home residence due to the high price of land, cost of materials, and the price of building components. Therefore, the government began to introduce the concept of Small Housing construction (RS) and Simple Healthy House (RSS) with the support from the national bank BTN under the government loan subsidy program (KPR) [2]. The housing construction has been initiated through a simple home program to reduce the price of production. The program was planned to accommodate the house with one bedroom, living room, dining room, kitchen, bathroom, with the total space area of 36 m² to 72 m² [4]. The purpose of this study is to figure out the principles used in the production of BTN housing program structure type 36 in Kendari.

2. Literature Review

2.1 The Principles of Building

Today, it is very essential to implement a good, reliable and fast work to construct a house, especially in foundation work [5]. Precast concrete structural system is one of the alternative way which supports time and energy efficiency [6]. Baduy tribe for instance, has been using this technology long time ago before high technology was known. the components of the house had been prepared to assembly later on by the homeowner so it takes a short time to construct a house. Generally, Baduy tribe took 5-10 days to construct home [7]. Considering the time issue in house construction, thus it is important to develop the idea of rapid house construction for a large scale community.

2.2 Principles of Economics in Buildings

A sufficient financial is a basic need for every household. If basic income of a family increases, the physical condition of the house will increase as well [9]. The building structure must meet the requirements of stability, balance, strength, aesthetics, and economical. In general, the construction of buildings must be economically viable. Based on the above it can be concluded that economic principles are needed in the building.

2.3 Simple Principles of Building

Architecture is not a matter of a luxury high-technology materials or the location of its building. Simple materials are even more capable to reflect the art of the building itself. A good architecture could also come from a simplicity [11]. The house is like a beautiful woman. No women from their appearance looks simple, but it has a very broad meaning of life. Thus, these women can be viewed as the beauty of the deeper value than just cosmetic [12]. Based on the above it can be concluded that these simple principles are needed in architecture.

3. Research Methods

This research was conducted in Kendari City. This research was done in qualitative method with case study approach to house type 36 in Kendari City. The selected architectural objects are the lower, middle and the upper structure of the type 36 house. The data collection method was done by observation and in-depth discussion with building workers. The analytical technique which used in this research was descriptive narrative analysis technique in the form of tabulation data on the structure of house type 36.

4. Results and Discussion

This research was conducted in public housing located in Wuawua Village, Wuawua District, Kendari. Houses in this housing there are two types. The first type is type 45 with total 57 units. The second type is type 36 with total of 139 units. The production cost of type 36 per unit is Rp. 50.000.000, - (Fifty million rupiah) with the cost of construction workers Rp. 14.000.000, - (Fourteen million rupiah) and development cost Rp. 36.000.000, - (Thirty-six million rupiah). The lower structure, the middle structure, and the upper structure of the house of type 36 are, as follows:

4.1. Bottom Structure

Bow plank was installed before digging the foundation hole. The tools which used in this work including hose, axe, and hammer. The materials which used in this work were are stacked wood, unused wooden boards, nails, and artisan rope. After the bow plank wood and rope has been installed, the next process was digging the foundation hole. There were two forms of foundation based on excavation pit in this housing. The first form of the foundation is the foundation located in the hole. The second form of the foundation is a partial foundation in the hole and partly above the ground. The height of the foundation in this housing may varies, as it adjusted the shape of the contours of the soil. The height of the foundation was about 20-80 cm. The width of the foundation head was 20-30 cm. The thickness of the foundation mortar head was 3 cm. The factors which affect the shape of the foundation head were the stirrup, thickness of plaster, and thickness board. The width of the best foundation head was 20 cm. The foundation formula was determined by the width of the foundation head, e.g. the width of the foundation head 25 cm, then the height of the foundation 25 x 2 = 50 cm, so the base foundation width would be equal to the height of the foundation. The foundation of the elbow was known by using an elbow tool, a cigarette pack (shown in figure 1), paper money, or using the formula 80: 60 = 100 by forming an elbow plane.

The foundation in this housing did not use sand dunes and empty stone pairs. The composition of the plaster ratio was 1: 4 for sloof and foundation. That means the top foundation was equipped with sloof, except foundation of terrace. The terrace foundation is made by using mountain rocks. The composition of the concrete mixture is 1: 4 or 1: 5 with sand material from the Pohara area and sand material from the Nambo area. The sand material which use for concrete mix design was material sand from Pohara area. This sand has the characteristic of coarse texture, solid and gray. The price for one truck was Rp. 550,000, - (five hundred fifty thousand rupiah). The other material which use was sand material from the Nambo area. The characteristics of this sand were white (if dried), red (when excavated), pebbly, and full of soil. The price of one truck was Rp. 350,000, - (three hundred fifty thousand rupiah). The composition of the mixture for the sloof concrete was 3 lorry of sand material from Pohara area, 1 lorry of sand material from Nambo area, 1 sack of cement, and water. The mix design for the ring balk is 1: 3; 1: 4; And 1: 5. The composition of the concrete mixture for the column is 1: 4. The cast thickness for sloof, column, and ring balk was 2 cm. The mixture composition for carport floor rebates was 1: 4 using sand material from the Nambo area and Pohara area, one cement sack. The floor rebate thickness was 8 cm.

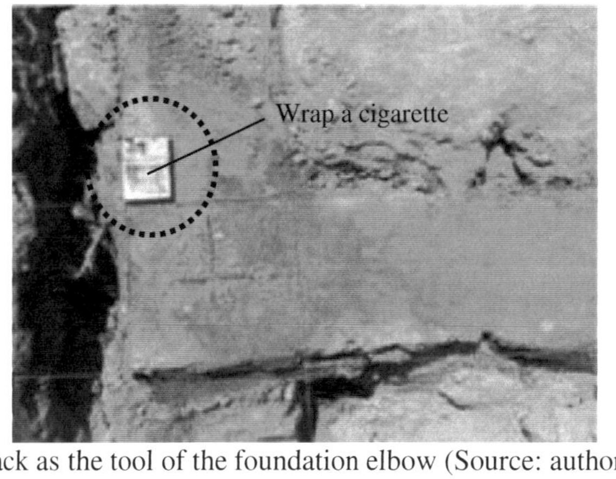

Figure 1. Cigarette pack as the tool of the foundation elbow (Source: author's collections, 2016)

The size of the iron stirrup for sloof was 8-12 cm high, 8-9 cm wide, 8 mm reinforcement steel, and 6 m long. The distance on each stirrup on the sloof refinery was 25-40 cm. The length of concrete binding wire was 35 cm. The length of the stirrup iron was cut by 42-45 cm. The amount of reinforcing steel in iron stirrup was four bars with an iron size of 8 mm. The reinforcing steel was bent over at the end of 3-5 cm long. After sloof was casted, the size becomes 9 x 10 cm and 12 x 15 cm. The tools which used for arranging iron sloof stirrup were including iron arches, iron fabricators, iron scissors, gauges, and iron locks. The materials were concrete binding wire, 8mm reinforcement steel, and iron for stirrup 6 mm. The dimensions of sloof stirrup needed to be taken into account with the dimensions of the brick. The brick dimensions which used in this housing was 19 cm long, 10 cm wide, and 5 cm high. The dimensions of sloof stirrup should be smaller than the width of the brick, so that the sloof and wall surfaces were flat.

Septic tank was located under the carport. Septic tank hole was dug as deep as one meter. Septic tanks were made into two wells. There were two types of sewer pipes in this housing, sewerage pipes from latrines and sewage pipes from the bathroom. Drainage was made as part of the foundation. Ground drainage was made from stack of rock mountain rocks and plastered. The soil type at this residential project location was waxy ground, this soil type is easy to melt when exposed to water and easily harden when exposed to the sun and has small swelling number.

4.2. Central Structure

Dimensions of stirrup for columns were 6-10 cm long and 6-10 cm wide. And it increased double in size as the concrete were casted. The distance between each column stirrup was 30-40 cm. Materials which used to make iron stirrup was a steel reinforcement with a size of 6 mm. The shape of the stirrup column should be smaller than the dimension of the brick. If the columns arouse between the brick, the concrete would get easily porous and broken. After the brick wall had been installed with a distance of one meter, then the casting of the column could be done. The columns were casted using sand material from the Nambo area and not sifted. The iron length for the column was 3.20 m. After floor installation, the column height becomes 2.70 m. The principle of steel casting was to make the stirrup distance on sloof, column, and ring balk is closer to the column and vice versa. The closest stirrup distance is 17-20 cm and the next distance 30-50 cm.

The composition of the mix ratio for wall plaster was 1: 4 with 1 cement sack, 3 pohara sand trays, and 1 Nambo sand tray. The sand material from the Pohara area was used as material for floor rebates and wall plaster. The sand material from the Nambo area tend to have less swelling number which were good material for plaster walls. This type of sand material was often used in public housing projects because of its economic price. The construction workers tend not to spray water on the walls before plastered and patched with mortar. The process of walls plastering started from 10 cm of the sloof head. Then, the plaster process was continued to the surface of the ring balk head. The walls were given skim coat to soften the walls with the composition of one bucket of skim coat mixed with half bucket of water or eight handfuls of skim coat mixed with half bucket of water. The process of skim coating was done after the wall has been completed plastered. The walls were cleaned using the unused cement bag then drawn the skim coat from the bottom by using two cape tools. Then applied the same process for the rest of the project.

The height of the walls in this housing varied from 3.00 m and 3.20 m. The wall height was measured from the sloof head to the ring balk head. The process of installing brick wall by weighing. The base of the yarn weigh was placed at the edges of the brick above the sloof. The position of the thread should be straight referring the reference point to erect the wall. If the wall was not weighed, then the wall could be potentially curved and collapsed. During the brick preparation, it is important to check its stability by touching the wall. if the brick wall was easy to shake when touched, the process should be stopped and wait until the mortar was dried. The wall might get collapsed if the span distance between them was too short and the wall height was more than 3 m. The wall stability was also determined by the sand material. Building workers prefer to put thick mortar in the installation of the brick, because the thicker the mortar, the faster the wall stands. The type of wood which used for the frame in this housing was the wood class type II.

6 mm iron was used for iron reinforcement and stirrup. The iron stirrup shape of the ring beam was a triangle (shown in figure 2). Distance between stirrup was 30-55 cm. Iron stirrup for ring beam were cut in 30 cm long, 7-8 cm height, and 7-8 cm width. The thickness of the cast beam ring was 1-2 cm. The reinforcement steel for ring beams fabrication shall be done in accordance with building area. The Nambo sand material was spread over a heap. The tiled floor was installed from the back corner of the door in the second row. The following floor installation follows the floor pattern. The tile floor needed to be soaked in water to permeate the mortar. Ceramic immersion was carried out in the lorry for 5-10 minutes. The mortar composition for the tile floor was 1: 4. The floor layers were made from ground embankment using Nambo san material, mortar 1: 4 with 5 cm thick.

Figure 2. The triangular stirring shape for the beam ring (Source: author's collections, 2016)

4.3. Top Structure
Ampig is a roof frame made of red stone. The height of ampig on the right side was 0.40 m. The height of the ampig on the left side was 0.20 m. The height of ampig was vary, due to the shape of the terraced roof. Mounted column was located at the center of the ampig to avoid instability. The steel reinforcement for ampig column was planted into a beam ring 0.10 m. The hand block on the ampig was not fitted with steel reinforcement rather be plastered. The concrete deck was printed on a terrace with the size of 0.80 x 2.00 m and a cast thickness of 0.10 m. The concrete deck used 6 mm steel woven reinforcement like a mat. Woven iron stirrup made of two layers and casted together with ring beam.

Working tools which used for assembling the lightweight steel roof truss were including hammer, waist bag, screw driver, self-drilling screw, steel scissors, meter, contacts stock, roll wire, and ladder. The materials which used for assembling the lightweight steel roof truss were light steel 72.35 mm with hat sections of 25.65 mm, screw, blue spandex roof, roof top camel, and lisplang. The details of the lightweight steel roof truss construction were including light steel stretched at its bottom which usually called the bottom chord with a span length of 3.40 m, triangle light steel as top chord (hand beam) with span length of 4.20 m, light steel which connected bottom and top chord (web or inner light steel). The next process was to cover the batten and spandex roof above ampig with red stone and mortar. Then, we arranged one layer of red stones on top of the hat sections then have it covered with mortar and plastered.

5. Conclusions and Recommendations
5.1. Conclusion
This study concluded that there are several principles of cost reduction which could be done in constructing public housing. Quick principles exist in constructing techniques such as using cigarette packs as a foundation pad, mortar usage for rapid wall standing, and the spacing of mortars could be done manually by using two fingers on a human hand. Economic principles could be used for material matters, such as eliminating the use of gravel for concrete, the use of sand material to contain the soil, the foundation does not use sand and empty stone, and the shape of the ring beam was made using triangle reinforcement. In addition, the use of simple working tools like handy strap, gauge, water hose, hammer, iron fabricator, and waist bag could also be a consideration in this study.

5.2. Suggestion
This research can be continued to examine the principles of precise and secure infrastructure for public housing buildings.

IOP Conf. Series: Materials Science and Engineering **267** (2017) 012025 doi:10.1088/1757-899X/267/1/012025

6. REFERENCES

[1] Nero I G T K, Amir S H and Puspawati I G A 2014 *Default Solved In Ownership Loan Agreement (KPR) On BTN Bank Branch Singaraja in Bali J. Student Legal.* p 2

[2] Akhmad A G 2011 Impact of Housing Area Development Simple Healthy House Against The Economy Farmers Suburbs In Palu *J. Space.* ***3*** *1*

[3] Nurfitriyanto H W 2013 *Economic. Lett.* **14**

[4] Warauw F 2014 Sustainable Design Approach In City Housing in Indonesia For Better Engineering *J. Media Matrasain.* ***11*** *2*

[5] Limanto S, Suwono J I, Wuisan D and Raharjo C 2013 *Conf. on National of Civil Engineering Context 7 (Surakarta: Sebelas Maret University)* p S-179

[6] Nurjannah S A 2011 *Conf. on The National Seminar AVoER 3rd (Palembang)* p 234

[7] Widyarti M, Setiawan B I, Arifin H S, and Yuwono A S 2012 Reconstruction of EcoHouse Structure In Baduy in Banten Province *J. Graduate Forum* ***35*** *69-78*

[8] Setiawan M I, Iskandar T, and Budiyanto 2015 *Conf. on National Seminar on Technology (Malang)* p. 2

[9] Atmaja J 2004 *Social Factors Economic Relations With Physical Condition Building Unhealthy In District Lubuk Alung J. R & B Scientific.* ***4*** *2*

[10] Wastika D N 2005 The Concept of Tri Hita Karana Application Planning for Housing in Bali *J. Settlement Natah.* ***3*** *262-105*

[11] MangunwijayaY B 2009 *Wastu Citra Introduction to Cultural Studies Architectural Forms and Their Philosophy Joints Practical Examples* (Jakarta: PTGramedia Pustaka Utama)

[12] Sing Y 2009 *Cheap Dream Home* (Jakarta: TransMedia)

IOP Conf. Series: Materials Science and Engineering **267** (2017) 012026 doi:10.1088/1757-899X/267/1/012026

Driver Performance Problems of Intercity Bus Public Transportation Safety in Indonesia

A Suraji[1], S Harnen[1], A Wicaksono[1], L Djakfar[1]

[1]Civil Engineering Department, Engineering Faculty, University of Brawijaya, Jl. MT Haryono 167 Malang, East Java, Indonesia

Email: ajisuraji@widyagama.ac.id

Abstract. The risk of an inter-city bus public bus accident can be influenced by various factors such as the driver's performance. Therefore, knowing the various influential factors related to driver's performance is very necessary as an effort to realize road traffic safety. This study aims to determine the factors that fall on the accident associated with the driver's performance and make mathematical modeling factors that affect the accident. Methods of data retrieval were obtained from NTSC secondary data. The data is processed by identifying factors that cause the accident. Furthermore, data processing and analysis used the PCA method to obtain mathematical modeling of factors influencing the inter-city bus accidents. The results showed that the main factors that cause accidents are health, discipline, and driver competence.

Keywords: driver, bus, safety, accident, traffic

1. Introduction

The safety of inter-city bus public transportation in Indonesia is still far from expectations. This can be seen from various accident incidents involving bus fleet in the last few years as well as data from NTSC and Indonesian Police [1] [2]. The incidents of bus accidents involving buses in Indonesia mostly include East Java, Central Java, West Java, DKI Jakarta, and other provinces. The incident is an accident with a lot of casualties. The incidents of the accident were at least more than 8 deaths, so the incident of the accident became the attention of the National Transportation Safety Committee (NTSC) to conduct an investigation.

As indicated by the party who has responsibility in the event of accident, either from the Police, Transportation Department / Ministry of Transportation, or NTSC, that the incident involving bus accidents other than caused by the factors of vehicles, roads, and environment also become the main factor Is the driver [3] [4]. The misbehaving driver behavior with high speed, as well as low traffic discipline is thought to be a major factor in road traffic accidents. In addition the health condition and stamina of the driver can also affect the concentration while driving on the highway [5] [6].

In the case of certain drivers who are still beginners or who are still a bit of experience in terms of driving the bus can also cause an accident. In general, driver performance issues are still many things that need to be addressed [7] [8]. Of course, in the improvement of public transport bus safety, more reliable identification is needed to determine the factors that caused the accident [9] 10] [11].

This research aims to determine the factors that cause accidents associated with the driver's performance as well as make mathematical modeling of factors that contribute to the cause of inter-city bus accidents. Thus, the safety of inter-city bus public transportation in Indonesia can be addressed from the driver behavior aspects found in this research.

2. Bus Accidents

Traffic Accidents are unexpected and unintentional incidents involving vehicles with or without other road users resulting in human casualties and / or property losses. While the opposite condition of the accident known as safety. Thus the notion of traffic safety is a situation where everyone is avoided from the risk of accidents during traffic caused by four main factors: human, vehicle, road, and / or

IOP Conf. Series: Materials Science and Engineering **267** (2017) 012026 doi:10.1088/1757-899X/267/1/012026

environment [12]. In public bus transportation, the human factor is more directed to the driver with various behaviors such as vehicle speed, traffic discipline, competence, health and stamina.

In general, inter-city bus public transportation has the main characteristics of vehicle operation on the highway such as long-distance and high-speed travel. Meanwhile, the driver also has different characteristics in driving. With the operational characteristics and the driver according to the research results of Taylor et al. and Yang, the driver tends to experience fatigue so that with inadequate vehicle control [6] [13].

Incidents involving bus accidents in Indonesia with varying degrees of severity and number of victims vary widely, ranging from minor accidents, serious accidents or fatal accidents. According to Police report records that accidents involving buses accounted for 4% of total accident incidents [2]. Since 2009 until 2015, the special fatal category and the center of public attention there are 25 events scattered throughout the territory of Indonesia. The fatal definition is very fatal in this case is the incidence of accidents involving public buses that resulted in the death of more than 8 people, and consequently the incident was investigated by the NTSC [1]. The incidence of fatal accidents involving bus public transport in Indonesia has been obtained data as in Table 1. While examples of bus conditions post-accident events as in Figure 1.

As shown in Table 1, the data has been detailed on the location and date of the accident, the incidence of accidents involving other vehicles, and the number of casualties. The table indicates that incidents involving bus accidents in Indonesia are very severe and require more focused attention by various parties, from the government, owners of bus companies, and the general public [1].

Figure 1. Example of Bus Condition after Accidents [1].

3. Driver Performance

The driver has a very important role in the safety of bus transportation, because the driver controls the vehicle in running the vehicle. Maneuvering of vehicles on the highway depends largely on the attitude and behavior of the driver. How the vehicle maneuver depends on the perception of the driver in maintaining traffic conditions on the highway. In addition the character of the driver's response style in maneuvering is also very influential on passenger comfort and safety [14] [15].

Much has to do with driver behavior both regarding the driver's personal condition and the behavior in the traffic interaction. A more personal aspect concerns health conditions, stamina, and competence / competence. While related to the interaction of traffic such as speed control and discipline in traffic. Of course when examined more in the above aspects can still grow again. However, according to Lin et al. Which has done the driver simulation in outline some of the aspects mentioned above is enough to describe some aspects related to driver behavior [3].

Table 1. The worst traffic accidents involving a bus in Indonesia [1].

No	Location/Date of incident	Bus accident	No. of victim (person)		
			Deaths	Serious injures	Minor injuries
1	West Jakarta /6 December 2015	Bus collision with train	19	4	0
2	Cirebon/14 Juliy2015	Single accident bus crashed into a pole	11	42	0
3	Subang1/17 June 2014	Bus collision with car	9	12	35
4	Bekasi/8 March 2014	Bus collision with train	0	38	0
5	Bogor1/21 August 2013	Bus collision with truck	20	34	0
6	Banyumas/10 August 2013	Bus VS Car and Motorcycle	15	6	6
7	Cianjur/27 February 2013	Single accident bus crashed into a cliff	17	26	38
8	Lampung/14 September 2012	Single accident bus crashed into a ravine	9	6	5
9	Sidoarjo/7 August 2012	Bus VS Car and Motorcycle	8	32	0
10	Simalungun Sumut/28 June 2012	Single accident bus crashed into a ravine	8	4	0
11	Hurau Sumut/1 May 2012	Bus fire	13	4	7
12	Bogor2/10 February 2012	Bus collision with car motorcycle	14	10	44
13	Sumedang/1 February 2012	Single accident bus crashed into a ravine	12	26	0
14	Indramayu/17 December 2011	Bus collision with car	8	22	0
15	Mojokerto/ 12 September 2011	Bus collision with bus	20	12	0
16	Tanapuli Sel/26 June 2011	Single accident bus crashed into a ravine	19	0	0
17	Madiun/22 May 2011	bus collision with truck	10	0	3
18	Temanggung/7 February 2011	Single accident bus crashed into a ravine	11	36	0
19	Batang/16 March 2010	Bus VS Trailer	10	12	21
20	Tuban1/1 November 2009	Bus collision with truck	8	4	2
21	Subang2/26 September 2009	Bus collision with car	9	14	0
22	Tuban2/24 September 2009	Bus collision with bus	7	2	1
23	Klaten/5 July 2009	Bus collision with train	7	2	1
24	Tanah Datar/31 May 2009	Single accident bus crashed into a ravine	13	12	13
25	Kediri/23 February 2009	Bus collision with train	9	19	6

4. Research Method

Research on safety-related driver behavior is based on the conceptual framework that there are many factors that affect the driver's performance. Poor driver performance may lead to a higher risk of accident risk. In this study, some driver performance variables are analyzed whether they have contributed to the effect of accidents. Driver factor variables include speed, stamina, health, discipline and driver competence. Based on previous research, many statistical analysis methods used to predict and determine the factors of accidents such as those conducted by Zairi and Khabiri, Suraji, and Sezhian et.al. [16] [17] [18].

The accident data was obtained from the National Transportation Safety Commission (NTSC). The scope of the incident area covers the whole of Indonesia over the past 6 years from 2009 to 2015. Causes of accident from the driver's aspect are identified which includes speed, stamina, health, discipline and competence. Furthermore, the data is codified by marking if there are factors causing the accident from the event given the code number 2, whereas if not the cause of the accident coded the number 1.

After codification, the data is processed and analyzed using SPSS software with Principal Component Analysis (PCA) method. This chosen statistical analysis method is similar to that done by Sezhian et.al [18]. In the early stages of analysis is done descriptive statistical test, Further data analysis is done PCA extraction to obtain mathematical equations. This mathematical equation is the result of accident modeling caused by the performance and behavior of the driver. To complete the understanding of the results of the analysis, also made biplot 2 dimension that shows the closeness between the locations of the accident with various factors that exist in the driver's aspect.

5. Data and Analysis

5.1. Descriptive statistics of driver performance

Driver's factor consists of five variables, namely: speed, stamina, health, discipline and competence. In each variable is divided into two categories. The first category is a condition related to accident prevention and the second category is an accident trigger.

Descriptive statistical results in Table 2 explain that the indiscipline and incompetence are two variables supporting accidents that are still often encountered in the driver. Accidents happen, 88% due to undisciplined and 60% due to incompetent driver. Furthermore, in PCA for the five variables in the factor will be extraction dimension.

Table 2. Descriptive statistics of driver performance.

Variable	Categories	Frequency	Percentage (%)
SPEED	Normal Speed	12	48,0
	High speed	13	52,0
	Total	25	100,0
STAMINA	Fresh	21	84,0
	Fatigue /Sleepy	4	16,0
	Total	25	100,0
HEALTH	Healthy	22	88,0
	Unwell	3	12,0
	Total	25	100,0
DISCIPLINE	Discipline	3	12,0
	Undisciplined	22	88,0
	Total	25	100,0
COMPETENCE	Competent	10	40,0
	Less competent	15	60,0
	Total	25	100,0

5.2. Mathematics model of PCA statistics method

Statistical analysis to determine the factors causing the accident used Principal Component Analysis (PCA) method. The data structure of the cause of the accident on the driver factor in the PCA can be extracted into 5 dimensions. The summary of the analysis results is shown in Table 3.

Table 3. PCA extraction of driver performance.

Variable	Dimension				
	1	2	3	4	5
SPEED (X_{11})	-0,342	0,805	-0,359	-0,235	0,224
STAMINA (X_{12})	0,159	0,709	0,606	0,296	-0,128
HEALTH (X_{13})	-0,844	-0,165	-0,029	0,459	0,222
DISCIPLNE (X_{14})	0,899	-0,073	0,153	0,094	0,393
COMPETENCE (X_{15})	-0,502	-0,177	0,759	-0,349	0,138
Eigen Value	1,915	1,216	1,096	0,484	0,289
Total of Varian (%)	38,30	24,31	21,92	9,67	5,78
Cumulative of Varian (%)	38,30	62,61	84,54	94,21	100,00

There are three dimensions that have Eigen value of more than 1, i.e. the first to third dimension with the cumulative total variance that can be explained is 84.54%. These three dimensions are sufficient to replace the extraction of the five variables present in the driver's factor. Thus the end result will be used extraction up to the third dimension. The cumulative total variant that can be explained up to the third dimension is 84.54%. The equations resulting from the three dimensions are found in Equations (1), Equations (2), and Equations (3).

$$PC_{11} = -0{,}342\ X_{11} + 0{,}805\ X_{12} - 0{,}359\ X_{13} - 0{,}235\ X_{14} + 0{,}224\ X_{15} \quad (1)$$

$$PC_{12} = 0{,}159\ X_{11} + 0{,}709\ X_{12} + 0{,}606\ X_{13} + 0{,}296\ X_{14} - 0{,}128\ X_{15} \quad (2)$$

$$PC_{13} = -0{,}359\ X_{11} + 0{,}606\ X_{12} - 0{,}029\ X_{13} + 0{,}153\ X_{14} + 0{,}759\ X_{15} \quad (3)$$

where:
PC_{13} : Principal component of driver performance
X_{11} : Variable of speed
X_{12} : Variable of stamina
X_{13} : Variable of health
X_{14} : Variable of discipline
X_{15} : Variable of competence

In the first dimension (PC_{11}) the great coefficients are in health (X_{13}), discipline (X_{14}) as well as competence (X_{15}) and the three are marked positive and negative. This first dimension explains a stronger relationship between the three variables. This dimension explains that the cause of accidents of the driver factor, 38.30% is explained by the health condition, the degree of discipline and competence. The results of this research are in line with research conducted by Taylor et al especially related to variable speed [6]. While the variable discipline of this research is similar to the research conducted by Symmons et.al [5]. From the research conducted this health variable is a newly discovered variable as the cause of the accident factor from the driver's condition.

The second dimension (PC_{12}) coefficient is large at speed (X_{11}) and stamina (X_{12}) and both coefficients are positive. This second dimension explains a stronger relationship between the two variables. This dimension explains that the cause of accidents of the driver factor, 24.31% is explained by the speed and stamina.

The third dimension (PC_{13}) coefficients are large in stamina (X_{12}) and prowess (X_{15}). The contribution of stamina has also been described in the second dimension. This dimension explains that the cause of accidents of the driver factor, 21.92% explained by stamina and competence. The results of this extraction give some conclusions that in many locations the driver's problems are strongly related to health and discipline. The results of this study also show there is no positive relationship between the conditions of the driver in good health with indiscipline.

In the second and third dimension results, the variables that are different from the first dimension are stamina, and the total variant for the two dimensions is smaller than in dimension 1. Therefore equation (1) as a sufficient result represents the modeling of accident factors from the behavioral aspect driver. Thus, the factors that cause accidents are influenced by poor health conditions, undisciplined, and incompetence of the driver. The results of this study for undisciplined behavioral variables and velocities are in line with research conducted by Symmons et.al, Taylor et al, and Suraji et al. [5] [6] [19] [20].

5.3. Biplot result of driver performance
Biplot driver factor is the visualization of the results of the analysis of all the events and the first two dimensions are generated. Biplot results as shown in Figure 2. The figure explains that more than half of accident incidents are always related to the five variables in the driver factor, namely variable speed, stamina, health, discipline, and competence.

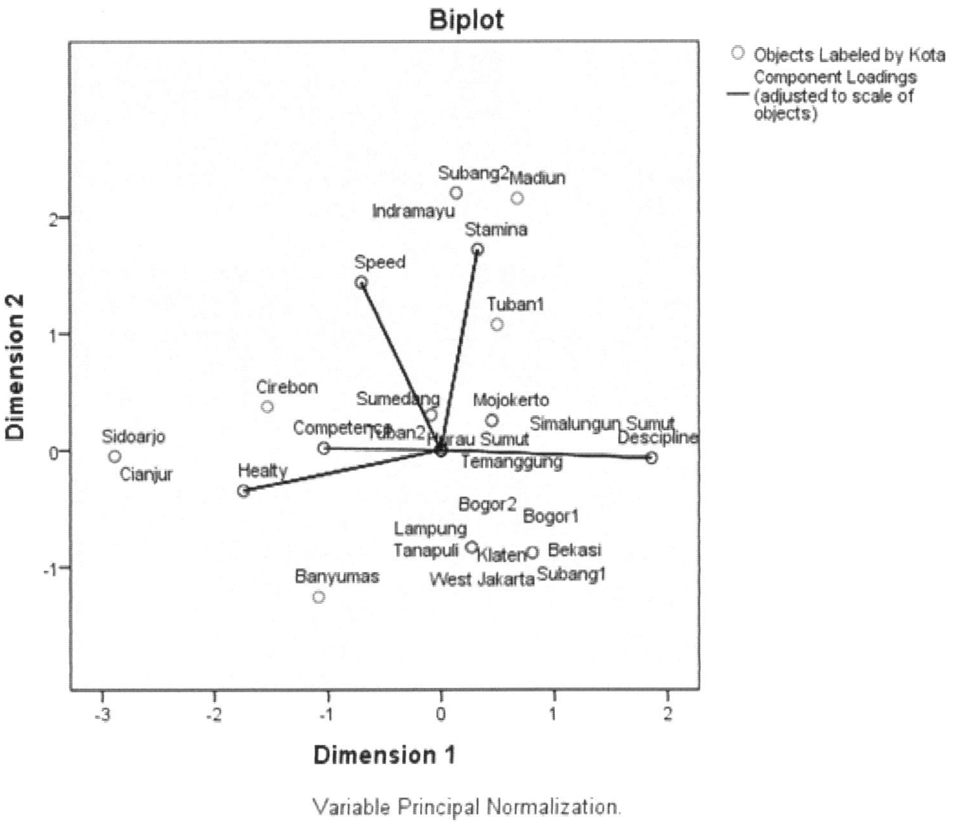

Figure 2. Biplot driver performance and accident locations

The results of research indicate that a number of accidents occur due to the condition of the driver with an unfavorable stamina so cannot control the speed. On the other side of the biplot picture also explains that the cause of accidents due to poor health conditions followed by the decline in proficiency. However, the results of this study also obtained information that the condition of drivers who are in poor health and less competent will always be careful in driving so tend to be more disciplined in bringing the vehicle. The results of this study are similar to the research done by Lin et.al, which used simulation method to know the driver behavior [3].

6. Conclusions

From the results of discussion and analysis of research can be concluded as follows: Mathematical equations modeling related to driver performance affecting bus inter-city bus accidents where accidents are affected by health condition, discipline rate, and driver's competence with equation $PC_{11} = -0{,}342\ X_{11} + 0{,}805\ X_{12} - 0{,}359\ X_{13} - 0{,}235\ X_{14} + 0{,}224\ X_{15}$. From the extraction results in statistical analysis shows that the problem of bus drivers of inter-city buses is very closely related to the health and discipline of the driver. Furthermore, the result of biplot behavior of the driver which is the visualization of all the resulting events shows that more than half of accident incidents are always associated with all the variables available are speed, stamina, health, and competence. A number of accidents occur due to the condition of the driver with good stamina that is not good enough to control the vehicle, poor health and less competent.

7. References

[1] National Transportation Safety Committee (NTCS) of Republik Indonesia 2016 *Road Accident Investigation Data*
[2] Police Department of Republik Indonesia 2013 *Accident Data Release Yearly on 2013*
[3] Lin T W, Hwang S L, and Green PA 2009 Effects of Time-Gap Settings of Adaptive Cruise Control (ACC) on Driving Performance and Subjective Acceptance in a Bus Driving Simulator *Safety Science* **47(5)** 620-625
[4] Salmon P M, Regan M A and Johnston I 2005 *Human Error and Road Transport: Phase One – A framework for an Error Tolerant Road Transport System* Accident Research Centre, Monash University: Australia
[5] Symmons S and Howarth N 2005 *Safety Attitudes and Behaviours in Work-Related Driving-Stage 1: Analyses of Crash* Data, No. 232 Accident Research Centre, Monash University: Australia
[6] Taylor M C, Baruya A and Kennedy J V 2002 The Relationship between Speed and Accidents on Rural Single-Carriageway Roads. *TRL Report* TRL **511**
[7] Cafiso S, Graziano A D and Pappalardo G 2013 Road Safety Issues for Bus Transport Management *J. of Accident Analysis and Prevention* **60** 324-333
[8] Chang H L, and Yeh C C 2005 Factors Affecting the Safety Performance of Bus Companies – the Experience of Taiwan Bus Deregulation *J. of Safety Science* **43(5)** 323-344
[9] Haworth N, Kowadlo N, and Tingvall C 2000 *Evaluation of Pre-driver Education Program*. Accident Research Centre, Monash University: Australia.
[10] Hoekstra T, and Wegman F 2011 Improving the Effectiveness of Road Safety Campaigns: Current and New Practices *IATTS Research* **34(2)** 80-86
[11] Mitsopoulos E, Regan M A, Anderson J, Salmon P M and Edquist J 2005 *Team Training for Safer Young Drivers and Passengers in the ACT: A Role for Crew Resource Management* Accident Research Centre, Monash University: Australia
[12] Laws of Republik Indonesia. No. 22/2009 *Traffic and Road Transportation*
[13] Yang C Y D 2007 Trends in Transit Bus Accidents and Promising Collision *The J. of Public Transportation* **10(3)** 119-136
[14] Assum T and Sorensen M 2010 Safety Performance Indicator for Alcohol in Road Accidents – International Comparison, Validity and Data Quality *J. of Accident Analysis and Prevention* **42(2)** 595-603
[15] Pulugurtha S S and Vanapalli V K 2008 Hazardous Bus Stops Identification *The J. of Public Transportation* **11(2)** 65-84
[16] Ziari H, and Khabiri M M 2006 Analysis Characteristics and Provide a Prediction Model of Public Bus Accident in Tehran *J. of Applied Science* **6(2)** 247-250
[17] Suraji A, and Tjahjono N 2012 A Confirmatory Factor Analysis of Accidents Caused By the Motorcycle Aspect in Urban Area *Int J. for Traffic and Transport Engineering* **2(1)** 60-69

[18] Sezhian M V, Muralidharan C, Nambirajan T, and Deshmukh S G 2011 Ranking of a Public Sector Passenger Bus Transport Company Using Principal Component Analysis: A Case Study. *Management Research and Practice* **3(1)** 60-71
[19] Suraji A, Harnen S, Wicaksono A, Djakfar L 2013 Accident Analysis Between Minibus and Intercity Bus on the By Pass Road Mojokerto KM 55.1 in the Kenanten Puri District, Mojokerto Regency, East Java. *Proc. of ATPW 2013 Seminar on Diploma Vocation, Institute of Technology Sepuluh November* Surabaya
[20] The Regulation of Transportation Minister, Republik Indonesia No. KM 35/2003 *The Regulation of Public Transportation on Road*

IOP Conf. Series: Materials Science and Engineering **267** (2017) 012027 doi:10.1088/1757-899X/267/1/012027

Road safety analysis on Achmad Yani frontage road Surabaya

Machsus[1,2], I Prayogo[1], Chomaedhi[1], D W Hayati[1] and A Utanaka[1]
[1]The Department of Civil Infrastructure Engineering, Institut Teknologi Sepuluh Nopember, Kampus ITS Manyar, Jalan Menur 127 Surabaya 60116, Indonesia

[2]E-mail: machsus@ce.its.ac.id

Abstract. This research discusses road safety analysis on the operation of frontage road on the west side of Achmad Yani Road Surabaya. This research began by conducting survey on secondary data of traffic accidents. In addition, primary data survey was conducted to obtain traffic data, geometric road data, and other supporting data at the study site along the west side frontage of Ahmad Yani Road Surabaya. Devices used in this research include camera, handy cam, speed gun, counters of vehicles, rolling meter, computer and others. In outline, the stages to conduct this research are divided into 4 stages, namely 1.the preparation stage, 2.data collection and processing, 3. analysis and discussion, and 4. conclusion. The results of this study showed that the accident characteristics of the frontage road are (i) 3 accidents occured per month, (ii) motorcycles was accounted for the largest proportion of accidents which amounted to 74.6 percent, (iii) there were 3 accident victims per month, and (iv) material losses per month worths 1.2 million. The accident rate in 2016 was 0.04 crashes per one million vehicle travels per kilometer, while during 2 months in 2017 it was 0.15 accidents per one million vehicle travels per kilometer. Black spot area of accident is located on Sta 2 + 800 to 2 + 900 which is in front of Graha Pena building and DBL Arena. The high rate of accidents is influenced by the speed of the vehicle which 85 percentile exceeds the speed limit of 40km per hour.

1. Introduction

Road safety is an important aspect in the issue of eco-friendly transportation or green transport. It becomes increasingly important as the phenomenon of the increase traffic accidents on the roads [1]. Traffic accidents can occur anytime and anywhere; these depend on the rider's condition, the technical condition of the vehicle and its roads, and the weather [2, 3, 4]. Recently, the number of traffic accidents has increased steadily. Based on the data of Land Transportation in Figures, the number of traffic accident incidents on the road from year to year did fluctuate, but the trend increased with an average growth of about 10% per year. Similarly, the number of vehicles involved, the number of victims, and material losses also increased [5, 6, 7].

Road traffic accidents in urban areas are interesting to assess, as the number of accidents in urban area is greater than in non-urban areas. Similarly, traffic accidents occured in major cities, such as Surabaya, are also estimated to be larger than those in small cities [7, 8]. The portrait of road traffic accidents in Surabaya also reflects the phenomenon in Indonesia. The number of road traffic accident casualties caused by motorcycles is higher than other types of vehicles [9].

[2] To whom any correspondence should be addressed.

The rapid economic growth in Surabaya needs to be supported by the provision of adequate road infrastructure. Currently, the road network in Surabaya has been too complex, the mobility of the population was very high as well. The road capacity which is unable to service the vehicle flow will potentially lead to congestions and traffic accidents [10, 11, 12]. Therefore, Surabaya Government have built frontage road on Jalan Achmad Yani, which is the access road in and out of Surabaya on the south side. Frontage road on the west side of Jalan Ahmad Yani Surabaya which is located at the southern end of Surabaya City is a busy and crowded highway because it is the entrance gate of vehicles from outside cities to Surabaya. Therefore, it makes frontage road on the west side of Ahmad Yani road Surabaya more prone to traffic accident.

The data obtained shows that the frequency of accidents is higher every year since frontage road on Jalan Achmad Yani was opened. There are 26 events in 12 months (2016) and 65 events in 5 months (2017). Due to a lot of accidents occur on the frontage road, it is particularly important to perform road safety analysis at that location.

2. Literature Review

The study of traffic accidents continues to grow. Several previous studies that examined road accidents and frontage roads include: Xie and Zhang [13] who examined roads in Toronto, Canada; Ackaah and Salifu [8] who examined roads in Ghana; and Polus and Cohen [14] who examined roads in Israel. In addition, Li, Lord, and Zhang [4] reviewed the frontage road in Texas and Machsus [9] reviewed the model of accident prediction on roads and intersections in Surabaya – Indonesia.

The accident rate is the number of traffic accidents compared to the volume of traffic and the length of the road. The accident rate expressed accidents per 100 million vehicles per kilomerter journey. This method is commonly used to calculate roads that have similar traffic types [15]. The annual accident rate is formulated as follows:

$$Tk = \frac{Fk \, x 10^8}{LHRt \, x \, n \, x \, L \, x \, 365} \quad , (100MVKT) \tag{1}$$

where:

Tk = Accident rate, 100 MVKT (Million Vehicle-Kilometre of Travel)
Fk = Accident Frequency on the road for n years of data
LHRt = Average of Traffic Volume
N = Number of years of data
L = Length of road, Km

The accident rate formula is used to analyze annual accident data. If it is necessary to analyze the monthly accident data, then the accident rate formula must be developed or transformed. The transformation results for the monthly accident rate are formulated as follows:

$$Tk = \frac{Fk \, x 10^6}{LHRt \, x \, n \, x \, L \, x \, 30} \quad , (MVKT) \tag{2}$$

where:

Tk = accident rate, MVKT
Fk = Frequency of accidents on the road for n data months
LHRt = Average of Traffic Volume (smp)
N = Number of months of data
L = Length of road (Km)
MVKT = Accident rate unit (number of accident / One Million Vehicle-Kilometre of Travel)

The fatality rate is the number of fatal traffic accidents compared to the volume of traffic and the length of the road. This fatality rate states the number of fatal accidents (causing death) per 100 million vehicle kilometres [15]. The annual fatality rate is formulated as follows:

$$\text{Tkf} = \frac{\text{Fk x}10^8}{\text{LHRt x n x L x 365}} \qquad , (100\text{MVKT}) \qquad (3)$$

where:
Tf = Fatality rate MVKT
Ff = Frequency of fatal accident on the road for n years of data
LHRt = Average of Traffic Volume (smp)
N = Number of years of data
L = Length of road, Km
100MVKT = Unit of accident level (number of accident / one hundred million vehicle-kilometre of travel)

The formula of the fatality rate is used to analyze the annual fatality data. If necessary to analyze monthly fatality data then the formula of the fatality rate must be developed or transformed. The transformation results for the monthly fatality rate are formulated as follows:

$$\text{Tkf} = \frac{\text{Fk x}10^8}{\text{LHRt x n x L x 30}} \qquad , (100\text{MVKT}) \qquad (4)$$

where:
Tf = Fatality rate, MVKT
Ff = Frequency of fatal accident on the road for n data months
LHRt = Average of Traffic Volume (smp)
N = Number of months data
L = Length of road, Km
100MVKT = Accident rate unit (number of accident / one hundred million travels-kilometre of travel)

Calculating the accident rate on each segment of the divided road segments per 100 meters is used as a method to determine black spot area. The highest accident rate calculation at a particular location indicates that the location is prone to accidents. At this accident which is prone location should be prioritized in mitigation to reduce traffic accidents rate [16, 17].

3. Material and Method

The location of this study is on the west side frontage road of Jalan Ahmad Yani Surabaya-Indonesia. Data used in this study are the number of traffic accidents, vehicle volume, vehicle speed, road geometry and land use. Data used in this research is the data which was collected since frontage road operate in 2016. Some devices used in this research are camera, handy cam, speed gun, vehicle counter, rolling meter, computer, stationery and others.

In outline, the stages to conduct this research are divided into 4 stages, namely 1. the preparation stage, 2. data collection and process, 3. analysis and discussion and 4. conclusion. Based on the result of traffic counting survey conducted on peak hour, it was obtained 2 important data which are the average daily traffic volume (ADT) and the degree of saturation (DS) of the road segment in this study location. Accident data obtained from Traffic Accident unit of Surabaya city resort is classified based on the number of accident incidents. Moreover, it is based on categories injuries: minor injury, heavy injury, and death. Then, these data used to calculate accident rate, fatality level, and black spot area. Based on the results of the survey speed, it can be calculated that the value of speed at the condition of 85 percentiles for each classification of vehicle types. Furthermore, it was conducted a correlation analysis between vehicle speed factor with accident rate. Finally, it was found solutions to reduce and avoid the occurrence of accidents and also formulated suggestions for this research, especially related to the utilization of research results and further research development.

Traffic flow data in this study is the data flow of incoming vehicles from and to the frontage road. While the types of vehicles surveyed are classified into: motorcycles (MC), light vehicles (LV), heavy vehicles (HV), and non-motorized vehicles (UM). Implementation of traffic surveys on frontage road was undertaken on weekdays. Survey time that was conducted during peak hour of the morning was at 6 am to 9 am. Surveys were conducted by using the camera recording method which used the following devices: digital camera, tripod, and memory card. The number of points to be surveyed is 11 points/locations. Point 1 located between STA 0+000 - 0+140, point 2 located between STA 0+140 - 0+160, point 3 located between STA 0+160 - 0+620, point 4 located between STA 0+585 - 0+690, point 5 located between STA 0+690 - 1+230, point 6 located between STA 1+230 - 1+640, point 7 located between STA 1+640 - 1+800, point 8 located between STA 1+800 - 2+800, point 9 located between STA 2+800 - 3+120, point 10 located between STA 3+120 - 3+800, point 11 located between STA 3+800 and 4+200. Speed survey was undertaken using electronic measurement method, namely speed gun.

Figure 1. Location of traffic survey (point 2).

4. Results and Discussions

This section deals with the results and discussion of the characteristics of traffic accidents, the accident rate and the level of fatality, the determination of black spot area, the evaluation of vehicle speeds, and the reduction of accident rates.

4.1. Characteristics of traffic accidents

The characteristics of traffic accidents can be obtained from field observations and accident statistics data during the frontage road operates. The construction of the road on the west side of Jalan Ahmad Yani Surabaya is intended as a frontage road facility for the movement of traffic from the south to the north. By operating frontage road, it increased the road capacity. Furthermore, it improved the performance of Jalan Ahmad Yani Surabaya. Nevertheless, the operation of frontage roads also has negatively impacted that increased the number of traffic accidents. The result of field observations indicated that it has no vehicle classification arrangement which regulates the permissibility of vehicles through fast lanes and or slow lanes. It means that mixed traffic occurs at the study site. It happens because riders are free to choose to pass the fast lane or slow lane (frontage road). These conditions will certainly affect on the high potential for accidents on a road [16, 18].

Traffic accidents statistics since frontage road operates, i.e. from January 2016 to the end of May 2017 can be read in Table 1 below.

Table 1. Characteristics of traffic accidents *in frontage road.*

No	Type	Unit	Year		Total	Percentage (%)	Monthly Average
			2016 (12mons)	2017 (5mons)			
1	Number o Accidents	accidents	26	65	91		6.5
2	Type of Vehicle						
	- Two-Wheeled Vehicles	unit	28	68	96	71.6	6.9
	- Four-Wheeled Vehicles	unit	4	6	10	7.5	0.7
	- Pedestrians	person	7	21	28	20.9	2.0

3	Accident Victims						
	- Death	person	2	6	8	8.8	0.6
	- Heavy Injury	person	3	5	8	8.8	0.6
	- Minor Injury	person	21	54	75	82.4	5.4
	Total of Accident Victims	person	26	65	91		6.5
4	Material Loss	million	12.1	23.1	35.2		2.5

Resource : State Police of the Republic of Indonesia, East Java Area, Directorate of Traffic and City Resort Police of Surabaya

Table 1 presents that the accident characteristic is shown by the frequency of accidents. There were 42 accidents since 17 months of frontage road operated or 6.5 accidents per month in average. In terms of the type of vehicle involved, the highest proportion of accident was caused by two-wheeled vehicles or motorcycle by 71.6%, while four-wheeled vehicles contributed 7.5% of accidents' number, and the rest by 20.9% were pedestrians or riding no vehicle. Furthermore, the numbers of accident victims were 91 people by 17 months or 6.5 people per month in average. While the classification of the victims died, serious injuries and minor injuries respectively were 8.8%, 8.8%, and 82.4%. Whereas, average material losses were 2.5 million per month.

4.2. Accident and fatality rates

Accident and fatality rates in each month in 2016 can be read in Table 2, while Table 3 shows the accident and fatality rates in January until May 2017.

Table 2. Number of accidents, number of casualties, accident rate and fatality rate in 2016.

No	Month	Number of Accidents	Number of Casualties			Accident Rate	Fatality Rate
			MD	LB	LR		
1	January	0	0	0	0	0.00	0.00
2	February	1	0	0	1	0.02	0.00
3	March	3	1	0	2	0.06	1.93
4	April	2	0	0	2	0.04	0.00
5	May	1	0	0	1	0.02	0.00
6	June	2	0	1	1	0.04	0.00
7	July	0	0	0	0	0.00	0.00
8	Augustus	1	0	0	1	0.02	0.00
9	September	2	0	1	1	0.04	0.00
10	October	5	1	1	3	0.10	1.93
11	November	0	0	0	0	0.00	0.00
12	December	9	0	0	9	0.17	0.00
	Amount		2	3	21		0.32
	Total Amount			26		0.04	
	%		7.7	11.5	80.8		

Table 3. Number of accidents, number of casualties, accident rate and fatality rate in 2017.

No	Month	Number of	Number of	Accident	Fatality

		Accidents	Casualties			Rate	Rate
			MD	LB	LR		
1	January	14	1	1	12	0.27	1.93
2	February	12	1	0	11	0.23	1.93
3	March	13	1	2	10	0.25	1.93
4	April	15	1	2	12	0.29	1.93
5	May	11	2	0	9	0.21	3.87
	Amount		6	5	54		2.32
	Total Amount			65		0.25	
	%		9.2	7.7	83.1		

Table 2 presents that the highest monthly accident rate in 2016 occurred in December by 0.17 accidents per one million vehicle-kilometre of travel, while for monthly fatality rates occurred in March and October by 1.93 accidents per one hundred million vehicle-kilometre of travel. The annual accident rate in 2016 was 0.04 accidents per one million vehicle-kilometre of travel, while the annual fatality rate in 2016 was 0.32 accidents per one hundred million vehicle-kilometre of travel.

Table 3 shows that the higher monthly accident rate in 2017 occurred in April, by 0.29 accidents per one million vehicle-kilometre of travel, while for monthly fatality rates occurred in May by 3.87 accidents / one hundred million vehicle-kilometre of travel. In terms of these two data, the annual accident rate in 2017 was 0.25 crashes / one million vehicle travels-kilometre of travel, while for the annual fatality rate was 2.32 accidents / hundred million vehicle-kilometre of travel.

The results of the accident and fatality rate calculation indicate that frontage road has a high accident and fatality rate since its first operation period. It is indicated that the presence of western side frontage has positive impact on the smoothness of traffic but it has negative impact as well. Frontage road increased the magnitude of accident and fatality rate.

4.3. Location of black spot area

The location of black spot area was obtained from the accident and the fatality rate on each segment road which is divided by 100 meters. The highest result of accident rate calculation at a particular location indicates that the location is one of black spot areas. Examples of accidents and fatalities rates are as follows:

- Accident rate on STA 2+800 to 2+900 in 2016

  $$\text{Tk} = \frac{\text{Fk x} 10^6}{\text{LHRt x n x L x 30}} = \frac{7 \times 10^6}{410189 \times 12 \times 0.1 \times 30} = 0.47$$

  Accident rate calculation results for other locations are presented in Appendix A.

- Fatality rate on STA 2+800 to 2+900 in 2016

  $$\text{Tk} = \frac{\text{Fkf x} 10^8}{\text{LHRt x n x L x 30}} = \frac{1 \times 10^8}{410189 \times 12 \times 0.1 \times 30} = 6.77$$

  Fatality rate calculation results for other locaions are presented in Appendix A.

Based on the calculation result of the accident and fatality rate, it was found that the area of STA 2 + 800 to STA 2 + 900 were the most vulnerable area to traffic accidents. Black spot areas which are in front of Graha Pena and DBL Arena buildings had the highest accident and fatality rate of 0.47 and

6.77. If the highest score between the accident rate and the fatality rate is different, then accident rate will be the priority to determine the black spot area.

The results of speed survey in peak hour of holidays and weekdays were calculated by 85 percentile of speed in each segment. Then, it was calculated the average of the four segments as shown in Table 4. The result of speed calculation, which is shown in the table 4, shows that the vehicle speed on holiday was higher than on the working day, with the largest average deviation reached 8 km / hour. This speed difference occurs in each type of vehicle. In addition, it happens to all peak hours ie morning, afternoon and evening. It means that traffic volume on holidays is less than on weekdays, so drivers can drive at higher speeds on holidays.

Table 4. Vehicle speed at peak hours on holidays and working days.

Peak Hour	Type of Vehicle	Instantaneous Speed (km / h)		
		Holiday	Weekday	Deviation
Morning	Light Vehicle (LV)	63.9	55.9	8.0
	Heavy Vehicle (HV)	55.9	49.6	6.2
	Motorcycle (MC)	63.7	58.2	5.6
Afternoon	Light Vehicle (LV)	61.1	57.8	3.3
	Heavy Vehicle (HV)	57.7	52.9	4.8
	Motorcycle (MC)	62.6	59.4	3.3
Evening	Light Vehicle (LV)	61.3	58.7	2.6
	Heavy Vehicle (HV)	50.6	46.6	4.0
	Motorcycle (MC)	63.6	60.2	3.4

If it is associated by regulation of maximum urban speed limit by 40 km / h, then the data in Table 4 shows that there is violation of the maximum speed limit on the frontage road. The maximum allowable speed limit is 40 km / hours. It is clearly visible on the maximum speed warning signs which is installed in Jalan Ahmad Yani as shown in Figure 1. Therefore, the high accident rates on the frontage road were closely related to the 85 percentile vehicle speed which exceeds 40 km / hours. Moreover, the geometric condition of frontage road of Jalan Ahmad Yani is flat with an average road width of 15 meters. These reasons encourage the riders to drive at high speed.

Speed exceeding the threshold could potentially increase the risk of traffic accidents. Someone who drives at speeds > 40 km / hours will have difficulty to controll the vehicle in urgent conditions such as the vehicle in front of it stop abruptly or it avoids damaged road conditions. The braking distance until the vehicle stops completely are directly proportional to vehicle speed. Vehicle speeding at 40 km / hours will be stopped optimally at a distance of about 20-100 meters after the braking process. However, the distance is also influenced by other factors, eg slippery road conditions will extend the required distance of vehicle to stop completely after the braking process [19].

4.4. Reduction of accident rate

Based on the discussion in this study, so the reduction of accident and fatality rate on the frontage road can be solved by some alternative as follows:

- Reducing mixed traffic by normalizing the frontage road function. Mixed traffic can be minimized by setting certain types of vehicles that are allowed to pass through the frontage road. For example, public vehicles and motorcycles must pass through frontage road and be forbidden to pass through fast lanes;
- Installation of rumble strips in black spot area. It is important to reduce the vehicle speed at black spot area.
- The addition of PCTL (Pedestrian Crossing Traffic Light) to reduce the potential of accidents between vehicles with pedestrians or road splinters;
- Painting road markings and completing signs at black spot area.

- Penalizing traffic violators, particularly for violations of speed limits.

5. Conclusion

- Characteristic of traffic accidents occurring along the west side frontage road of Jalan Ahmad Yani shows that there are mixed traffic. Since 17 months of frontage road operates, there were 6.5 accidents per month in average. In terms of the type of vehicle involved, the highest proportion of accident was caused by two-wheeled vehicles or motorcycle by 71.6%, while four-wheeled vehicles contributed 7.5% of accidents's number, and the rest by 20.9% were pedestrians or riding no vehicle. Furthermore, the number of accident victims were 91 people by 15 months or 6.5 people per month in average. While the classification of the victims died, serious injuries and minor injuries respectively were 8.8%, 8.8%, and 82.4%. Whereas, average material losses were 2.5 million per month.
- The annual accident rate in 2016 was 0.04 accidents per one million vehicle-kilometre of travel, while for 5 months in 2017 was 0.25 accidents per one million vehicle-kilometre of travel.
- Black spot areas located along west side of frontage road on Jalan Ahmad Yani which is on STA 2 + 800 to 2 + 900, located in front of Graha Pena and DBL Arena building, with accident rate by 0.47 accidents per one million vehicle travels per kilometre, and a fatality rate by 6.77 million accidents per kilometre.
- The high rate of accidents is influenced by the vehicle speed factor that is 85 percentiles which exceeds the allowable speed limit 40 km per hour. The vehicle speed at the 85th percentile condition that crossed the frontage road segment on the west side of Jalan Ahmad Yani for LV classification is 50 km per hour, the HV classification is is 47.5 km per hour, and the MC classification is 67 km per hour.
- Alternatives of counter measures to reduce accident rates include: (i) reducing mixed; (ii) installating of rumble strips at the spot black spot area, (iii) installing PCTL (Pedestrian Crossing Traffic Light); (iv) painting road markings; and (v) penalizing traffic violators, particularly for violations of speed limits.

6. References

[1] Petrov A. and Petrov D., 2016, *Assessment of Spatial Unevenness of Road Accidents Severity as Instrument of Preventive Protection from Emergency Situations in Road Complex*. in *7th International Scientific Practical Conference "Innovative Technologies in Engineering", IOP Publishing, IOP Conf. Series: Materials Science and Engineering* **142** 012116, doi:10.1088/1757-899X/142/1/012116

[2] Sakhapov RL. et al. 2016, *Risk management model in road transport systems.* In *5th International Conference on Mathematical Modeling in Physical Sciences (IC-MSquare 2016), IOP Publishing, Journal of Physics: Conference Series* **738** 012008, doi:10.1088/1742-6596/738/1/012008

[3] Li, X. G., Lord, D., Zhang, Y., & Xie, Y. C. 2008. *Predicting Motor Vehicle Crashes Using Support Vector Machine Models.* Accident Analysis and Prevention, **40(4)**, p.1611–1618.

[4] Li, X. G., Lord, D., & Zhang, Y. 2011. *Development of Accident Modification Factors for Rural Frontage Road Segments in Texas Using Results from Generalized Additive Models.* ASCE Journal of Transportation Engineering, **137(1)**, 74–83.

[5] Direktorat Jenderal Perhubungan Darat, 2015. Perhubungan Darat Dalam Angka (PDDA). Jakarta.

[6] Machsus, Rachmad Basuki, and Amalia F. Mawardi, 2015. *Generalized Additive Models for Estimating Motorcycle Collisions on Collector Roads.* In *the 5th International Conference of Euro Asia Civil Engineering Forum (EACEF-5)*, doi: 10.1016/j.proeng.2015.11.105, ISSN: 1877-7058, September 15-18, 2015, Surabaya, Indonesia.

[7] Machsus, Harnen S., Wicaksono A., and Djakfar L., 2013. *The Prediction Models of Motorcycle Accidents on Surabaya Arterial Roads Using Generalized Linear Models*. Middle-East J. Sci. Res., **18 (12)**: 1859-1866, 2013. ISSN 1990-9233, © IDOSI Publications.

[8] Ackaah W. and Salifu. M. 2011. *Crash prediction model for two-lane rural highways in the Ashanti region of Ghana*. IATSS Research, **35**, p.34-40

[9] Machsus, Harnen Sulistio, Achmad Wicaksono and Ludfi Djakfar. 2014. *The Effect of Access Points on Motorcycle Accident Rates on Surabaya Arterial Roads*. Australian Journal of Basic and Applied Sciences (AJBAS), **8 (10)**, p.38-43

[10] Ispas N. and Trusca D., 2016, *Urban planning and traffic safety at night*. 7th International Conference on Advanced Concepts in Mechanical Engineering, IOP Publishing, IOP Conf. Series: Materials Science and Engineering **147** (2016) 012130, doi:10.1088/1757-899X/147/1/012130

[11] Nambuusi et.al. 2008. *A review of accident prediction models for road intersections*. Steunpunt Mobiliteit & Openbare Werken – Spoor Verkeersveiligheid. RA-MOW-2008-004

[12] Karacasu Murat et al. 2016, *Traffic Perception in Eskişehir Province*. World Multidisciplinary Earth Sciences Symposium (WMESS 2016), IOP Publishing, IOP Conf. Series: Earth and Environmental Science **44** 052060, doi:10.1088/1755-1315/44/5/052063

[13] Xie, Y., & Zhang, Y. 2008. *Crash Frequency Analysis Using Generalized Additive Models*. Transportation Research Record, **2061, 39–45**.

[14] Polus A. and Cohen M. 2011. *A new, non-canonical Poisson regression model for the prediction of crashes on low-volume rural roads*. IATSS Research, **Volume 35, Issue 2**, March 2012, pages 98-103.

[15] National Highway Traffic Safety Administration (NHTSA). 2012. Traffic Safety Facts 2012, A Compilation of Motor Vehicle Crash Data from the Fatality Analysis Reporting System and the General Estimates System, U.S. Department of Transportation, Washington, DC 20590. https://crashstats.nhtsa.dot.gov/Api/Public/ViewPublication/812032

[16] Zhang, Y. et al. 2012. *Crash frequency analysis of different types of urban roadway segments using generalized additive mode*. Journal of Safety Research **43**, 107–114

[17] Akalin KB. et al. 2016, *Curve Estimation of Number of People Killed in Traffic Accidents in Turkey*. World Multidisciplinary Earth Sciences Symposium (WMESS 2016), IOP Publishing, IOP Conf. Series: Earth and Environmental Science **44** (2016) 052060, doi:10.1088/1755-1315/44/5/052060

[18] Mansur A., and Nasution MI.. 2016, *Identification of Behavior Based Safety by Using Traffic Light Analysis to Reduce Accidents*. ICET4SD, IOP Publishing, IOP Conf. Series: Materials Science and Engineering **105** 012033, doi:10.1088/1757-899X/105/1/012033

[19] Reynolds CCO. et al. 2009. *The impact of transportation infrastructure on bicycling injuries and crashes: A Review of the Literature. Environmental Health*

Acknowledgements

We wish to thank to the reviewers for the constructive remarks. We also would like to acknowledge the financial support provided by Institut Teknologi Sepuluh Nopember (ITS) Surabaya through Laboratory Research Grant 2017.

Appendix A. Black spot area based on accident rate per road segment

No	Station	Number of Accidents 2016 (12 mons)	Number of Accidents 2017 (5 mons)	LHR_T (smp)	T_k 2016 (12 mons)	T_k 2017 (5 mons)
1	0+000 - 0+100	0	0	410189	0.00	0.00
2	0+100 - 0+200	2	2	410189	0.14	0.33

IOP Conf. Series: Materials Science and Engineering **267** (2017) 012027 doi:10.1088/1757-899X/267/1/012027

3	0+200 - 0+300	0	0	410189	0.00	0.00
4	0+300 - 0+400	1	3	410189	0.07	0.49
5	0+400 - 0+500	1	1	410189	0.07	0.16
6	0+500 - 0+600	0	0	410189	0.00	0.00
7	0+600 - 0+700	0	0	410189	0.00	0.00
8	0+700 - 0+800	2	0	410189	0.14	0.00
9	0+800 - 0+900	4	2	410189	0.27	0.49
10	0+900 - 1+000	0	1	410189	0.00	0.16
11	1+000 - 1+100	0	0	410189	0.00	0.00
12	1+100 - 1+200	0	3	410189	0.00	0.49
13	1+200 - 1+300	0	0	410189	0.00	0.00
14	1+300 - 1+400	0	2	410189	0.00	0.33
15	1+400 - 1+500	0	0	410189	0.00	0.00
16	1+500 - 1+600	0	0	410189	0.00	0.00
17	1+600 - 1+700	0	1	410189	0.00	0.16
18	1+700 - 1+800	0	0	410189	0.00	0.00
19	1+800 - 1+900	0	1	410189	0.00	0.16
20	1+900 - 2+000	0	0	410189	0.00	0.00
21	2+000 - 2+100	0	0	410189	0.00	0.00
22	2+100 - 2+200	2	1	410189	0.14	0.16
23	2+200 - 2+300	0	0	410189	0.00	0.00
24	2+300 - 2+400	0	0	410189	0.00	0.00
25	2+400 - 2+500	0	5	410189	0.00	0.81
26	2+500 - 2+600	0	4	410189	0.00	0.65
27	2+600 - 2+700	0	0	410189	0.00	0.00
28	2+700 - 2+800	0	0	410189	0.00	0.00
29	**2+800 - 2+900**	**7**	**13**	**410189**	**0.47**	**2.11**
30	2+900 - 3+000	2	1	410189	0.14	0.16
31	3+000 - 3+100	0	0	410189	0.00	0.00
32	3+100 - 3+200	0	0	410189	0.00	0.00
33	3+200 - 3+300	1	2	410189	0.07	0.33
34	3+300 - 3+400	0	0	410189	0.00	0.00
35	3+400 - 3+500	4	13	410189	0.27	2.11
36	3+500 - 3+600	0	4	410189	0.00	0.65
37	3+600 - 3+700	0	0	410189	0.00	0.00
38	3+700 - 3+800	0	0	410189	0.00	0.00
39	3+800 - 3+900	0	5	410189	0.00	0.81
40	3+900 - 4+000	0	0	410189	0.00	0.00
41	4+000 - 4+100	0	0	410189	0.00	0.00
42	4+100 - 4+200	0	0	410189	0.00	0.00

Appendix B. Black spot area based on fatality rate per road segment

No	Station	Death 2016 (12 mons)	Death 2017 (5 mons)	LHR_T (smp)	T_f 2016 (12 mons)	T_f 2017 (5 mons)
1	0+000 - 0+100	0	0	410189	0.00	0.0
2	0+100 - 0+200	0	1	410189	0.00	16.3

3	0+200 - 0+300	0	0	410189	0.00	0.0
4	0+300 - 0+400	0	1	410189	0.00	16.3
5	0+400 - 0+500	0	0	410189	0.00	0.0
6	0+500 - 0+600	0	0	410189	0.00	0.0
7	0+600 - 0+700	0	0	410189	0.00	0.0
8	0+700 - 0+800	0	0	410189	0.00	0.0
9	0+800 - 0+900	0	0	410189	0.00	0.0
10	0+900 - 1+000	0	0	410189	0.00	0.0
11	1+000 - 1+100	0	0	410189	0.00	0.0
12	1+100 - 1+200	0	0	410189	0.00	0.0
13	1+200 - 1+300	0	0	410189	0.00	0.0
14	1+300 - 1+400	0	0	410189	0.00	0.0
15	1+400 - 1+500	0	0	410189	0.00	0.0
16	1+500 - 1+600	0	0	410189	0.00	0.0
17	1+600 - 1+700	0	0	410189	0.00	0.0
18	1+700 - 1+800	0	0	410189	0.00	0.0
19	1+800 - 1+900	0	0	410189	0.00	0.0
20	1+900 - 2+000	0	0	410189	0.00	0.0
21	2+000 - 2+100	0	0	410189	0.00	0.0
22	2+100 - 2+200	0	0	410189	0.00	0.0
23	2+200 - 2+300	0	0	410189	0.00	0.0
24	2+300 - 2+400	0	0	410189	0.00	0.0
25	2+400 - 2+500	0	0	410189	0.00	0.0
26	2+500 - 2+600	0	0	410189	0.00	0.0
27	2+600 - 2+700	0	0	410189	0.00	0.0
28	2+700 - 2+800	0	0	410189	0.00	0.0
29	**2+800 - 2+900**	**1**	**1**	**410189**	**6.77**	**16.3**
30	2+900 - 3+000	1	0	410189	6.77	0.0
31	3+000 - 3+100	0	0	410189	0.00	0.0
32	3+100 - 3+200	0	0	410189	0.00	0.0
33	3+200 - 3+300	0	0	410189	0.00	0.0
34	3+300 - 3+400	0	0	410189	0.00	0.0
35	3+400 - 3+500	0	1	410189	0.00	16.3
36	3+500 - 3+600	0	2	410189	0.00	32.5
37	3+600 - 3+700	0	0	410189	0.00	0.0
38	3+700 - 3+800	0	0	410189	0.00	0.0
39	3+800 - 3+900	0	0	410189	0.00	0.0
40	3+900 - 4+000	0	0	410189	0.00	0.0
41	4+000 - 4+100	0	0	410189	0.00	0.0
42	4+100 - 4+200	0	0	410189	0.00	0.0

IOP Conf. Series: Materials Science and Engineering **267** (2017) 012028 doi:10.1088/1757-899X/267/1/012028

Comparison study on flexible pavement design using FAA (Federal Aviation Administration) and LCN (Load Classification Number) code in Ahmad Yani international airport's runway.

S E Santoso[1], D Sulistiono[2], A F Mawardi[2]
[1] Civil Infrastructure Department. Institut Teknologi Sepuluh Nopember, Menur 127, Surabaya Indonesia.
[2] Lecturer at Civil Infrastructure Department. Institut Teknologi Sepuluh Nopember, Menur 127, Surabaya Indonesia

E-mail: shelvysantoso@gmail.com

Abstract. FAA code for airport design has been broadly used by Indonesian Ministry of Aviation since decades ago. However, there is not much comprehensive study about its relevance and efficiency towards current situation in Indonesia. Therefore, a further comparison study on flexible pavement design for airport runway using comparable method has become essential.
The main focus of this study is to compare which method between FAA and LCN that offer the most efficient and effective way in runway pavement planning. The comparative methods in this study mainly use the variety of variable approach. FAA code for instance, will use the approach on the aircraft's maximum take-off weight and annual departure. Whilst LCN code use the variable of equivalent single wheel load and tire pressure. Based on the variables mentioned above, a further classification and rated method will be used to determine which code is best implemented. According to the analysis, it is clear that FAA method is the most effective way to plan runway design in Indonesia with consecutively total pavement thickness of 127cm and LCN method total pavement thickness of 70cm. Although, FAA total pavement is thicker that LCN its relevance towards sustainable and pristine condition in the future has become an essential aspect to consider in design and planning.

1. Introduction

Pavement structure is a structure consisting of one or more layers of processed materials. A pavement consisting of a mixture of bituminous material and aggregate placed on high quality granular materials is referred to as flexible pavement [1]. Airfield pavement is intended to provide a smooth and safe all weather riding surface that can support the weights of such heavy objects as aircraft on top of the natural ground base. Airfield pavements are typically designed in layers, with each layer designed to a sufficient thickness to be adequate to ensure that the applied loads will not lead to distress or failure to support its imposed loads [1].

[1] Corresponding author. Tel: +62 856 48904890
Email: shelvy13@mhs.ce.its.ac.id (S E Santoso)
[2] Corresponding author. Tel: +62 812 3212190, +62 812 34082635
Email: djoko_sulistiono@ce.its.ac.id (D Sulistiono), amaliafmawardi@gmail.com (A F Mawardi)

IOP Conf. Series: Materials Science and Engineering **267** (2017) 012028 doi:10.1088/1757-899X/267/1/012028

2. Research significance

By comparing two methods, it is very possible to project which method has significant benefits and drawbacks. A smaller number in pavement's thickness and heavy load carriage could yield into a cheaper and more efficient design. In another hand, a thick and less load carriage pavement could be a significant draw back that people should avoid. Therefore, knowing each code in detail will help the society to determine the best way to design Airport pavement.

3. Pavement design codes which are compared in this study.

There are two main codes which are being compare in this study. Those codes are FAA and LCN pavement design codes. Each code has its distinguish approach to determine the runway's total pavement thickness which will be explain on the next section.

3.1. FAA (Federal Aviation Administration) Code Design.

FAA code is originally generated by the US Ministry of Aviation. This code has a specific approach on the aircraft's annual departure, main landing gear configuration, maximum take-off weight (MTOW) and number of tires in order to determine the total thickness of runway pavement. Each approach represents a significant factor which contributed to the runway's design [1]. In order to calculate the pavement's thickness, FAA code use this following equations:

$LogR1 = Log\ R2 \left(\frac{W2}{W1}\right)^{\frac{1}{2}}$ .... equation (3.1)

Log R1: $Log\ R_2 . \sqrt{\frac{W2}{W1}}$ .... equation (3.2)

Log R2: $Log\ R_2$ .... equation (3.3)

R1 : $10^{Log\ R1}$ .... equation (3.4)

R2 : Annual Departure x tire configuration converted factor into *dual wheel gear*.. equation (3.5)

W2 : $MTOW . 0.95 . \frac{1}{Number\ of\ landing\ gear}$ .... equation (3.6)

W1 : Maximum value of W2 column.. equation (3.7)

Where R1 = Equivalent annual departures by the design aircraft.
R2 = Annual number of departures by an aircraft in terms of design aircraft landing gear configuration.
W1 = Wheel load of the design aircraft.
W2 = Wheel load of the aircraft being converted.

In order to get all aircraft's tire configuration converted in one single form, below (Table.1) is the table guide for factors for converting annual departures by aircraft to equivalent annual departures by design aircraft:

Table 1. Factors for Converting Annual Departures by Aircraft to Equivalent Annual Departures by Design Aircraft [3]

To Convert from	To	Multiply Departure By
Single Wheel	Dual Wheel	0,8
Single Wheel	Dual Tandem	0,5
Dual Wheel	Dual Tandem	0,6
Double Dual Tandem	Dual Tandem	1,0
Dual Tandem	Single Wheel	2,0
Dual Tandem	Dual Wheel	1,7
Dual Wheel	Single Wheel	1,3
Double Dual Tandem	Dual Wheel	1,7

3.2. LCN (Load Classification Number) Code Design

LCN code is originally generated by the British Air Ministry Directorate of General Work. This code has a specific approach on the aircraft's annual departure, main landing gear configuration, equivalent single wheel load, tire pressure and number of tires in order to determine the total thickness of runway pavement. Each approach represents a significant factor which contributed to the runway's design [1]. In order to achieve the proper strength during the service period, the LCN number of the runway's pavement need to be greater or at least equal than the LCN number of the aircraft itself.

4. Data

There are two main data which provided by PT. Angkasa Pura I Semarang, the government owned aviation corporation in Indonesia. Those data including the aircraft's annual departure from 2011 up to 2016 (Table. 2) along with the runway's cross section data which also show the detail CBR percentage of certain layer of the pavement (Figure 1.)

Table 2. Aircraft Annual Departure from 2011 to 2016

Domestic Flight															
No.	Aircraft Type	2011		2012		2013		2014		2015		2016		Total	
		Arrive	Depart	Arrive	Depart	Arrive	Depart	Arrive	Depart	Arrive	Depart	Arrive	Depart	Arrive	Depart
1	A 320	150	149	657	657	886	886	974	972	1319	1319	2579	2579	6,565	6,562
2	ATR 72-600	1314	1313	1557	1558	1759	1759	2408	2407	3138	3138	3754	3754	13,930	13,929
3	B 738	2697	2696	3240	3239	3291	3291	3728	3730	4834	4836	4939	4940	22,729	22,732
4	B 739	1658	1657	2827	2827	3371	3372	3237	3236	2716	2715	2378	2377	16,187	16,184
5	C 172	519	519	1601	1590	1345	1327	1506	1473	2274	2285	1866	1875	9,111	9,069
Total														22,729	22,732

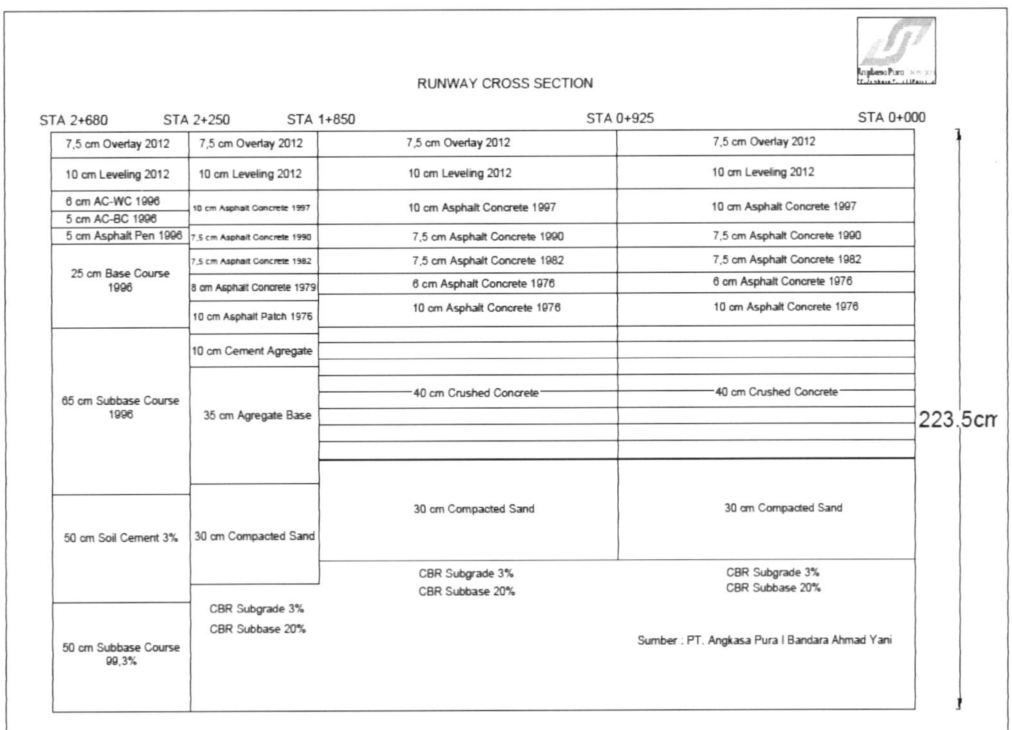

Figure 1. runway's cross section

		International Flight													
No.	Aircraft Type)	2011		2012		2013		2014		2015		2016		Total	
		Arrive	Depart	Arrive	Depart	Arrive	Depart	Arrive	Depart	Arrive	Depart	Arrive	Depart	Arrive	Depart
1	A 319	118	118	0	0	0	0	45	45	95	95	30	30	288	288
2	A 320	50	51	396	396	521	521	581	582	468	468	556	556	2,572	2,574
3	B 733	31	30	13	13	0	0	0	0	0	0	0	0	44	43
4	B 734	0	0	0	0	15	15	0	0	0	0	0	0	15	15
						Total								2.919	2.920

5. Methodology

The method that are being used to determine which design code suit the current circumstances best is using the evaluation framework with the three planning objectives described in Table 1. Individual strategies are rated by the author according to the degree they support each objective, using a four-point scale, from 4 (best), 1 (worst). Ratings in this paper are based on the authors' judgment (the "Notes" columns in tables 6-7 provide brief explanations of these ratings). This framework is reasonably comprehensive but not too complex. Of course, it could be expanded by adding more objectives (such as cost estimation), by weighting the objectives, and by using a panel of experts and stakeholders to rate impacts, but such elaborations are not necessary for this level of analysis, which is primarily illustrative. Except for total pavement thickness, this objective is directly related to code option. All else being equal, a change in annual aircraft departure causes a similar change in thickness of the pavement, thickness of the subgrade and surface layer although the relationships are not exactly proportional.

Table 3. Planning Objectives

Objective	Description
Total Pavement Thickness	Overall thickness of the pavement. Measure from the surface ,base course, subbase course and subgrade.
Subgrade Thickness	The thickness of the very bottom layer in the pavement structure.
Surface Thickness	The thickness of the very top layer in the pavement structure.
Cost estimation	The cost that will be spent to construct the whole pavements.

This table describes the planning objectives used in this evaluation framework.

6. Analysis

The analysis in this study is done based on each code references and different approach. FAA code will have the analysis based on the aircraft's annual departure, main landing gear configuration, maximum take-off weight (MTOW) and number of tires in order to determine the total thickness of runway pavement. While LCN code will have the aircraft's annual departure, main landing gear configuration, equivalent single wheel load, tire pressure and number of tires in order to determine the total thickness of runway pavement.

6.1 FAA Codes Design Analysis

FAA design code has a specific approach on the aircraft's annual departure, main landing gear configuration, gross aircraft weight (lbs.) and number of tires in order to determine the equivalent annual departures by the design aircraft. Then the calculated of the equivalent annual departures by the design aircraft will be used to design the total thickness of runway pavement [1]. This approach has been calculated and put into an equation (Equation. 3.1) which is broadly addressed in this following table (Table.3)

Table 4. FAA equivalent annual departures by the design aircraft analysis.

Aircraft's Type	Main Landing Configuration	Gross Aircraft Weight (lbs.)	Annual Departure	Number of Tires	Factor for Converting Dual Wheel	W2	R2	W1	Log R2	Log R1	R1
ATR 72-600	Dual Wheel	50,705	13,929	4	1.0	12,043	13,929	44,698	4.14	2.15	142
B 737-800 (B-738)	Dual Wheel	174,700	22,732	4	1.0	41,491	22,732	44,698	4.36	4.20	15,757
B-727-900 (B-739)	Dual Wheel	188,200	16,184	4	1.0	44,698	16,184	44,698	4.21	4.21	16,184
C 172	Dual Tandem wheel	2,552	9,099	8	1.7	303	15,468	44,698	4.19	0.34	2
A 320	Dual Tandem wheel	158,730	2,574	8	1.7	18,849	4,376	44,698	3.64	2.36	231
B 737-300 (B-733)	Dual Tandem wheel	140,000	43	8	1.7	16,625	73	44,698	1.86	1.14	14
B 737-400 (B-734)	Dual Tandem wheel	150,500	15	8	1.7	17,872	26	44,698	1.41	0.89	8
Total											32,382

Right after the calculation of the equivalent annual departures by the design aircraft. Then the result will be plotted into the following curve in order to acquire the total pavement of the runway.
Prior to 2008, the FAA's standard method for flexible pavement design was known as the CBR method. The CBR method was based on approximation charts that factored in the CBR value of the subgrade and the number and gross weight of equivalent annual departures of the design aircraft. Separate approximation charts were provided by the FAA for different generic aircraft landing gear configurations, and for aircraft greater than 300,000 lbs. maximum gross weight, specific individual aircraft [1]. Figure 2 provides an illustrative example of the CBR method. The example nomograph found in Figure 2 represents the historical method of estimating the total thickness of flexible pavement for a Boeing 737-900. The arrow within the nomograph represents the example for a subgrade with CBR value of 3 (Figure 1.), a 188,200 –lbs. aircraft gross weight (Table.4), and 32,382 annual equivalent departures (Table.4), resulting in a required total pavement of 50 in thickness.

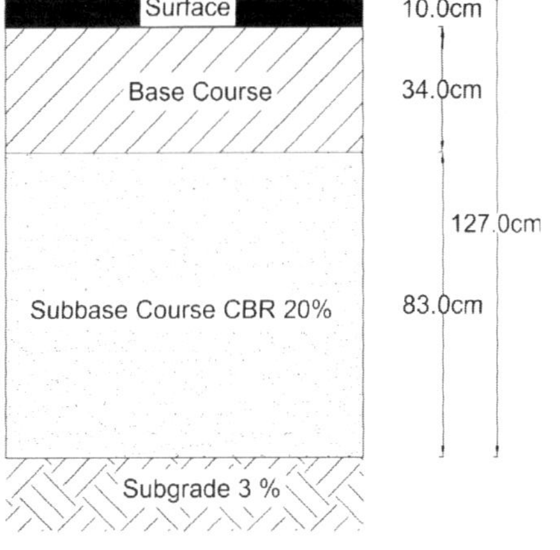

Figure 2. Cross section of FAA design

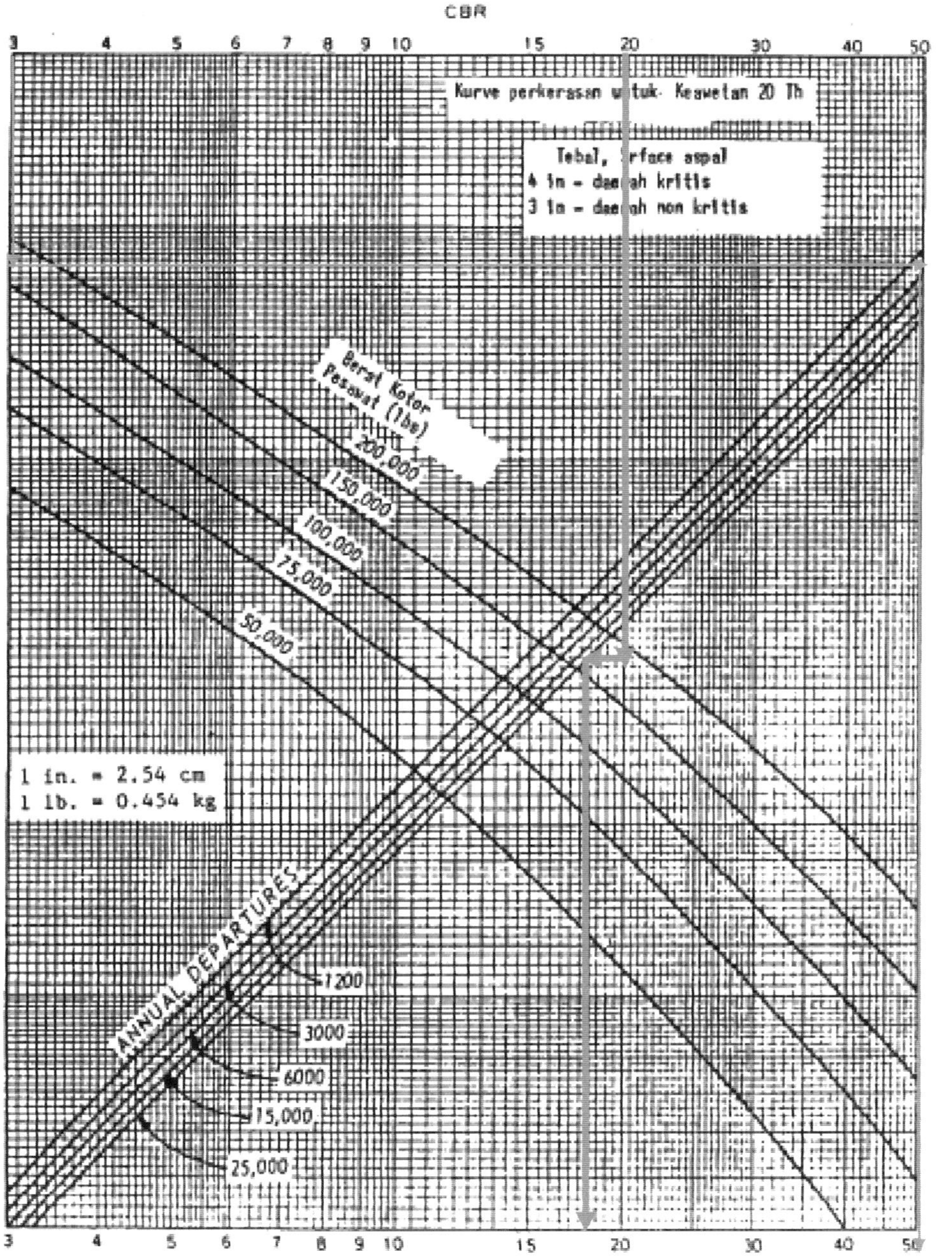

Figure 3. Approximation chart, CBR method of flexible pavement design pavement

The nomograph above also provides the necessary thickness for the surface layer, at 4 in thick for critical areas and 3 in thick for noncritical areas, such as pavement shoulders. The same process applied to determine the thickness of the subbase course. The arrow within the nomograph show the value of subbase course with CBR value of 20 (Figure1.), a 188,200 –lbs. aircraft gross weight (Table.4), and 32,382 annual equivalent departures (Table.4), resulting in a required subbase course of 18 in thickness.

IOP Conf. Series: Materials Science and Engineering **267** (2017) 012028 doi:10.1088/1757-899X/267/1/012028

6.2 LCN Codes Design Analysis

6.2.1 LCN for Aircrafts

According to procedure by the British Air Ministry Directorate of General Work, LCN design code has a specific approach on the aircraft's annual departure, main landing gear configuration, equivalent single wheel load, tire pressure and number of tires in order to determine the equivalent annual departures by the design aircraft. Then the calculated of the equivalent annual departures by the design aircraft will be used to design the total thickness of runway pavement [4]. This approach will be broadly addressed in this following table (Table.5)

Table 5. LCN data analysis based on the approach variable.

Aircraft's type	Main Landing Configuration	MTOW (lbs)	Wheel Number	Tire Pressure (Psi)	Gear Loads (lbs)	Tire's Contact Area(inchi^2)
ATR 72-600	Dual Wheel	50,705	4	55	12042	219
B 737-800 (B-738)	Dual Wheel	174,700	4	204	41491	203
B-727-900 (B-739)	Dual Wheel	188,200	4	204	44698	219
A 319	Dual Tandem wheel	141,095	8	200	16755	84
A 320	Dual Tandem wheel	158,730	8	200	18849	94
B 737-300 (B-733)	Dual Tandem wheel	140,000	8	166	16625	100
B 737-400 (B-734)	Dual Tandem wheel	150,500	8	177	17872	101
	Maximum			204	44698	219

IOP Conf. Series: Materials Science and Engineering **267** (2017) 012028 doi:10.1088/1757-899X/267/1/012028

In order to determine the pavement thickness, the LCN number has to be found first. The arrow within the nomograph represents the calculation of the equivalent single wheel load and aircraft's tire pressure. Above is the nomogram (Figure 4.) shown that the adequate LCN number which suit the needs of all aircraft type is 40. Due to a safety reason the result should be multiple by 1.5. Therefore, the final LCN number is 60.

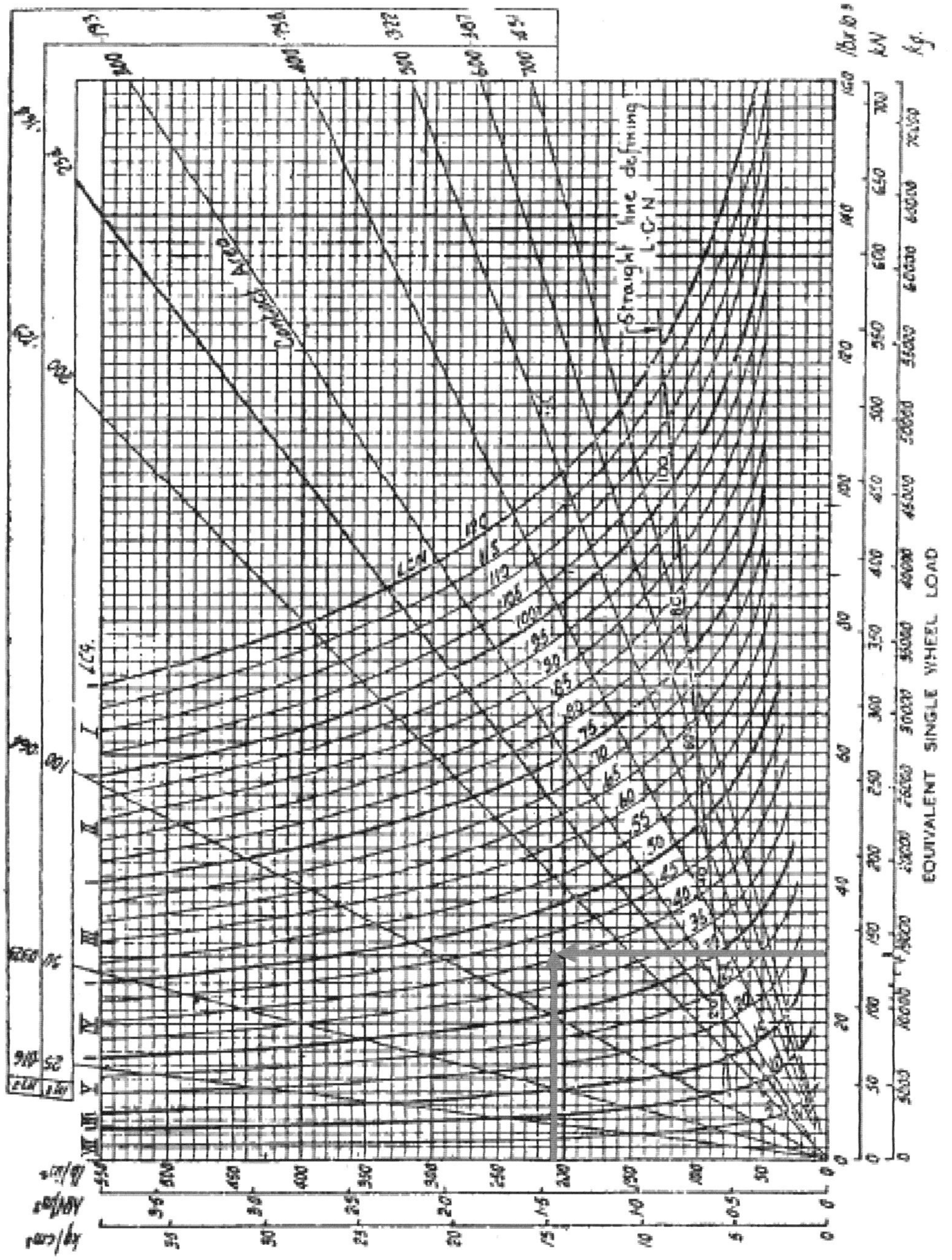

Figure 4. Curve to determine LCN number

6.2.2 LCN for Pavement

According to the LCN method which use soil parameter as the main approach, the method was based on approximation charts that factored in the CBR value of the subgrade and LCN category of the aircraft [4]. Figure 5 provides an illustrative example of the method. The example nomograph found in Figure 5 represents the historical method of estimating the total thickness of flexible pavement. The arrow within the nomograph represents the calculation for a subgrade modulation (K) of 300 Pci and flexural strength 350 Psi, resulting in a required total pavement of 700mm of thickness (Figure 6.). The nomograph also provides the necessary thickness for the surface layer, at 100mm thick, base course layer of 500mm and subbase course layer of 100mm. According to this data, it is then concluded that the LCG of the pavement was categorized as the LCG III.

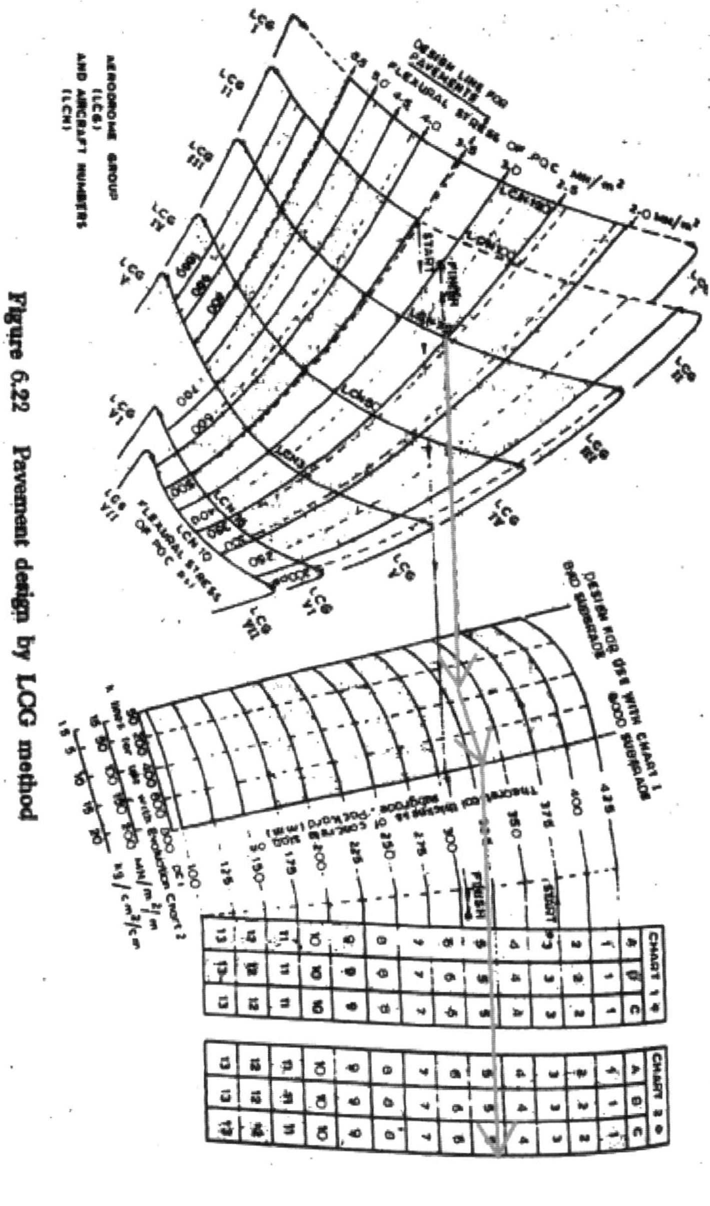

CHART 1

Recommended Aircraft Pavements			
S. No	A	B	C
1	425 100	100 250 100	100 700 100
2	425 100	100 240 100	100 650 100
3	400 100	100 220 100	100 600 100
4	375 100	100 200 100	100 550 100
5	350 100	100 180 100	100 500 100
6	325 100	100 160 100	100 450 100
7	300 100	100 140 100	100 400 100
8	275 100	100 120 100	100 350 100
9	250 100	100 100 100	100 300 100
10	225 100	100 100 100	100 250 100
11	200 100	100 100 100	100 200 100
12	175 100	100 100 100	100 150 100
13	150 100	100 100 100	100 150 100

Figure 5 Flexible Pavement design curve using LCN method

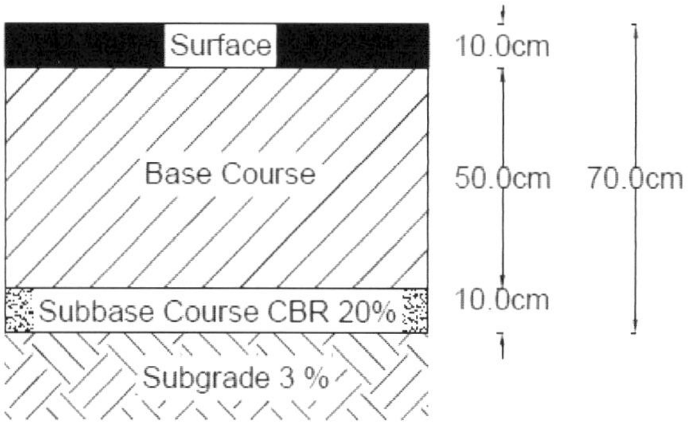

Figure 6. Cross section of LCN design pavement

Pleasse bear in mind that the LCN pavement should surpass the LCN of aircrafts in order to well serve the heavy load carried from the airlines traffic. Therefore, the check of this matter has become essential. Based on the previous calculation, it was noted that the LCG of the pavements was categorized as LCG III.

Table 6. LCN and LCG Correlation

LCG	LCN *Range*
I	101-120
II	76-100
III	51-75
IV	31-50
V	16-30
VI	11-15
VII	10 and under

Source: Robert Horonjeff *Planning and Design of Airports 5th Edition*

Based on (Table 6.) it is concluded that the pavement LCN is in 51-75 range. Therefore, the design pavement has surpassed the LCN of aircraft of 60. In this case the design pavement successfully met the requirements.

7. Results and Discussion

The comprehensive study presented in this paper allowed us in the first hand to develop a performing methodology to offer an applicative solution towards major problem in pavement code decision. The analysis that we offer will help save materials, money, time and also a revenue generator for airport holdings. Having all the evaluation framework system will also increase the understanding of both ideas which will serve the best solution for everybody.

Table 7. FAA (Federal Aviation Administration) code of pavement analysis.

Objective	Rating	Notes
Cost Estimation	2	The overall pavement thickness is 127cm. Which is considerably thick.
Method Result Precision	4	The variable approach used in this method is well represented and close to the actual condition.
Technical Calculation	4	The way to calculate the total thickness of the pavement is quiet easy. The supporting data is easy to find as well.
Application Method	3	The field construction is vary depends on the orders and procedures of every country.
Total Score	13	

Rating from 4 (best) to 1 (worst).

Table 8. LCN (Load Classification Number) code of pavement analysis.

Objective	Rating	Notes
Cost Estimation	3	The overall pavement thickness is 70cm. Which is considerably thin.
Method Result Precision	2	The variable approach used in this method is moderately represented the actual variety of the airport environment.
Technical Calculation	3	The way to calculate the total thickness of the pavement is slightly complicated. It need more control variable which need to be verified and checked first. The supporting data is rather easy to find.
Application Method	3	The field construction is vary depends on the orders and procedures of every country.
Total Score	11	

Rating from 4 (best) to 1 (worst).

According to the above analysis, it is clear that FAA method is the most effective way to plan runway design in Indonesia with consecutively total pavement thickness of 127cm and LCN method total pavement thickness of 70cm. Although, FAA total pavement is thicker that LCN its relevance towards sustainable and pristine condition in the future has become an essential aspect to consider in design and planning.

8. Acknowledgement

The author would like to express her gratitude to the department Civil Infrastructure Engineering of Institut Teknologi Sepuluh Nopember (ITS) and the head of the department Dr. Machsus for funding the conference.

9. Reference

[1] R. Horonjeff and F. Mc Kelvey, Planning and Design of Airports, vol. 5th Edition, United States: United States, 2010.

[2] H. Basuki, Merancang dan Merencana Lapangan Terbang, Bandung: P.T Alumni, 1986.

[3] Boeing, "Boeing 737 Airplane Characteristics for Airport Planning," 2013. [Online].

[4] S. Khanna and M. Arora, Airport Planning and Design, Roorkee: Roorkee Press, 1979.

[5] FAA, "Airfield Pavement Design and Evaluation," *Advisory Circular AC 150/5320-6E,* 2008.

[6] FAA, "Airport Pavement Analysis," [Online]. Available: http://www.faa.gov..

[7] I. C. A. Organization, "Aerodrome Design Manual Part1: Runway," International Civil Aviation Organization, Montreal, 1980.

[8] ICAO, "Aerodrome Design Manual, Annex 14 to the Convention on International Civil Aviation.," International Civil Aviation Organization, Montreal, 1976.

[9] FAA, "Advisory Circular AC 150/5335-5:Standardized Method of Reporting Airport Pavement Strength-LCN," US Department of Transportation., Washington, D.C., 1983.

IOP Conf. Series: Materials Science and Engineering **267** (2017) 012029 doi:10.1088/1757-899X/267/1/012029

Experiment Design for Determining the Minimum Sample Size for Developing Sample Based Trip Length Distribution

H Suprayitno[1], V Ratnasari[2] and N Saraswati[3]
[1]Civil Engineering Department
[2]Statistics Department
[3]Master Student, Civil Engineering Department
Institut Teknologi Sepuluh Nopember (ITS), Surabaya, Indonesia

E-mail: suprayitno.hita@gmail.com

Abstract. In transportation modelling, the origin and destination matrix is essential. To develop an origin-destination matrix, trip length distribution pattern is needed. Thus, it is important to develop a representative trip length distribution based on data sample. This paper present an attempt to develop experiment design for investigating the behaviour of curve pattern acceptance and error value acceptance due to the sample size variation. The experiment design indicate that, in general, smaller sample size produce higher error value. The experiment also indicate that even if the curve pattern is consider accepted, the error value can be unacceptable. Since the experiment design development is merely based on a particular small sample size, the research result can only give an indication yet the calculation result value can not be taken as a proper dictum. The experiment design can be considered as appropriate to be used for any further assessment.

1. Introduction

Transport modelling is needed for transportation planning. The main role of transport modelling is to calculate the trip desired lines and the trip flow assignment in the network of modelling area. These two products are calculated based on origin destination matrix (OD Matrix). The base OD Matrix can be refined and corrected through validation-and-calibration process in the end. The future prediction model is done by calculating the future OD Matrix. Thus, OD Matrix is an essential part of the transport modelling [1-5].

In conventional modelling method, the OD Matrix is a product of mathematical operation among Trip Generation, Generalized Cost Matrix and Deterrence Function. The Deterrence Function itself is developed based on Trip Length Distribution (TLD). Meanwhile the TLD is gotten from a sample. Hence, collecting a good Sample of TLD is important. A Special Conventional Transport Model for a New BRT Line Passenger Demand Prediction has been developed and it is not incorporate the validation-and-calibration process. Thus, having an accurate Sample Based TLD is absolutely crucial for this modelling. One of the basic sampling requirements is to get a minimum number of samples. However, guidelines on number of samples for developing Sample Based TLD is very weak [1-10].

Sample based TLD Quality Measure has ever been formulated before. The main point is having the same curve pattern as the reference's with having an acceptable error, both at a certain confidence level. It has been mentioned before that deriving formula to express the minimum sample size to get a good Sample Based TLD can not be done. Therefore, the minimum sample size requirement can only be gotten from an extensive experiment [8-9].

This paper presents the result of an attempt to develop an Experimental Design for determining the Minimum Sample Size. If this design can be proved appropriate, then it can be used to execute further extensive experiment.

2. Research Method

The research was conduted by following these steps: research objective definition, recalling the Sample Based TLD Quality measurement method, experiment design, case study determination, calculation and analysis, and finished by a conclusion.

3. Sample Based TLD Quality Measure

Sample based TLD quality measures follows the principal of Maximum Acceptable Error at a Certain Confidence Level (MAE at CCL). For this case, it has two main measures : having a curve pattern which considered the same as the reference's curve pattern and having an accepted error, both are measure at a certain confidence level. This pairs is similar to the Zonal Regression Trip Production model quality measure which are : accepted coefficient of determination value with accepted span of prediction value. This principle is conform to the statistics basic principle of sample quality : "MAE at CCL" [8-9,11-12].

The Curve Pattern Acceptance (CPA) test is the statistical goodness of fit test based on χ^2 value. The test is as follows [9,11].

H_0 : $\chi^2 \le \chi_0^2$, curve pattern is considered the same as the reference's.
H_1 : $\chi^2 > \chi_0^2$, curve is not part of the reference's curve pattern.

$$\chi^2 = \Sigma \{(Y_i - Y^0_i)^2 / Y^0_i\} \quad (1)$$
$$\chi_0^2 = \chi^2_{(\upsilon,\beta)} \quad (2)$$
$$\upsilon = n - k - 1 \quad (3)$$

where :

χ^2	=	sample's chi square value
$\chi^2_{(\upsilon,\beta)}$	=	chi squre value et a certain degree freedom and confidence level value
Y_i	=	the ith sample distribution value, measured in percentage
Y^0_i	=	the ith reference distribution value, measured in percentage
β	=	one tail confidence level (1-α, where α is one tail significance level)
υ	=	degree of freedom of χ^2
n	=	number of distribution points
k	=	number of observed parameter

The Error Value Acceptance (EVA) test follows the principal of "MAE at CCL". The error used is the Mean Absolute Error. The Confidence Level value used is the same as for measuring CPA. The test is as follows [11].

H_0 : $|\bar{e}| \le E$, the case is accepted
H_1 : $|\bar{e}| > E$, the case is rejected

$$|\bar{e}| = \Sigma |e_i| / n \quad (4)$$
$$|e_i| = |Yi - Y^0_i| / Y^0_i \quad (5)$$

where :

E	=	accepted error value, measured in percentage.
$\|e_i\|$	=	absolute error value at each distribution points, measured in percentage.
$\|\bar{e}\|$	=	mean of absolute error values, measured in percentage.

n	=	number of distribution points.
Y_i	=	the ith sample distribution value, measured in percentage
Y^0_i	=	the ith reference distribution value, measured in percentage

4. Analysis and Discusion

4.1. Research Objective

Having a good Sample Based TLD is crucial in Transportation Modelling. The Sample Based TLD Quality Measurement Method does not incorporate Sample Size Formula for ensuring good sample quality. An experiment to determine Sample Size must be done. Thus, the Research Objective is to design an experiment for investigating sample's χ^2 value - Curve Pattern Acceptance variation and Error value - Error Value Acceptance variation due to the Sample Size variation.

4.2. Experiment Design

To execute this experiment, a case must be taken, from which the population or the reference sample data were collected. The reference sample can be considered as 'population'. If the population is taken, the experiment is a 'real experiment', but if a reference sample is used, the experiment is called a 'simulation experiment'. Then, from the 'population' (real population of reference sample) take several Samples Groupes of various Reference Sample Percentage. Therefore, the Reference Sample must be big enough, so that the minimum sample size still consists of at least 30 individus. For each Sample Groups construct several Samples with the same sample size but consists of different sample individu composition. Then, the "MAE at CCL" value must be set. It consists of : acceptable error value (%) and confident level value (%). Then for each sample, calculate the TLD. Finally make comparaison between the sample against the 'population', in terms of the Curve Pattern Acceptance and Error Value Acceptance.

4.3. Case Study and Reference Sample

The BRI Branch Office Kertajaya was taken as the Case Study. Its total number of staff is 78 persons. The BRI Branch Office is located in Jl. Kertajaya 24 Surabaya, the center area of Surabaya.

A Reference Sample of 50 individus was taken from BRI Kertajaya. As this is designed for Working Trip TLD, so the TLD is developed based on the distance between home to office. The employee adresses were noted. To construct the TLD, the trip length is classified into each 3 km incremental distances.

The developed TLD seems very reasonable, the distribution value is low at short distance, then arrive at peak at second distance category, then gradually decrease. The TLD is presented in Table 1, while the TLD Graph is presented in Figure 1.

Tabel 1. Reference Sample TLD.

No	Trip Length	Distribution	
	km	*person*	%
1	0-3	20	40
2	3-6	21	42
3	6-9	8	16
4	9-12	1	2
	Total	50	100

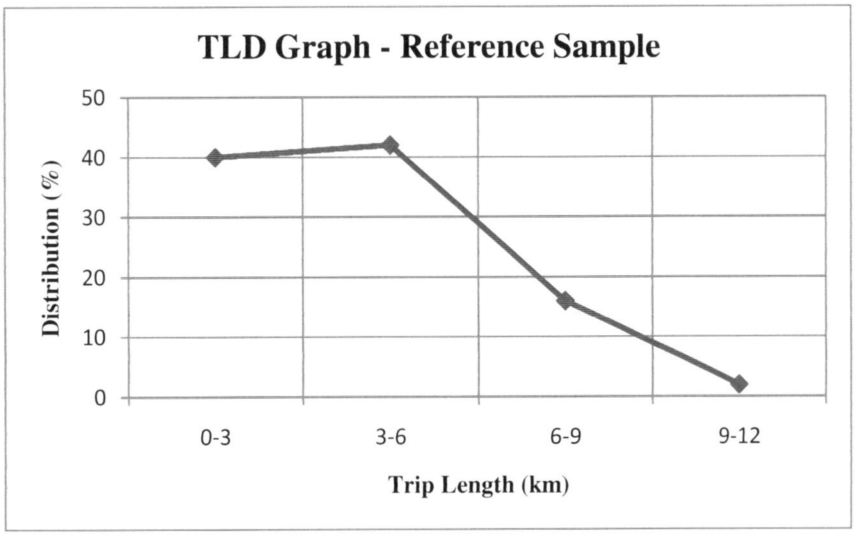

Figure 1 . TLD Graph of Reference Sample.

4.4. Analysis of Sample Size Variation

4.4.1. Sample Size Variation. A Reference Sample was taken, this experiment is consider as 'Population', consists of 50 individuals. This number can be considered as fair, since this number is close enough to the real population number of 78. Reference Sample is 67,5% of Population and the 60% of the Reference Sample is 30 individus.

Four Sample Sizes were taken at 90%, 80%, 70% and 60% of the Reference Sample. Three samples of each Sample Size were constructed. For the development of this experiment design, the samplings were not taken yet as random sample. Each three samples were taken from the first n%, the middle n% and the late n% individuals. Each sample parameters were calculated, i.e. each TLD, each Curve Pattern Acceptance and each Mean Absolute Error Acceptance, all in a certain Confidence Level. The "MAE at CCL" value has 10% Error at 90% Confidence Level. Later on, in the real experiment, sampling must be done in random situation.

4.4.2. Analysis of 90% Sample.
The 90% TLD Sample are all in normal form. The calculated χ^2 values are 0.031, 0.288 and 0.344. These all are far less than $\chi_0^2 = 5.99$. This means that all curve pattern is the same as the Reference Curve Pattern. The calculated error values are 0.5%, 3.3% and 2.9%. These all are well below the MAE = 10%. It means that the 90% Sample is very good.

The 90% TLD Sample calculation result is presented in Table 2, the TLD Graph is presented in Figure 2, while the Curve Pattern acceptance and the Error Value acceptance calculation is presented in Table 3 below.

Table 2. TLD Calculation of the 90% Samples.

No	Trip Length	Person Distribution			Percentage Distribution		
		90A	90B	90C	90A	90B	90C
	km	*person*	*person*	*person*	%	%	%
1	0-3	18	17	17	40.0	37.8	37.8
2	3-6	19	20	19	42.2	44.4	42.2
3	6-9	7	7	8	15.6	15.6	17.8
4	9-12	1	1	1	2.2	2.2	2.2
	Total	45	45	45	100.0	100.0	100.0

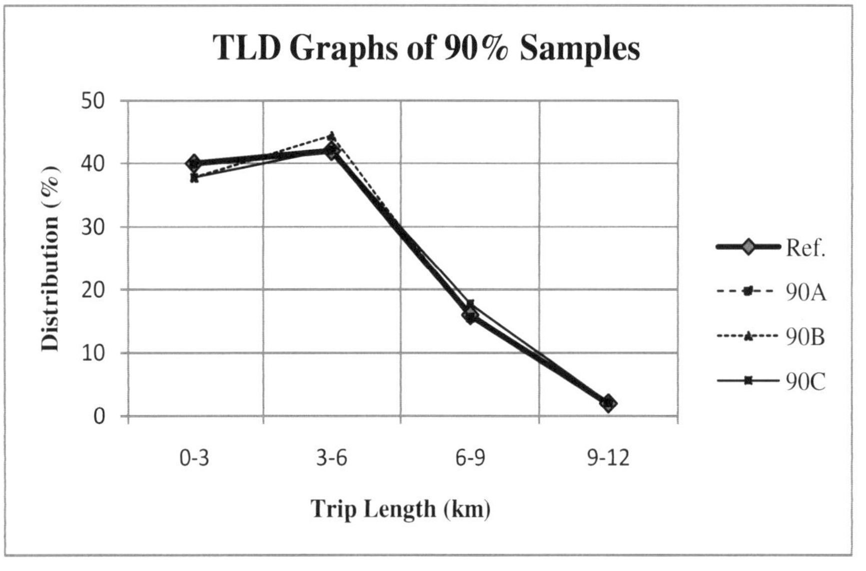

Figure 2. TLD Graphs of 90% Samples.

Table 3. Calculation of Curve Pattern Acceptance and Error Value of 90% Samples.

No	Trip Length	Percentage Distribution				χ2 Calculation			|ē| Calculation		
		Ref	90A	90B	90C	χ^2Ai	χ^2Bi	χ^2Ci	|ē|A	|ē|B	|ē|C
	km	%	%	%	%				%	%	%
1	0-3	40	40.0	37.8	37.8	0.000	0.121	0.121	0.0	5.5	5.8
2	3-6	42	42.2	44.4	42.2	0.001	0.137	0.001	0.5	6.0	0.5
3	6-9	16	15.6	15.6	17.8	0.010	0.010	0.203	1.0	1.0	4.8
4	9-12	2	2.2	2.2	2.2	0.020	0.020	0.020	0.5	0.5	0.5
	Total	100	100	100	100						
CPA		$\upsilon = 2$				**0.031**	**0.288**	**0.344**			
		$\chi^2_{(2,95\%)} = 5{,}99$				**ok**	**ok**	**ok**			
EVA		Maximum Accepted Error Value = 10%							**0.5**	**3.3**	**2.9**
									ok	**ok**	**ok**

4.4.3. Analysis of 80% Sample.

Calculation of the three Samples gives the following χ^2 value : 3.006, 0.428 and 1.376. All of these χ^2 values are less than χ_0^2 =5.99. The error calculation give error values of 7.5%, 2.8% and 6.0%. Those are less than MAE of 10%. It can be concluded that the Curve Patterns and the Errors are all accepted. The 80% Sample is acceptable.

The calculation and the calculation result are presented in the following Table 4, Figure 3 and Table 5 below.

Table 4. TLD Calculation of 80% Samples.

No	Trip Length	Person Distribution			Percentage Distribution		
		80A	**80B**	**80C**	**80A**	**80B**	**80C**
	km	*person*	*person*	*person*	%	%	%
1	0-3	18	15	15	45.0	37.5	37.5
2	3-6	17	17	16	42.5	42.5	40.0
3	6-9	4	7	8	10.0	17.5	20.0
4	9-12	1	1	1	2.5	2.5	2.5
		40	40	40	100.0	100.0	100.0

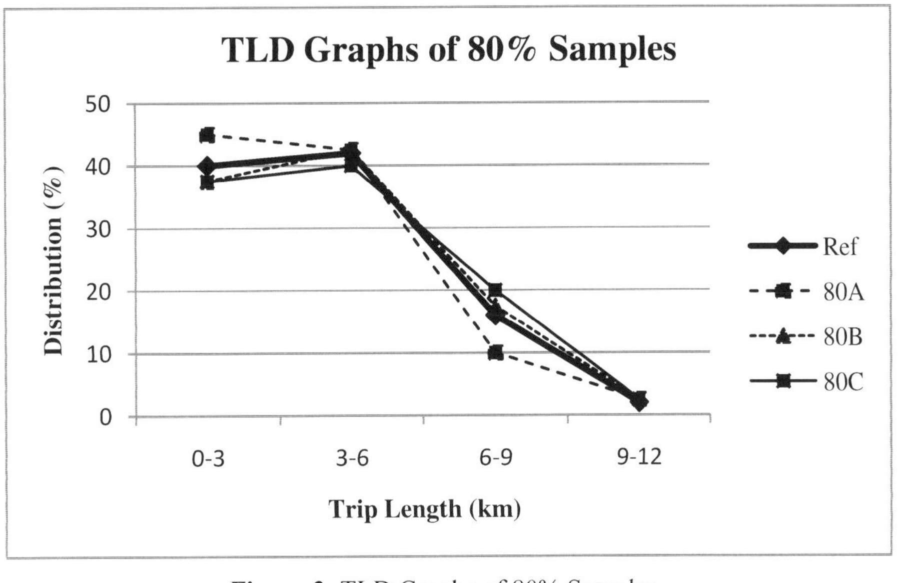

Figure 3. TLD Graphs of 80% Samples.

IOP Conf. Series: Materials Science and Engineering **267** (2017) 012029 doi:10.1088/1757-899X/267/1/012029

Table 5. Calculation of Curve Pattern Acceptance and Error Value of 80% Samples.

No	Trip Length	Percentage Distribution				χ^2 Calculation			|ē| Calculation		
		Ref	80A	80B	80C	$\chi^2 A_i$	$\chi^2 B_i$	$\chi^2 C_i$	|ē|A	|ē|B	|ē|C
	km	%	%	%	%				%	%	%
1	0-3	40	45.0	37.5	37.5	0.625	0.156	0.156	12.5	5.6	6.7
2	3-6	42	42.5	42.5	40.0	0.006	0.006	0.095	1.3	1.1	5.3
3	6-9	16	10.0	17.5	20.0	2.250	0.141	1.000	15.0	3.3	10.7
4	9-12	2	2.5	2.5	2.5	0.125	0.125	0.125	1.3	1.1	1.3
	Total	100	100	100	100						
CPA		$\upsilon = 2$				3.006	0.428	1.376			
		$\chi^2_{(2,95\%)} = 5,99$				**ok**	**ok**	**ok**			
EVA		Maximum Acceptable Error = 10%							**7.5**	**2.8**	**6.0**
									ok	**ok**	**ok**

4.4.4. Analysis of 70% Sample.

The analysis of the 70% Sample gives the following χ^2 value : 6.670, 2.053 and 1.170. Two of them are less than $\chi_0^2 = 5.99$, while one value is slightly more than χ_0^2. This mean that 2 samples have curve pattern which can be considered the same as the reference curve pattern, and one of them is not. The calculation gave these error values : 11.8%, 5.4% and 6.1%. One sample has an error which is slightly bigger than the acceptable value. Not all of them are really good. The 70% Sample is very close to be acceptable.

The calculation and analysis result together with TLD graph are presented in the following Table 6, Figure 4 and Table 7.

Table 6. TLD Calculation of 70% Samples.

No	Trip Length	Person Distribution			Percentage Distribution		
		70A	70B	70C	70A	70B	70C
	km	*person*	*person*	*person*	%	%	%
1	0-3	15	14	13	42.9	40.0	37.1
2	3-6	17	16	14	48.6	45.7	40.0
3	6-9	3	4	7	8.5	11.4	20.0
4	9-12	0	1	1	0.0	2.9	2.9
		35	35	35	100.0	100.0	100.0

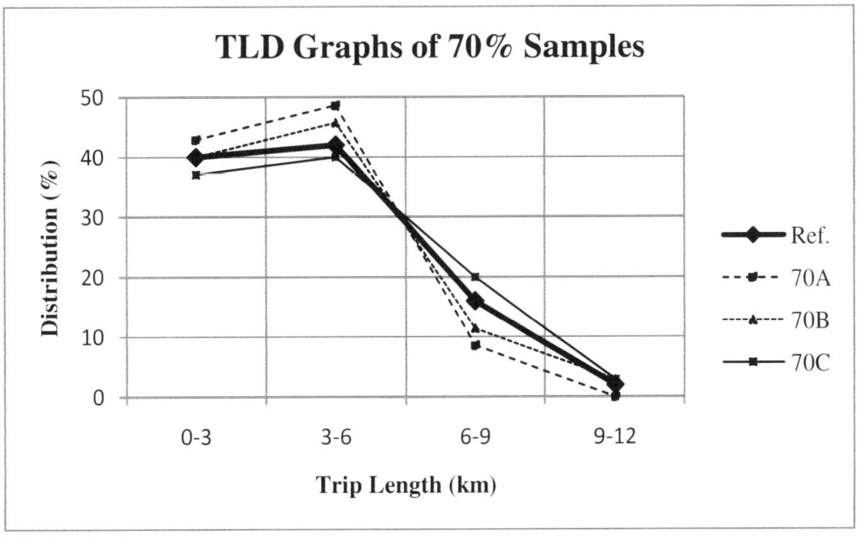

Figure 4. TLD Graphs of 70% Samples.

Table 7. Calculation of Curve Pattern Acceptance and Error Value of 70% Samples.

No	Trip Length	Percentage Distribution				χ^2 Calculation			$\lvert\bar{e}\rvert$ Calculation		
		Ref	70A	70B	70C	$\chi^2 A_i$	$\chi^2 B_i$	$\chi^2 C_i$	$\lvert\bar{e}\rvert$A	$\lvert\bar{e}\rvert$B	$\lvert\bar{e}\rvert$C
	km	%	%	%	%				%	%	%
1	0-3	40	42.9	40.0	37.1	0.210	0.000	0.210	7.3	0.0	7.3
2	3-6	42	48.6	45.7	40.0	1.037	0.326	0.095	16.5	8.6	5.0
3	6-9	16	8.6	11.4	20.0	3.423	1.323	1.000	18.5	10.7	10.0
4	9-12	2	0.0	2.9	2.9	2.000	0.405	0.405	5.0	2.1	2.3
	Total	100	100.1	100	100						
	CPA	$\upsilon = 2$				**6.670**	**2.053**	**1.710**			
		$\chi^2_{(2,95\%)} = 5,99$				**x**	**ok**	**ok**			
	EVA	Maximum Acceptable Error = 10%							**11.8**	**5.4**	**6.1**
									x	**ok**	**ok**

4.4.5. Analysis of 60% Sample.

The analysis of the 60% Sample gives the following χ^2 value : 5,412, 1.803 and 3.142. It means that all sample's curve patterns are accepted. On the other hand, error calculation gave error values of : 39.3%, 25.0% and 28.7%. This means that all samples are very bad.

The calculation, trip length distribution curve and analysis result are presented in the following Table 8, Figure 5 and Table 9.

Table 8. TLD Calculation of 60% Samples.

No	Trip Length	Person Distribution			Percentage Distribution		
		60A	**60B**	**60C**	**60A**	**60B**	**60C**
	km	*person*	*person*	*person*	%	%	%
1	0-3	14	13	10	46.7	43.3	33.3
2	3-6	13	12	13	43.3	40.0	43.3
3	6-9	3	4	6	10.0	13.3	20.0
4	9-12	0	1	1	0.0	3.3	3.3
		30	30	30	100.0	100.0	100.0

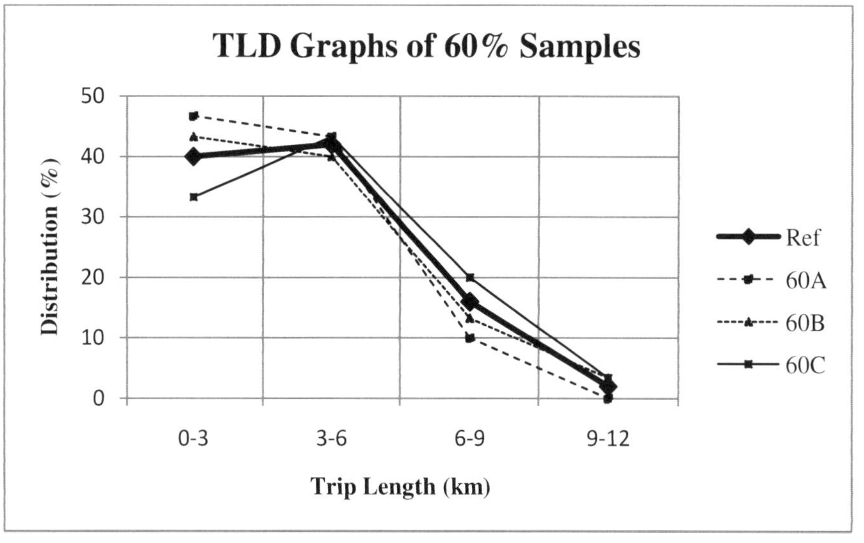

Figure 5. TLD Graphs of 60% Samples.

Table 9. Calculation of Curve Pattern Acceptance and Error Value of 60% Samples.

No	Trip Length	Percentage Distribution				χ^2 Calculation			$\lvert\bar{e}\rvert$ Calculation		
		Ref	**60A**	**60B**	**60C**	$\chi^2 A_i$	$\chi^2 B_i$	$\chi^2 C_i$	$\lvert\bar{e}\rvert$A	$\lvert\bar{e}\rvert$B	$\lvert\bar{e}\rvert$C
	km	%	%	%	%				%	%	%
1	0-3	40.0	46.7	43.3	33.3	1.122	0.272	1.122	16.8	8.2	16.8
2	3-6	42.0	43.3	40.0	43.3	0.040	0.095	0.040	3.1	4.8	3.1
3	6-9	16.0	10.0	13.3	20.0	2.250	0.456	1.000	37.5	16.9	25.0
4	9-12	2.0	0.0	3.4	3.4	2.000	0.980	0.980	100.0	70.0	70.0
	Total	100	100	100	100						
	CPA	$\upsilon = 2$				5.412	1.803	3.142			
		$\chi 2_{(2,95\%)} = 5{,}99$				**ok**	**ok**	**ok**			
	EVA	Maximum Acceptable Error = 10%							**39.3**	**25.0**	**28.7**
									x	**x**	**x**

4.5. Analyse of Curve Pattern Acceptance and Error Value Acceptance Variation

Different Sample Size gives different χ^2 and error values, thus different Curve Pattern Acceptance Level and Error Acceptance Level. The experiment also indicates that the higher the Sample Size Percentage give the better Curve Pattern Acceptance level and the smaller Error Value. This experiment indicates that all of the samples Curve Patterns are accepted except the 70% sample. The 90% Samples gave a range of Error Value between 0.5% – 2.9%. The 80% Samples gave a range of Error Value between 2.8% – 7.5%. The 70% Samples have a range of Error Value between 5.4% - 11.8%. While the 60% Samples gave an exceptionally high Error Value of more than 20%. The variation of χ^2 value, the Curve Pattern acceptance and the Error Value accross different Sample Sizes is presented in Table 10.

Table 10. Variation Curve Acceptances and Error Acceptances.

Sample	χ^2			Curve Accept.			Error			Error Accept.		
Size	A	B	C	A	B	C	A	B	C	A	B	C
%							%	%	%			
90	0.031	0.288	0.344	ok	ok	ok	0.5	3.3	2.9	ok	ok	ok
80	3.006	0.428	1.376	ok	ok	ok	7.5	2.8	6.0	ok	ok	ok
70	6.670	2.053	1.710	x	ok	ok	11.8	5.4	6.1	x	ok	ok
60	5.412	1.803	3.142	ok	ok	ok	39.3	25.0	28.7	x	x	x

The χ^2 value and curve pattern acceptance for each samples are presented in Figure 6 and Figure 7 below. It can be seen that, in general, the higher sample size give lesser χ^2 value and contrary the less sample size give higher χ^2 value. It means that higher sample value gives higher curve pattern acceptance level. The curve pattern of 70% Sample is rejected.

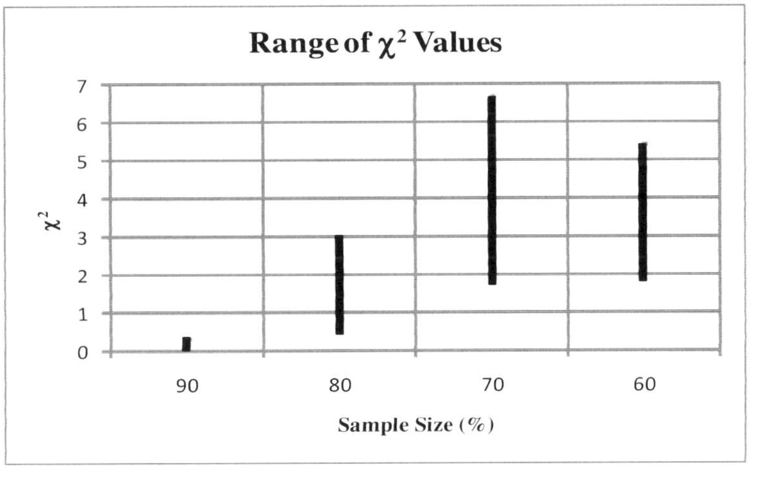

Figure 6. Range of χ^2 Values across Different Sample Sizes.

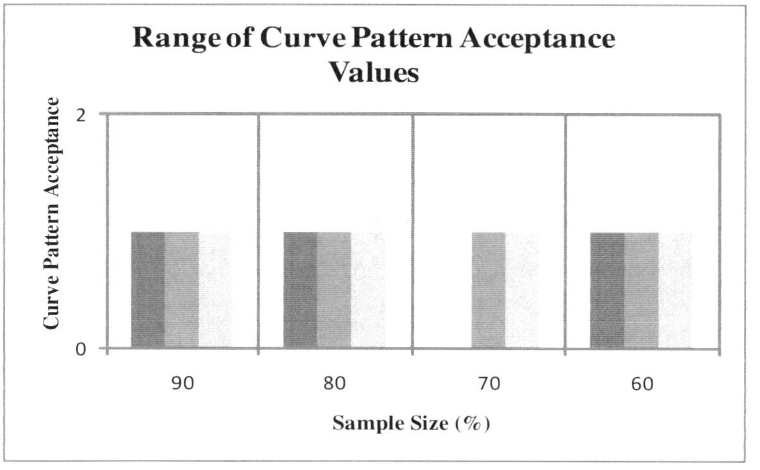

Figure 7. Range of Confidence Level Acceptance across Different Sample Sizes.

The range of Error Value across different sample are presented in Figure 8 below. It can be seen easily that higher sample size give smaller Error Value. The 90% Samples and the 80% Samples produced acceptable Error Values. The 70% Samples produced 2 acceptable error values and 1 slightly more than 10% error value. While the 60% Samples produced very high unacceptable Error Values.

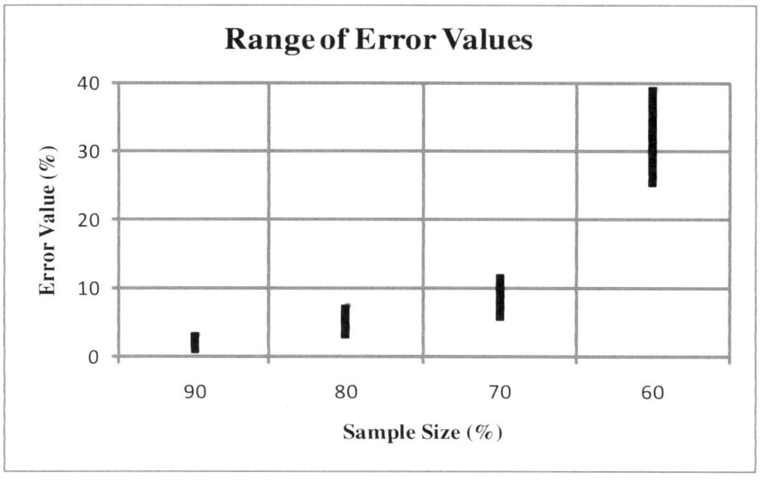

Figure 8. Range of Error Value across Different Sample Sizes.

To complete the research, a Correlation between χ^2 value and Error Value is calculated and presented in Figure 9 below. It can be seen easily that higher χ^2 Value produce higher Error Value. For 60% Sample, especially, all Error Values are extra ordinarily very high, more than 25%. Acceptable χ^2 Values do not always give Acceptable Error Values. Meanwhile, if the Error Value is accceptable, the χ^2 Value is also acceptable.

The correlation between χ^2 Value against Error Value can be formulated by a Linear Regression formula as follows. The correlation is not strong as indicated by a very low R^2 value.

$$E = 3.702\,\chi^2 + 3.505 \quad (6)$$
$$R^2 = 0.385$$

where :

E	:	error value
χ^2	:	chi square value

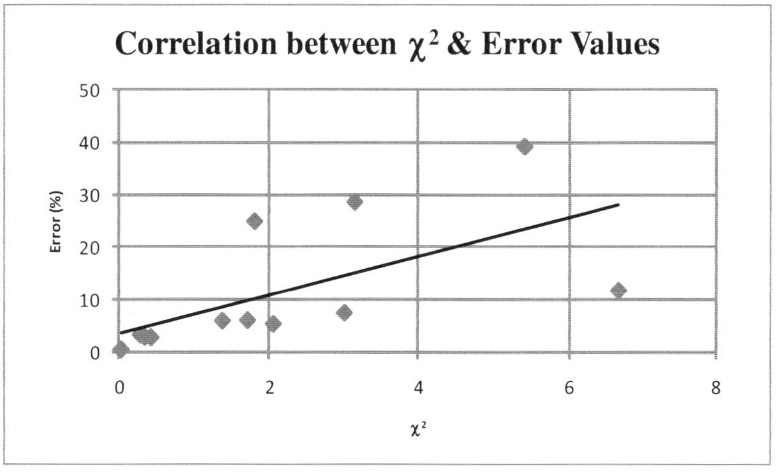

Figure 9. Correlation Between χ^2 and Error Values.

4.6. Remarks

The Experiment Design was succesfully developed. The experiment is capable to indicate acceptable sample size, by indicating sample's Curve Pattern acceptance and Error Value acceptance.

This experiment calculation result indicates that higher sample size give better result, in terms of Curve Pattern Acceptance level and Error Value Acceptance level. The 90% sample and 80% sample are accurate enough to present the population. The 70% sample and the 60% sample can not be used. This Experiment indicates that accepted curve pattern does not always give acceptable error, but on the other hand accepted error value always gives acceptable curve pattern.

It must be noted that this experiment has very limited number of data. The data was taken not from a whole population rather from one small office. The principal objective of this research is to develop an experiment design for determining appropriate sample size. Hence, this experiment result can not be taken as a dictum, the experiment result can only give an indication of the Curve Pattern Acceptance level variation and Error Value variation across different Sample Size.

5. Conclusions

The research objective has been succesfully attained. Several principal conclusions can be written as follows :

- The experiment design has been finished and can be used to define the Minimum Sample Size to develop a Sample Based TLD. The experiment design is capable to produce χ^2 Value, Error Value, Curve Pattern Acceptance, Error Value Acceptance and Corelation between χ^2 and Error Value.
- In terms of experiment result, indicative conclusions can be presented as follows :
  - In general, higher sample size will give better result : lower χ^2 value, higher level of Curve Pattern acceptance, lower Error Value, higher level of Error Value acceptance.
  - In terms of sample size, it can be noted that 90% Sample and 80% Sample are good sample sizes. On the other hand, 70% Samples has a fair quality sample size and 60% Sample is unacceptable sample size.

- Accepted Curve Pattern does not automatically gives Accepted Error Value. On the other hand, Accepted Error Value always give Accepted Curve Pattern.
- This experiment result can not be taken as a dictum. The experiment result can only give an indication of the χ^2 value, the Error value, the Curve Pattern acceptance level and the Error Value acceptance level variation across different Sample Size.

In order to be easily mentioned, this experiment is called "Experiment for Determining Sample Size for Sample Based Trip Length Distribution Development".

This research need a further action to investigate different type of morning trips and other related cases.

6. Acknowledgements

This research paper is a part of the main research in trying to find the Trip Length Distribution Quality variation across different Sample Sizes. The data are collected by Nina Saraswati as part of her thesis work.

7. References

[1] Tamin OZ 2008 *Perencanaan dan Pemodelan Transportasi* Edisi Kedua (Penerbit ITB. Bandung).

[2] Ortuzar JD & Willumsen LG 2004 *Modelling Transport* Third Edition (John Wiley & Sons Ltd. Chichester).

[3] O'Flaherty et al. 2006. *Transport Planning and Traffic Engineering*. (Elsevier. Amsterdam).

[4] Suprayitno H 2015. "Penyusunan Metoda Perhitungan Model Distribusi Perjalanan Berbasis Data Volume Lalu Lintas pada Kasus Pembebanan All-or-Nothing". *Seminar Nasional Aplikasi Teknik Preasarana Wilayah 2015.*

[5] Suprayitno H 2017. "Manual Validation and Calibration Method for All-or-Nothing Traffic Assignment". *IPTEK Journal of Proceeding Series, Issue 2, Year 2017, pages 29-36.*

[6] Suprayitno H & Upa VA 2016. "Mamminasata BRT User Trip Characteristics for the Design of Demand Modelling Method for a New BRT Line". *IPTEKS – Journal for Technology and Science, Vol. 23, No. 3, December 2016.*

[7] Suprayitno H & Upa VA 2017. "Special Conventional Transport Model for a New BRT Line Passenger Demand Prediction – The General Modelling Method". *Proceedings of International Conference on Technology and Social Science, May 2017, Kiryu, Japan.*

[8] Suprayitno H, Saraswati N & Fajriani CP 2016. "Developing a Method for Measuring the Quality of a Sample Based Trip Length Distribution for Urban Trip". *Jurnal Rekayasa Teknik Sipil, Semester Juli-Desember 2016*, page 252-258.

[9] Suprayitno H & Ratnasari V 2017. "Sample Based Trip Length Distribution Quality based on χ^2 and Mean Absolute Error Values". *IPTEKS The Journal of Technologies and Sciences – (submitted on April 2017).*

[10] Jonker NJ 2016. Modelling The Trip Length Distribution of Shopping Trips from GPS Data. *Master of Engineering Dissertation*. Faculty of Engineering. University of Pretoria.

[11] Blank L 1982 *Statistical Procedures for Engineering, Management and Science* (McGraw-Hill International Book Company. New York)

[12] Suprayitno H & Ratnasari V 2017. "Reflexion on Linear Regression Trip Production Modelling Method for Ensuring Good Model Quality". *ICONBUILD 2017 – International Conference on Construction and Building Engineering, Sriwijaya University, Palembang, August 2017.*

IOP Conf. Series: Materials Science and Engineering **267** (2017) 012030 doi:10.1088/1757-899X/267/1/012030

The Effect of Drying-Wetting Cycle's Repetition to the Characteristic of Natural and Stabilization Residual Soils Jawa Timur - Indonesia

M Muntaha[1]
Department of Civil Engineering, Institut Teknologi Kalimantan, Balikpapan, and Institut Teknologi Sepuluh Nopember, Surabaya Indonesia

E-mail: mohamad_m74@ce.its.ac.id

Abstract. Indonesia, which located in tropical region, continuously undergoes wetting and drying cycles due to the changeable seasons. An important role in activating the clay minerals on tropical residual soils is the main factor that affects the static and dynamic properties, such as: volume change, soil suction and dynamic modulus. The purpose of this paper is to evaluate the effect of drying-wetting cycles repetition on volume change, soil suction and mechanical characteristics of natural and stabilization of residual soils from Jawa Timur - Indonesia. The natural undisturbed and stabilized residual soil sample was naturally and gradually dried up with air to 25%, 50%, 75%, and 100 % of the initial water content. The wetting processes were carried out with the gradual increment water content of 25 %($w_{sat} - w_i$), 50 %($w_{sat} - w_i$), 75 %($w_{sat} - w_i$), up to 100 %($w_{sat} - w_i$). The Direct Shear test is used to measure the mechanic properties, and Whatman filter paper No. 42 is used to measure the soil suction. The drying-wetting processes were carried out for 1, 2, 4, and 6 cycles. The laboratory test results showed that, the void ratio decreased, the unit weight, cohesion and the internal friction angle were increasing due to stabilization. Drying-wetting cycle repetition reduces void ratio, negative pore-water pressure, cohesion and internal friction angle of natural and stabilized soils. Briefly, the decreased of mechanical soil properties was proven from the physical properties change observation.

1. Introduction

Indonesian geological condition induces residual soil formation over the mountains for more than 53% land coverage region [1]. Due to its proper mechanical soil properties, many types of construction could be built on residual soil [2]. On the other hand, usage of land due to human requirements transforms agriculture area to residential area. Moreover, many residual soil slope instabilities occur rapidly, for example in Tawang Mangu area at Karang Anyar district-Manting, Mojokerto district, and other regions in Indonesia.

Several slope instability shows that residual slope failure occurred during raining or afterwards, for example slope failure in Pacet, Mojokerto, Panti and Arjasa, Jember; failure at several sections in Purbaleunyi toll road, Bandung. Recently, infiltration capacity reaches minimum value since soil pores have already been filled by the water, therefore soil loses its capacity that is usually strengthened by

[1] To whom any correspondence should be addressed.

the negative pore water pressure [3]. Uncontrolled construction plan activity such as highway, toll road, train railway and dams trigger the residual soil slope clearing, therefore many geotechnical engineering face the residual soil properties problem. Slope clearing also affects the soil which is initially located in the bottom ground become into surface soil and in the following time the drying wetting cycle happens due to changeable season.

Many soil stabilizations have been conducted in order to improve the residual soil characteristics, for instance treated with cement, shotcrete and geogrid. In line with the time, residual soil experiences cyclic drying-wetting due to changeable dry and rainy seasons. This cyclic drying-wetting is predicted to influence the soil volume, soil shear strength and soil negative pore water pressure changes [4].

The purpose of the research is to assess the effect of the drying-wetting cycles on the physic properties, mechanical soil properties (cohesion and internal friction ratio), the negative pore water pressure of natural and stabilized residual soil.

2. Research Methodology

An experimental research was conducted at field and laboratory on Mojokerto residual soils Jawa Timur - Indonesia. Soil sampling was conducted on field and the physical, mechanical soil properties were investigated in laboratory. These investigations were conducted by the standard equipment at laboratory. The flowchart of research methodology could be seen in **Figure 1**.

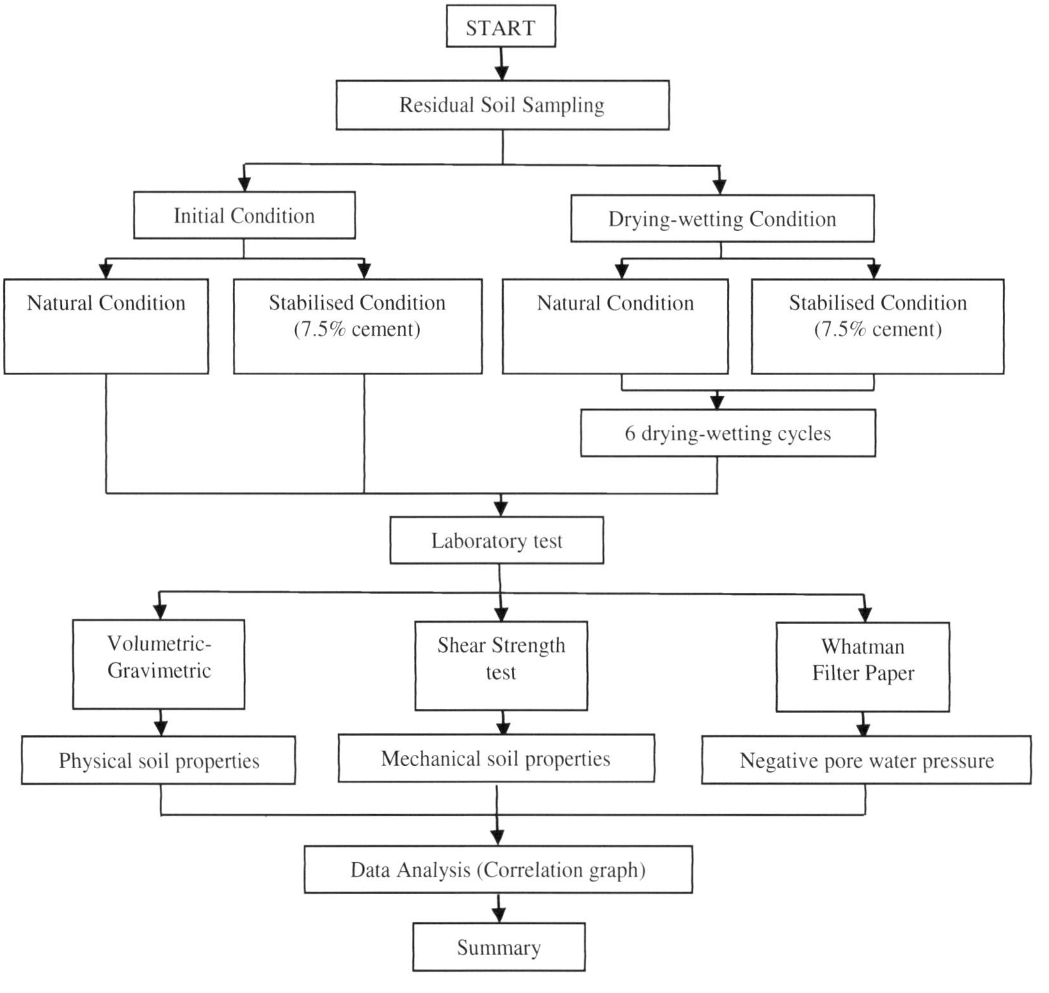

Figure 1. Research flow chart.

Drying-wetting process involves laboratory works to dry and wet the soil samples in cycle generation, and it would be continued with laboratory tests. To determine the drying and wetting model step in each cycle, there is an upper and lower limit of water content range. The upper limit is the saturated soil water content, whilst lower limit is the water content when the sample is dried. Then from the range, step is created by dividing the range into certain value. The degree of saturation would increase during saturation process and decrease during desaturation process [5].

The soil sample starts in initial condition (1st cycle), then it is dried until the lower limit (W_{dry}) and wetted back until the initial condition, this drying-wetting path is regarded as the 1st cycle. Then the next action is wetting the soil sample from initial until upper limit condition (W_{sat}) and continue to dry until its initial condition, this path is regarded as the 2nd cycle. The similar process followed until 3rd, 4th, 5th and 6th cycle [6]. The soil properties tests would be conducted on each step.

Sets of laboratory investigation are then performed to obtain the physical and mechanical soil properties at initial condition and drying-wetting condition (cyclic repetition from 1st cycle to 6th cycle). Physical and mechanic soil properties are proposed to be obtained by performing tests as follows:

- Volumetric-gravimetry (ASTM D 2216-71; ASTM D 854-72).
- Negative pore water pressure by using Whatman paper no. 42 (ASTM D5298-03).
- Triaxial test (ASTM D4767-04), or direct shear test (ASTM D3080).

Laboratory investigation on both initial condition and drying-wetting condition are only conducted at the initial condition point. Therefore, the comparison curves of soil properties are analyzed based on the initial condition point at each cycles.

3. Results and Discussions

3.1. Initial condition

The result of physical, mechanical and negative pore water pressure on natural and stabilized residual soils is presented on Table 1. In order to compare with the initial condition in which the soil was sampled, only the initial condition point was observed in each drying wetting cycles. Then the comparison curves of soil properties could be analyzed at initial condition point during the cycles.

Table 1. Physical, mechanical soil properties at initial condition.

Type of test		Natural	Stabilised
1. Water content	(w, %)	49.67	25.41
2. Degree of saturation	(S_r, %)	97.51	57.45
3. Moisture unit weight	(γ_t, kN/m^3)	15.45	16.23
4. Specific gravity	(G_s)	2.432	2.624
5. Void ratio	(e)	0.963	0.905
6. Cohesion	(c, kPa)	6.5	14.7
7. Internal friction	(ϕ, ^{0})	10.9	14.75
8. Suction	(kPa)	5.85	1815

From Table 1 shows that natural and stabilized soil undergo significant physical soil characteristics changing. Soil moisture unit weight increased 10.33 %, which is followed with specific gravity for about 7.31 %. As well as water content the void ratio also decreased as much as 38.8 %. Similar to mechanical soil and suction properties which the soil cohesion increased from 65 kPa into 147 kPa (55.7 % increment); meanwhile the soil suction increased 99.5 %. Regarding to the analysis, it shows that stabilized soil has higher density rather than natural soil. This phenomenon is predicted due to cement addition into soil that enhance the weight and soil grain volume (W_s and V_s). In line with the soil shear strength, the pore volume reduction affected the lower void ratio, hence the soil effective stress increased. In accordance with results that investigated by Yusdiantoro [7], there is an enhancement of soild grain composition in clay that changes soil physical properties due to soil stabilization.

3.2. Cyclic drying-wetting influence

Laboratory experiments has been conducted in order to assess the influence of cyclic drying-wetting on natural and stabilized soil (initial condition point). **Figure 2** shows the soil-water characteristic curve (1st cycles), **and Figure 3** shows the influence of drying-wetting cycle's to the moisture unit weight and void ratio and soil suction that classified as soil physical properties; mechanical soil properties (cohesion).

3.2.1. Inserting superscripts to link names and addresses.

Figure 2. shows the soil-water characteristic curve (SWCC). At the natural condition, initial water content of the soil specimen was 49.67%. The initial void ratio, e 0.963, moisture unit weight, γ_t 15.45 kN/m^3 and the spesific gravity, G_s 2.432. Stabilised soils shows the initial water content soil was 25.41%. The initial void ratio, e 0.905, moisture unit weight, γ_t 16.23 kN/m^3 and the spesific gravity, G_s 2.624.

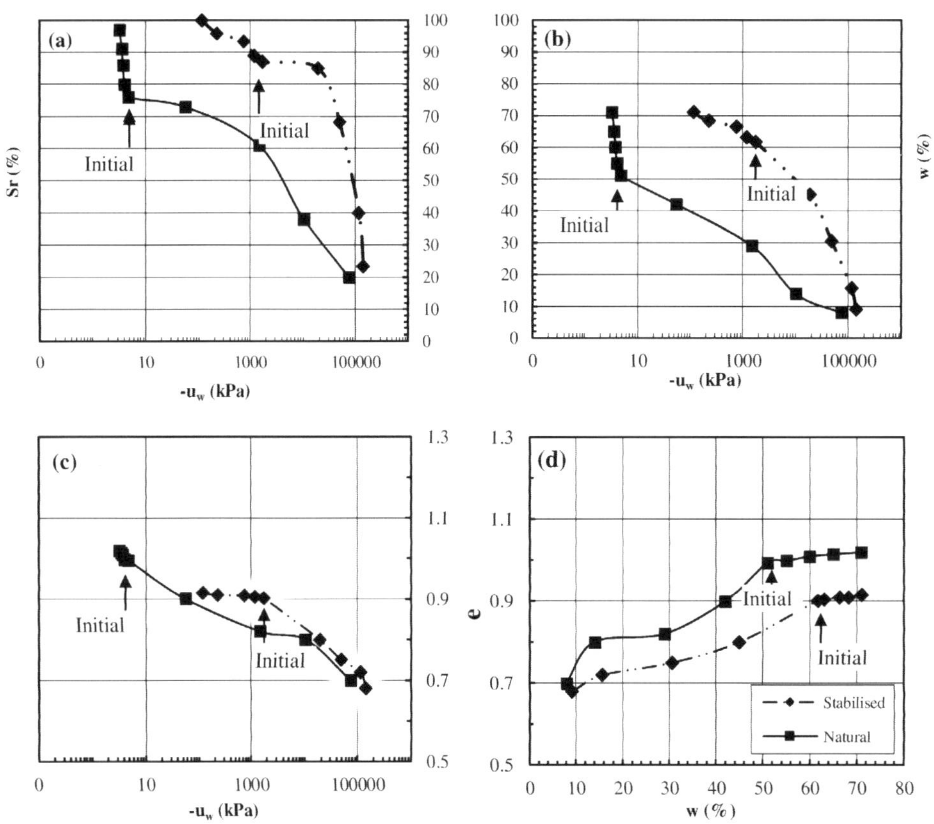

Figure 2. Soil Water Characteristic Curve (SWCC).

During wetting process, for natural and stabilised soils, as water introduced to a relatively nearly saturated soil, it was assumed to be directly taken into the voids. As additional water became available to the soil, the adsorption forces to the clay were attenuated and the voids were gradually and fully filled; resulted in the relatively step increased the water content and the degree of saturation. From the relationship between negative pore water pressure and degree of saturation showed that at the same degree of saturation, the stabilised soil has a value negative pore water pressure higher than natural soils. This happens because the stabilized soil has better density.

Upon drying, for natural and stabilised soils, the water phase within the soil was continued until air entered the soil-water system and the air entry value was reached. On approaching the air entry value, it was assumed that progressively smaller pores were emptied and, were present, the hydrated water layer enveloping clay minerals reduced. It can be obviously seen on the degree of saturation-negative pore pressure curve. Soil water continuity progessively diminished over this range of negative pore pressure as air filled channels began to form within the soil voids.

3.2.2. Cyclic influence on soil physical properties (density).

Figure 3A. shows the relationship between soil moisture unit weight and number of cycles. For instance, in natural soil, soil moisture unit weight increases as much as 5.79 % during cycle that started from initial condition into 1st cycle; meanwhile soil moisture unit weight increased as much as 1.57 % in stabilized soil. From the evidence showed that higher soil moisture unit weight changes occurred in natural soil since the cyclic undergo the higher density.

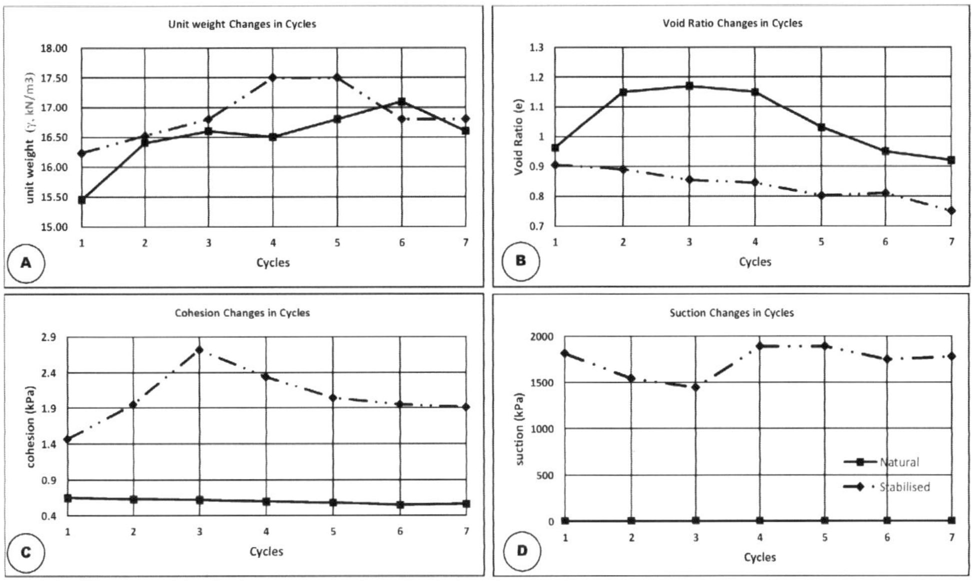

Figure 3. Relationship between void ratio, soil unit volume, cohesion, suction and number of cycles (initial condition point).

Application of the 1st cycle until the 6th cycle on natural soil affected the soil moisture unit weight where γ_t increases as much as 9.6 %; while stabilized soil increased up to 3.39 %. Figure 2A shows that natural soil initial density condition is more affected by the cyclic, compared with the stabilized soil.

Stabilized soil was less affected by the cyclic drying-wetting condition due to higher density. Asmaranto [8] had found that the fly ash added into clay soil induces hydration that triggers the increase of soil volume grain and soil particle orientation as well.

Figure 3B. shows the effects of cyclic drying-wetting of natural and stabilized residual soil on soil void ratio. **Figure 3B.** shows that the 1st cycle until the 6th cycle application induces the decrease of void ratio as much as 25.01 %; while void ratio decreases 18.66 % on stabilized soil. It could be concluded that cyclic drying-wetting induces the void ratio decrease. Muntaha [3] had found that the cyclic drying-wetting induces water between inter particular space and free water in between inter aggregate that comes out from the pore space, hence the soil pore volume will be reduced.

The other phenomenon revealed that void ratio changes on stabilzed soil is tend to be constant in each of cycle. In natural soil, void ratio change for 1st, 2nd, 4th and 6th cycle in consecutive way are 16%, 15,6%, 15,6% and 16,2%, while for stabilized soil are 33,9%, 33,2%, 32,3% and 32,3%. It is caused by the first drying-wetting cycle that the soil re-composed its structure so it is denser and more stable to reduce the soil deformation.

3.2.3. Cyclic influence on soil mechanical properties.

Figure 3.C. shows the relationship curve between soil cohesion and cycle for both natural and stabilized soils. It could be seen that due to cyclic drying-wetting, cohesion for both soils either tend to be constant in the same cycle or decrease in the cycles generation. For example, in natural soil, the soil cohesion consecutively for 1st, 2nd, 4th and 6th cycle are 63, 62, 58 and 56 kPa. These analyses inappropriate with the conventional soil mechanics where due to drying-wetting cyclic generation would reduce the water content and void ratio, it is correlated with the denser and stiffer soil structure. The soil mechanics theory states that as the density and stiffness increase, the soil shear strength will be higher. On the other hand, in this research, the generation of drying-wetting cycle induces soil cohesion reduction for boyh of natural and stabilized soils.

This phenomenon is predicted to be affected by the water content and void ratio reduction that followed by increase of the dry unit weight. However, the degree of saturation is getting higher, it indicates that there is an apparent dry density due to high degree of saturation. The high degree of saturation mostly influences the reduction of soil shear strength. This analysis supports the research conducted by Miyakita [9].

Figure 3.D. reveals that in same cycle and same initial condition point, negative pore water pressure of stabilized soil is higher than natural soil. In same negative pore water pressure during wetting, stabilized soil undergoes increase of solid grain and reduction of pore volume as the cement and water hydration process. As consequency, the soil effective stress will be increased as well. If the cyclic drying-wetting being analogyzed as the loading-unloading during consolidation process, hence the increase of soil effective stress would affect the reduction of void ratio. As it was stated by Yusdiantoro [7], void ratio of stabilized soil with fly ash is much lower than natural soil. Therefore, the stabilized soil with fly ash possesses higher negative pore water pressure rather than natural soil. Paulus [10] stated that lime addition into natural clay soil induces reduction of void ratio, so the degree of saturation increases and negative pore water pressure increases respectively.

4. Conclusion

Test results showed that soil stabilization increased both of the soil density and soil shear strength shown from the reduction of void ratio. Cyclic drying-wetting generation induced the density of natural and stabilized soil. However, the soil density increase is not followed by the soil shear strength increase in stabilized soil. Moreover, the soil density is predicted in the form of apparent density which possesses high degree of saturation. The high degree of saturation mainly induces the decrease of soil shear strength when the density is high.

5. References

[1] Saroso, B.S. (2002). "Geology and Landslides in Indonesia", National Conference in Slope, Bandung 2002, pp 1-13.

[2] Wesley, L.D. 2005. "Residual and Volcanics Soils for University Parahyangan Presentation". Seminar Nasional Parahyangan 2005. Bandung.

[3] Muntaha, M. (2015). "The static and dynamic characteristic of undisturbed residual soils under drying-wetting cycle's repetition", International Conference ARC-15, Fukuoka-Japan.

[4] Soemitro, R.A. dan Saing, Z. (2001), "Drying-Wetting Cyclic Repetition Influence on Expansive Soil Compression Strength", Proceeding National Conference - ITS, Surabaya.

IOP Conf. Series: Materials Science and Engineering **267** (2017) 012030 doi:10.1088/1757-899X/267/1/012030

[5] Vanapalli, S.K.; Fredlund D.G.; Pufahl, D.E. and Clifton, A.W.1996. "Model for the Prediction of Shear Strength with respect to Soil Suction", Canadian Geotechnical Journal, Vol. 33, pp. 379-392.
[6] Muntaha, M. 2013. "The Static and Dinamic Characteristic of Undisturbed Residual Soils under Drying-Wetting Cycle's Repetition", Dissertation, Institut Teknologi Sepuluh Nopember.
[7] Yusdiantoro, Y. (2001), Drying-Wetting Cyclic Repetition Influence on Dynamic Soil properties of Unsaturated Expansive Clay Treated with Fly Ash by using Resonance Column Test", Master thesis, Post-graduate program of ITS.
[8] Asmaranto, R. (2001), "Drying-Wetting Cyclic Repetition Influence on Dynamic Soil properties of Unsaturated Expansive Clay Treated with Fly Ash by using Cyclic Triaxial", Master thesis, Pascasarjana ITS, Surabaya.
[9] Miyakita, K. 1991. "Effect of Repetition of Drying-Wetting on Mechanical Characteristics of Diatomaceous Mudstone", Japanese Society of Soil Mechanics and Foundation Engineering, PP. 117-133.
[10] Paulus, N. 2007, "Pengaruh Air Garam pada Proses Pembasahan-Pengeringan terhadap Kuat Geser Tanah Lempung yang Distabilisasi dengan Fly Ash", Tesis, Pascasarjana ITS, Surabaya
[11] ASTM D 2216-71. "Standard Test Method for Laboratory Determination of Water (Moisture) Content of Soil and Rock by Mass"
[12] ASTM D5298-03. "Standard Test Method for Measurement of Soil Potential (Suction) Using Filter Paper"
[13] ASTM D 854-72. "Standard Test Methods for Specific Gravity of Soil Solids by Water Pycnometer"

IOP Conf. Series: Materials Science and Engineering **267** (2017) 012031 doi:10.1088/1757-899X/267/1/012031

Effect of Density and Resistivity Measurement for Foam Flooding Propagation in Static Condition

S Irawan [1], **K A Permatasari** [1(*)] **and R Bayuaji** [2]
[1]Department of Petroleum Engineering
Universiti Teknologi PETRONAS, Seri Iskandar, Perak, Malaysia
[2]Department of Civil Engineering
Institut Teknologi Sepuluh Nopember, Surabaya, East Java, Indonesia

E-mail: karina_g03560@utp.edu.my

Abstract. Foam flooding is one of the method that is currently used to optimize the amount of oil in sub surface. Foam flooding is used to overcome the weakness of water injection, gas injection, and water alternating gas injection. Foam is the mixture of gas, water and surfactant. Foam propagation plays an important part in porous medium, foam propagation is related to intrinsic properties of the fluids such as density and viscosity. Resistivity of the fluids should be taken also because in the end of the experiment, method that is used to track foam propagation in porous medium is resistivity measurement. The objectives of this experiment are to obtain fluid characterization and to understand the phenomena that is possibly occurred during foam injection in porous medium, specifically, to obtain the correlation between density and resistivity measurement. Foam is generated with several concentration of solutions which are various concentration of brine, various concentration of MFOMAX, oil, and distilled water. Several tests should be taken to observe the intrinsic properties of the fluids such as density test, viscosity test, resistivity and conductivity test. Based on the experiment, correlation between density of brine and brine concentration towards resistivity is inversely proportional. Correlation between density of fluid mix with MFOMAX and MFOMAX concentration towards resistivity is proportional.

1. Introduction

Enhanced Oil Recovery (EOR) is one of the method for optimizing oil in sub surface. The most recent method for producing large amount of oil is foam flooding injection. Foam Flooding is introduced to overcome water injection, gas injection and water alternating gas injection. Foam is used to increase oil production in sub surface by reducing the mobility of gas that is injected, increase sweep efficiency and is able to travel in both high and low permeability due to higher viscosity of foam compared to gas. [1] However, foam also has negative side which are easily rupture and the velocity of foam is slower than viscosity of gas.

Foam flooding propagation is influenced by properties of each fluid that is mix with other fluids. Based on the negative side of foam, properties of each fluid should be known to determine foam behaviour, velocity and flow rate of foam when foam is injected in porous medium (core sample). Properties of foam that is related to velocity and flow rate of foam are density and viscosity of foam. Another properties of foam are resistivity and conductivity; these properties are related to the method of tracking foam propagation inside the porous medium (core sample). The method that is chosen for tracking foam propagation is resistivity measurement.

2. Theory and Literature Review

2.1.1 Concept of Foam

Foam is made by the mixture of gas and liquid. Foam is formed in a liquid if bubbles of gas are injected faster than liquid. Gas is dispersed in a continuous liquid phase as expressed in Figure 1. Foam disintegrate when the bubbles coalesce as soon as the liquid between the bubbles has drained away [2].

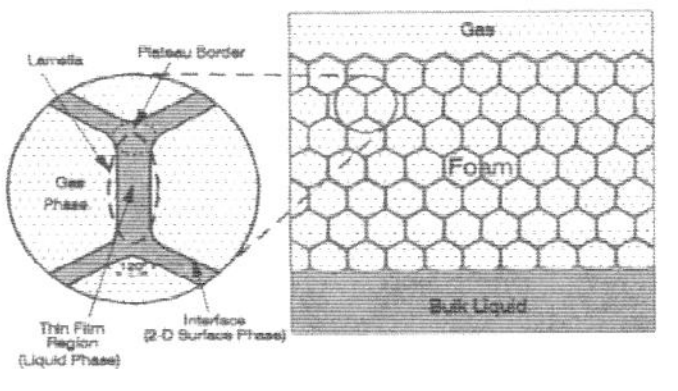

Figure 1. Foam Structure. [2]

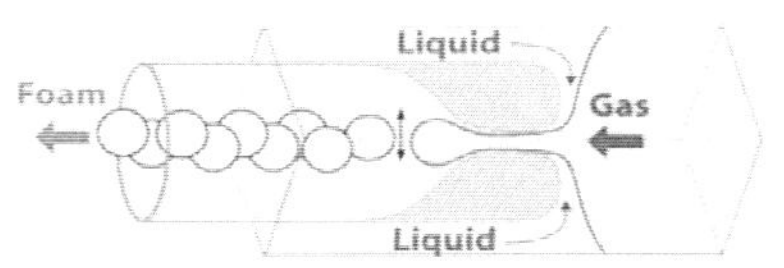

Figure 2. Foam Generated. [3]

Based on figure 2, lamellae is defined as the region that encompasses the thin film, the connection of lamellae (120°) and contains two gas-liquid interfaces separated by a layer of fluid thin film.[2]

Foam has electrical properties; surface potential is a charged interface influences the distribution of nearby ions in a polar medium. Surface potential is related to electrical forces which are repulsive forces and attractive forces. Attractive forces which are the opposite of ion charge (counter ions) are attracted to the surface, while those charge (co-ions) which are repelled are called repulsive forces. The Electric Double Layer (EDL) consists of the charged surface and a neutralizing excess of counter ions over co-ions, distributed near the surface. The EDL can be viewed as being composed of two layers, an inner layer that may include adsorbed ions and a diffuse layer where ions are distributed according to the influence of electrical forces and thermal motions. [4]

An inner layer exists because ions are not really point charges, and an ion can approach a surface only to the extent allowed by its hydration sphere. The Stern model incorporates a layer of specifically adsorbed ions bounded by a plane, the Stern plane. [2]

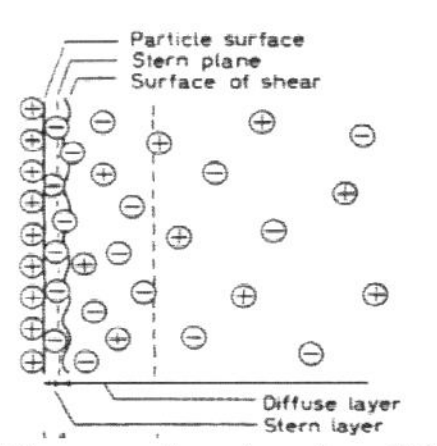

Figure 3. A simplified illustration of the Electrical Double Layer [2].

2.1.2 Density and Viscosity.

Foam propagation in porous medium (core sample) plays an important part and it is related to velocity and flow rate of foam. Velocity and flow rate of foam is related to intrinsic properties in each solution that will be mixed together in one solution. Properties that is influenced by how foam propagate inside the porous medium (core sample) are density and viscosity. In this chapter, density and viscosity will be explained further.

- Density

The mass density or density of a material is defined as its mass per unit volume. The symbol which mostly used for density is ρ (rho). [6] The Mathematical formulation of density can be expressed as below.

$$\rho = \frac{m}{v} \tag{1}$$

Where m is mass, v is volume. From this equation, density should have units of a unit of mass per unit of volume. Density unit in this experiment is g/cm^3.

- Viscosity

Viscosity is a measurement of the resistance of a fluid which is being deformed by either shear stress or tensile stress. In general terms, viscosity is related with the thickness and internal friction. The fluids which have thicker solution make viscosity higher and lower its ease of movement (fluidity). [7] Viscosity of the fluid that is used is related with Newton's Law and stated that the shear stress between adjacent fluid layers is proportional to the negative value of the velocity gradient between them as shown as equation (2) and (3) below. [7]

$$Shear\ stress = \tau = \frac{Force}{Area} \tag{2}$$

Based on the equation (2) shear stress is the ration of force divide by surface area of the plate.

$$Shear\ stress = \tau_{y,x} = -\mu \frac{d\mu_x}{dy} \tag{3}$$

Where $\tau_{y,x}$ is the shear stress in y axis at distance y from the origin and x is the direction of the velocity, μ is the viscosity, negative sign expresses the shear stress that is applied from higher velocity to a lower velocity.

- Resistivity

Resistivity is an electrical resistance measured between opposite faces of a unit cube of a substance at a specified temperature. Conductivity is the reciprocal of resistivity. Ohm's Law is the basic theory to predict foam propagation along the porous medium. Ohm's Law can be expressed as equation (1) and (2).[8]

$$r = \frac{V}{I} \tag{4}$$

$$R = \frac{V}{I} \cdot \frac{A}{L} = \frac{r \cdot A}{L} \tag{5}$$

$$C = \frac{1}{R} \tag{6}$$

Where r is the resistance in ohms (Ω), V is the potential difference in volts (V), and I is the current in amperes (A), R is the resistivity in ohm meter (Ωcm), A is the cross sectional area of core plug (cm^2) and L is the length of the particular zone in meters (cm), C is conductivity in S/cm.

3. Research Methodology

This experiment used several concentration of solutions. As mentioned earlier foam is the mixture of water, surfactant and gas. Each fluid has number variation of concentration, the fluids that is used is mentioned below.

- Distilled Water
- Brine (1 wt%, 2 wt%, 3wt%)
- MFOMAX (0.05 wt%, 0.1 wt%, 0.5 wt%, 1 wt%)
- Oil (7.5 ml)

All of the solutions will be mixed together after characterize has been done for each solution. The solutions will be placed in 250 ml of beaker. After all the solutions is ready, several test will be applied on each solutions in ambient condition (25°). Several test are density test, viscosity test and resistivity test.

3.2 Density Test

1. Each solution is placed on the bottle solution.

IOP Conf. Series: Materials Science and Engineering **267** (2017) 012031 doi:10.1088/1757-899X/267/1/012031

2. Bottle solution is placed inside the density meter.

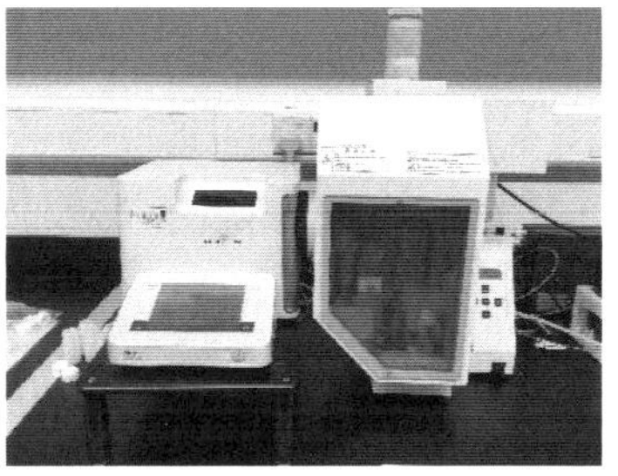

Figure 4. Density Meter.

3. Equipment is ready to operate. First, choose "Methods/products" on the screen then choose "Methods".
4. Choose "Measure 25°" due to ambient condition measurement then click "Start".
5. Sample ID should be added to name for each solution then click "Start" for starting the measurement.
6. The needle starts to move inside the bottle solution and suck the solution into the "U" tube.
7. When equipment starts to measure the solution, there are several process while running equipment. First stage is filling the solution in "U" tube in 90 s, measuring the solution in 74 s, draining the solution from "U" tube in 5 s, rinsing using solvent in 6 s, rinsing using acetone in 8 s, drying the "U" tube in 116 s, calculating the result and reporting the result.
8. Density and specific gravity result is obtained by choosing "Result" on the screen, choosing sample ID that has been measured, and choosing "Calculated Results".
9. After all the process has been done, take out the bottle solution from density meter.

3.3 Viscocity Test

1. Each solution is placed on the small tank.
2. Small tank and spindle is placed on the rheometer until the status in the screen is "Ready".

Figure 5. Rheometer.

3. Small tank and spindle is placed on the rheometer until the status in the screen is "Ready".
4. Open rheometer software in PC then click "Rheometer" in setup measuring profile, set lift position of spindle in 140 mm and measuring position of the spindle in 1 mm, set temperature is 25°, click "Measurement Position" to shift the spindle into measurement position then click "OK".
5. Click "Edit Shear Rate" in setup measuring profile, set measurement points: 20, time unit: s, interval: 120 s, profile : ramp, initial: 1 and final 1000 then click "OK".

6. Click "Start The Test" in setup measuring profile, name the sample ID then click "Start The Test Now".
7. The graph and the data will be shown in the diagram 1: time test single point frame.
8. After data and graph has been obtained, click "Rheometer" and click "Lift Position" to pull out the spindle then take out and clean the small tank.

3.4 Resistivity and Conductivity Test

1. Solution which is placed in the small container is provided to proceed resistivity and conductivity test using conductivity meter.

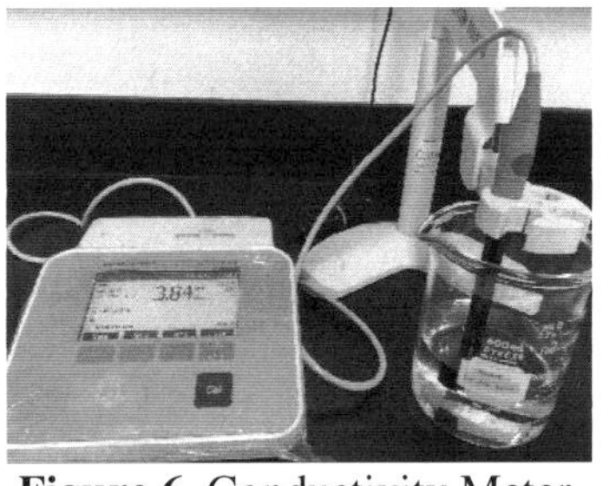

Figure 6. Conductivity Meter.

2. The electrodes are immersed into solution and press "Read".
3. Conductivity meter is start to measuring the solution and wait in few seconds until the reading is stable.
4. The measurement is taken five times and average the result.
5. Parameter conductivity and resistivity can be obtained at the same time. Parameter result is possibly changed by pressing "Mode" in the conductivity meter.

3. Result and Dicussion

There are several results that had been obtained from experiment such as correlation between density of brine and brine concentration towards resistivity, correlation between density of the mixture of MFOMAX and distilled water towards resistivity, correlation between density of the mixture of various MFOMAX concentration and various brine concentration towards resistivity, correlation between density of the mixture of various MFOMAX concentration, various brine concentration, and oil towards resistivity.

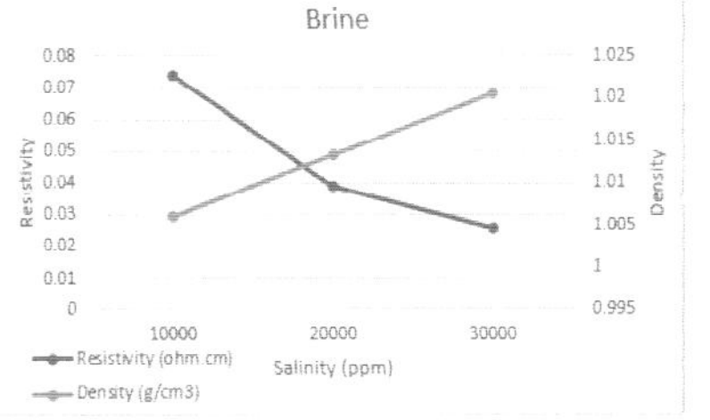

Figure 7. Correlation between Density of Brine and Brine Concentration towards Resistivity.

Normally, higher concentration of fluid, density will be higher. It occurred because of smaller size of molecules that was dissolved in water which was higher based on the equation (1). Theory of density and equation (1) is only applicable for density in various brine concentration. Density and resistivity of brine 1 wt% is 1.006 g/cm^3 and 0.074 Ωcm, brine 2 wt% is 1.0132 g/cm^3 and 0.0388 Ωcm, brine 3 wt%

IOP Conf. Series: Materials Science and Engineering **267** (2017) 012031 doi:10.1088/1757-899X/267/1/012031

is 1.0206 g/cm^3 and 0.0255 Ωcm. Based on the figure 7, correlation between density of brine and resistivity of brine is inversely proportional. As can be seen, NaCl has positive (+) and (-) negative charge. The higher concentration of NaCl that is dissolved in water, mass of the NaCl increase and positive (+) (-) negative charge which is carried out by NaCl is also increasing. However, when and positive (+) and (-) negative increase means conductive, in other words lower resistivity.

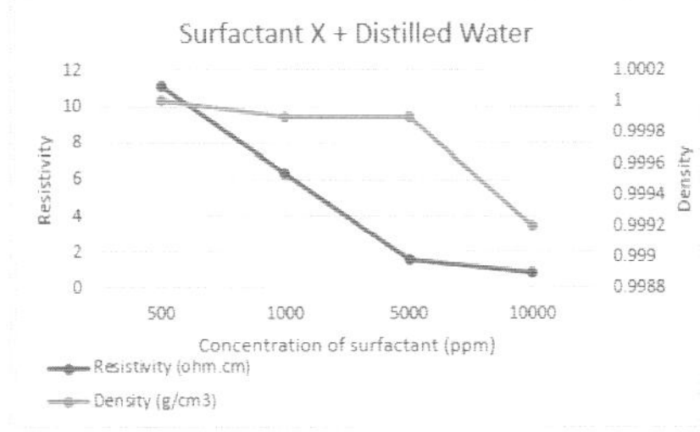

Figure 8. Correlation between Density of the mixture of MFOMAX and Distilled Water towards Resistivity.

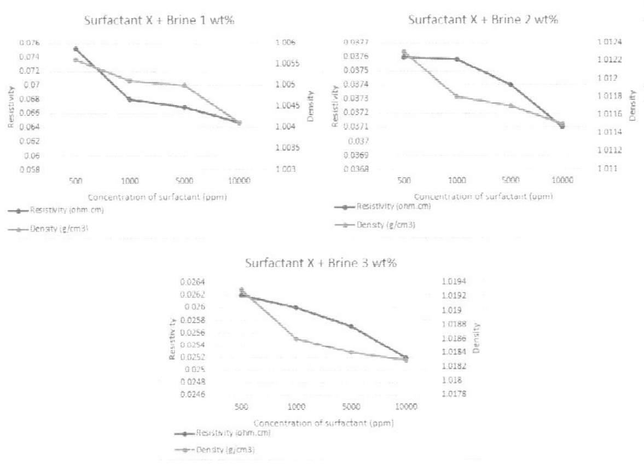

Figure 9. Correlation between Density of the mixture of Various MFOMAX Concentration and Various Brine Concentration towards Resistivity.

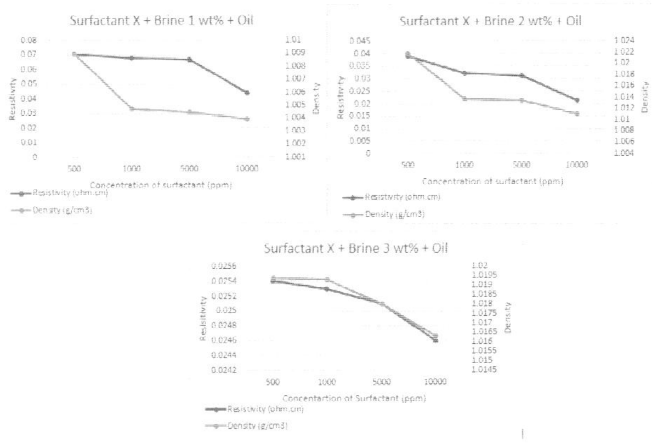

IOP Conf. Series: Materials Science and Engineering **267** (2017) 012031 doi:10.1088/1757-899X/267/1/012031

Figure 10. Correlation between Density of the mixture of Various MFOMAX Concentration, Various Brine Concentration, and Oil towards Resistivity.

When MFOMAX is added to the solution, theory of density and equation (1) is not applicable. Density tend to reduce while concentration of MFOMAX is increasing. It occurs because MFOMAX has bigger molecule in the same volume. Bigger molecules in solution will affect the density because bigger molecules will not be able to occupy the empty spaces between other molecules which derive from another fluid.

Based on the EDL theory, conductivity result occurs when positive (+) charge of the fluid and negative (-) charge of the ion of the electrodes is attracted each other or vice versa. Resistivity result occurs when positive (+) charge of the fluid and positive (+) charge of the ion of the electrodes is repelled each other or vice versa. All figures are expressed correlation between resistivity and fluid concentration. Generally, resistivity for all the solution is decreased. It means, all the liquid is conductive. In other words, all the solutions are capable to transmit the current and voltage. Resistivity between the mixture of MFOMAX with various brine concentration and the mixture of various MFOMAX concentration, various brine concentration, and oil is not much different. This issue occurs because when concentration increased, the amount of brine and MFOMAX also increased but the amount of brine that is added into solution is more dominant than the amount of MFOMAX and oil. Conductivity of solution depends on the concentration of all the ions present in the fluid, the greater concentrations, the greater conductivity. MFOMAX is the surfactant that has (+) and (-) charge, in other word MFOMAX can transmit current by itself without any conductive fluid added because MFOMAX is made from the mixture of anionic (-) and zuitorionic (+) and (-). Based on the mixture, MFOMAX has high conductivity.

Theoretically, there is no direct correlation between density and resistivity because density is related with atomic packing, mass and volume of the molecules inside the fluid and resistivity is related with positive (+) and negative (-) charge inside the fluid (electronic structure). Based on the result, the higher concentration of the fluid, density and resistivity will be reduced. It is possibly occurred because when density decrease, means it has bigger molecules that is not fully dissolved in water and reducing the resistivity due to the increasing positive (+) negative (-) charge that is carried out from the solution.

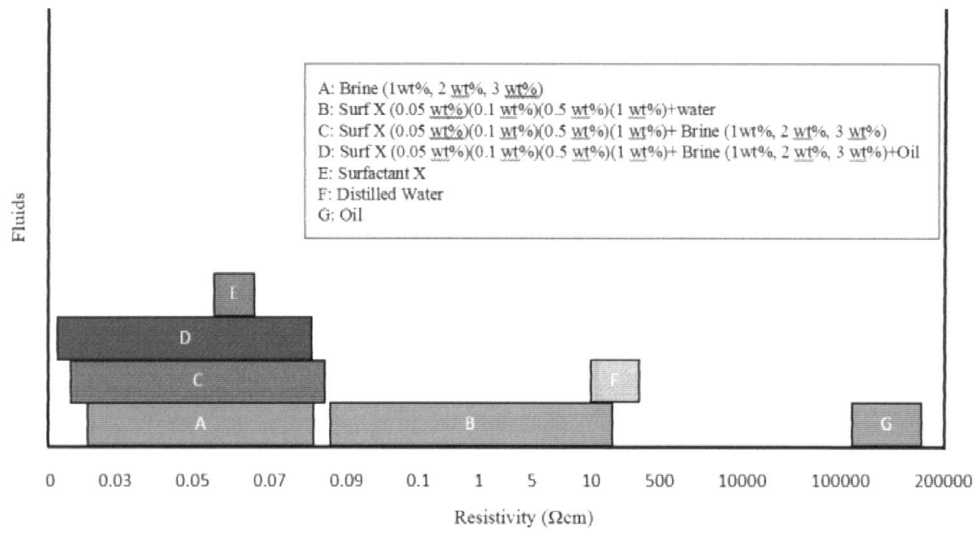

Figure 11. Range of Fluid Resistivity.

As can be seen in figure 11, the smallest resistivity is MFOMAX mix with brine and oil with various concentration. The highest resistivity is oil. Based on figure 11, range of resistivity is quite similar and the resistivity tend to lower. It occurs because the amount volume of brine is more dominant than the amount volume of MFOMAX and oil added in the solution.

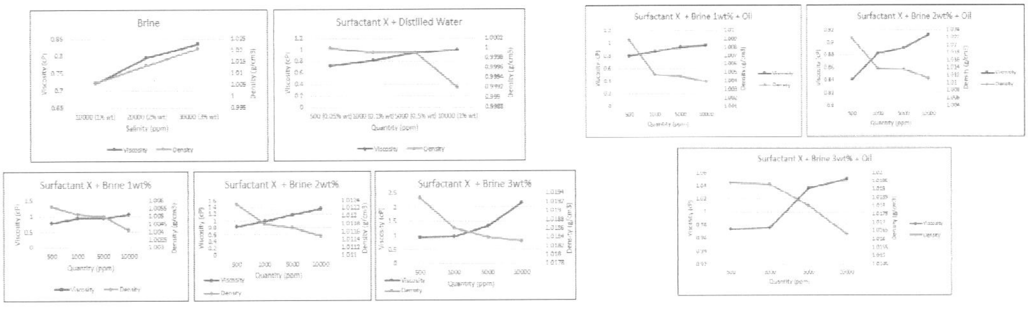

Figure 12. Correlation between Density and Viscosity.

Theoretically, greater number of fluid concentration, density and viscosity will be higher. This theory is only applicable for brine with various concentration. Correlation between density and viscosity towards brine concentration was proportional. Density and viscosity of brine 1 wt% was 1.006 g/cm^3 and 0.7213 cP, brine 2 wt% was 1.0132 g/cm^3 and 0.7954 cP, brine 3 wt% was 1.0206 g/cm^3 and 0.835 cP. For another solution, specifically solution which has MFOMAX, correlation between density and viscosity towards fluid concentration was inversely proportional.

4. Conclusion

In this experiment, several conclusions can be obtained based on the phenomena that occurred during the experiment. All the results can be concluded as stated below:

1. Correlation between density of brine and brine concentration towards resistivity was inversely proportional.
2. Correlation between density of fluid mix with MFOMAX and MFOMAX concentration towards resistivity was proportional.

Acknowledgment

The authors would like to thank to UTP-DELFT-SHELL-PETRONAS for the funding and also for Petroleum Department – Universiti Teknologi PETRONAS for assistance in this study.

5. References

[1] Sunmonu, R. M., Onyekonwu, M. (2013). Enhanced Oil Recovery Using Foam Injection; A Mechanistic Approach. SPE 167589.

[2] Schramm. L. L, Turta. A. T, Novosad. J. J. (1993). Microvisual and Core flood Studies of Foam Interaction with a Light Crude Oil. SPERE, 201.

[3] Schramm, L. L. (1994). Foams: Fundamentals and Applications in the Petroleum Industry. American Chemical Society: Washington, DC. doi :10.102

[4] Collin, M. (2014). Core Analysis: A Best Practice Guide, Malaysia.

[5] Jones, F. E., Harris, G. L., (1992). ITS-90 Density of Water Formulation for Volumetric Standards Calibration. Journal of Research of the national Institute of Standards and Technology, Vol 97.

[6] Lecture Note. The concept of Viscosity. University of Columbia.

[7] Glover, P. W. J. (2000). Petrophysics MSc Course Notes. UK: University of Aberdeen.

IOP Conf. Series: Materials Science and Engineering **267** (2017) 012032 doi:10.1088/1757-899X/267/1/012032

Discharge Fluctuation Effect on Meandering River Bed Evolution

Kuntjoro[1]
[1]Civil Infrastructure Engineering Department, Vocational Faculty of Institut Teknologi Sepuluh Nopember Surabaya
E-mail: kuntjoro_rivers@yahoo.co.id

Ismail Saud[2], Didik Harijanto[2]
[2]Civil Infrastructure Engineering Department, Vocational Faculty of Institut Teknologi Sepuluh Nopember Surabaya

Abstract. This research was based on some considerations: first discharge fluctuation argued that none rivers with constant discharge and second meandering river bed evolution with considering that none of rivers in a steady state without bed change. This research developed to get formulation the relationship between fluctuations discharge with the evolution of the bottom of river considering discharge, parameter rivers and parameter sediment. In the span of daily discharge data 1997-2011 and cross section monitoring annual results 1997-2011 evolution of bottom of a river subjects obtained: Formula 1 is the relationship between discharge fluctuations with rate of sedimentation (*S*) and Formula 2 is the relationship between discharge fluctuations with rate of erosion (*E*). Thus formula have higher prediction accuracy than other published formulas and it is applicable to predict Brantas River bed evolution approximate with the real conditions. Further analysis from the output KUN-QArSHOV formula produces: Erosion equation $\Delta S = 25{,}167e^{0{,}0034\,\Delta Q}$, on constant discharge, sedimentation value occur is $25.167.10^{-5}$ meter. Sedimentation equation $\Delta E = 8{,}3455e^{0{,}0075\,\Delta Q}$, on constant discharge, erosion value occure is $8.3455.10^{-5}$ meter. Critical point between sedimentation and erosion happened on discharge between 269 m^3/second and 270 m^3/second

1. Introduction

No constant discharge in the real river conditions. In the notes of discharge rivers resulting from automatic water level a recorder (AWLR), river discharge is always changing. It means there are always fluctuations in discharge in all the time. Erosion and sedimentation in the river depend on the discharge in the river. Thus discharge fluctuations will have an influence to the formation and change the bottom of a river.

In general, arch is undesirable in open channels, because they friction would increase loss and lead to danger due to levee of the river by erosion and sliding caused helical flow [1],[2]. In erodible channels, the action of helical flow will develop a configuration in the bed [4]. In an alluvial bend it seems that the configuration of the channel cross section is defined more or less in accordance with certain natural laws [18], [19]. The form of bottom in arch river cross section deeper approaching cliff beyond according to Ripley and V.T. Chow [4]. The river bottom evolution in this research was

[1] To whom any correspondence should be addressed.

erosion and sedimentation quantities analysed based on the discharge quantity Q that happen all the time. Thus obtained change of bottom vertical (ΔV) and horizontal (ΔH) in any change of discharge (ΔQ).

2. Form of River Bottom in Arch of the River

The form of bottom in arch river cross section were deeper approached by cliff beyond according to Ripley and V.T. Chow expressed in Figure 1[4]

$$y = 6{,}35D\left(\sqrt{0{,}437 - \frac{x^2}{T^2}} - 0{,}433\right)\left(1 + \frac{xK}{r_o}\right) \qquad (1)$$

Where:
y = river depth (feet)
x = abscise (feet)
D = hydraulic depth (feet)
r_o = outer arc radius (feet)
K = 17,52 (Chow, 1959)

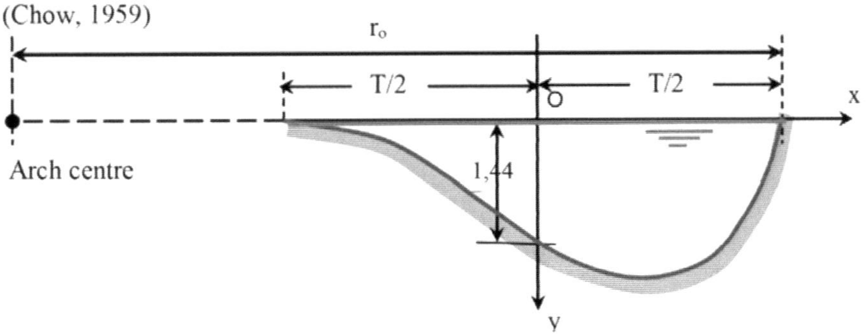

Figure 1. The form of river cross section in arch condition.

3. Meandering River Evolution and Method for Predicting Scouring and Sedimentation in Meandering River

3.1. Meandering River Evolution

Meander evolution depends on initial geometric condition and upstream meanders that was reviewed. Thus can be estimated the movement of meanders evolution direction [4], some possible direction of meander evolution as shown in Figure 2. [6],[7],[9]

Type A describe evolution meanders with low amplitude with the rate of evolution of sluggish. Type B is meander evolution that occurs along flood plains in the narrow flow of river. Type C is meander evolution that occurs in unstable levee of the river. Type D is occurs on highly meandering stream, when the meandering bend becomes too large, secondary meanders were created along the existing loop. Type E is quite similar to Mode D, but cut off is expected in this type. Type F and G were occured along locally braided sinuous or meandering stream. [8],[10]

IOP Conf. Series: Materials Science and Engineering **267** (2017) 012032 doi:10.1088/1757-899X/267/1/012032

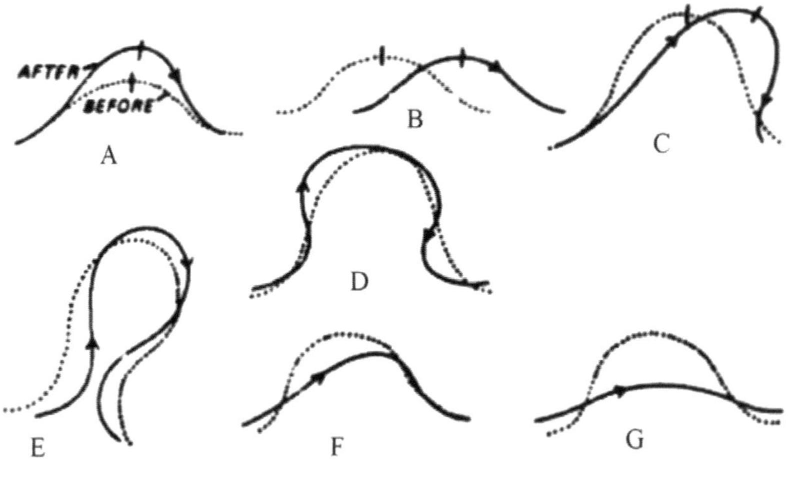

Figure 2. Type of Meander Loop Evolution [3]

3.2. Predicting Scouring and Sedimentation in Meandering River

River meandering geometry changes is a natural process [12], for long time will change the river geometry [13]. Erosion and Sedimentation in Meandering River position are illustrated in Figure 3. then predicted by KUN-QArSHOV equation [14], [15], [16].

Erosion in toe of bank of the river cause sliding and other process in the river [17], these can be harmful to the river bank and hydraulic structure near closely with the river.

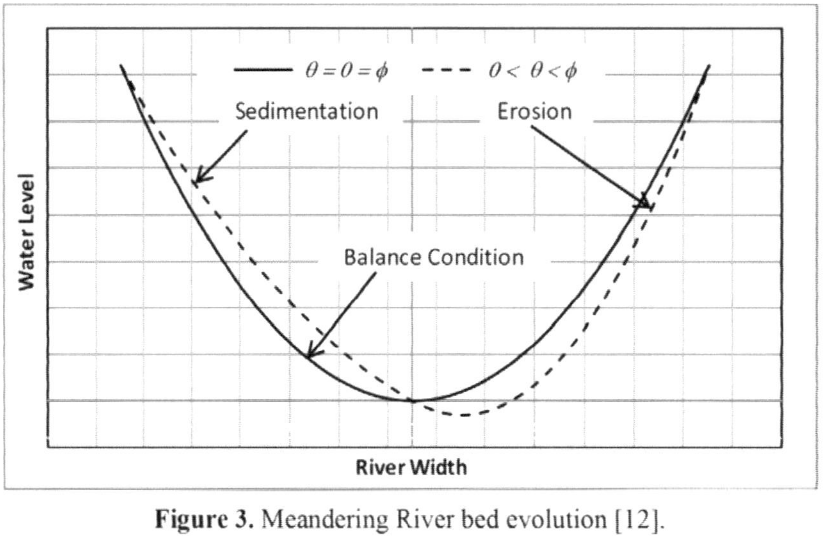

Figure 3. Meandering River bed evolution [12].

4. Study area and data collection

4.1. Meandering and Geometry of Brantas River

Plan form of Brantas River looked from Google Earth shown in Figure 4. And cross section in two points of observation show in Figure 5 and Figure 6.

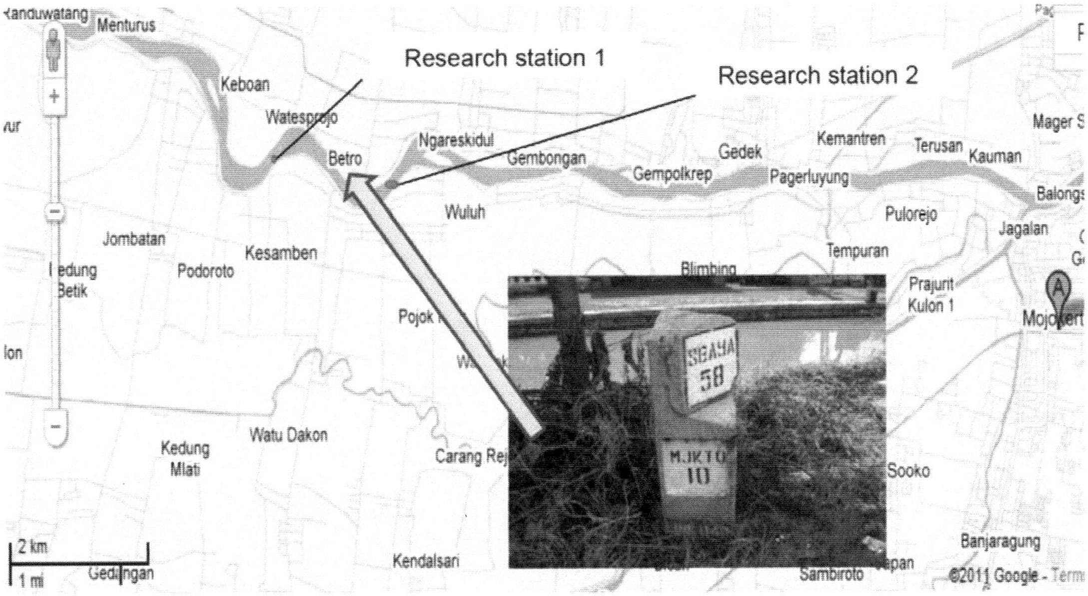

Figure 4. Two points of observation in Meandering of Brantas River.[12],[16]

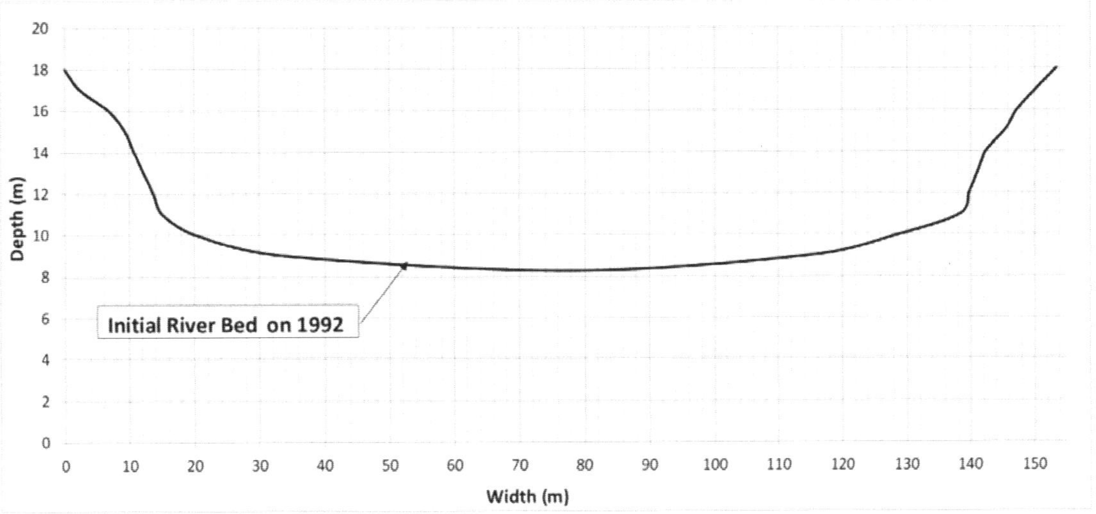

Figure 5. Cross Section of the river in Research Point 1.

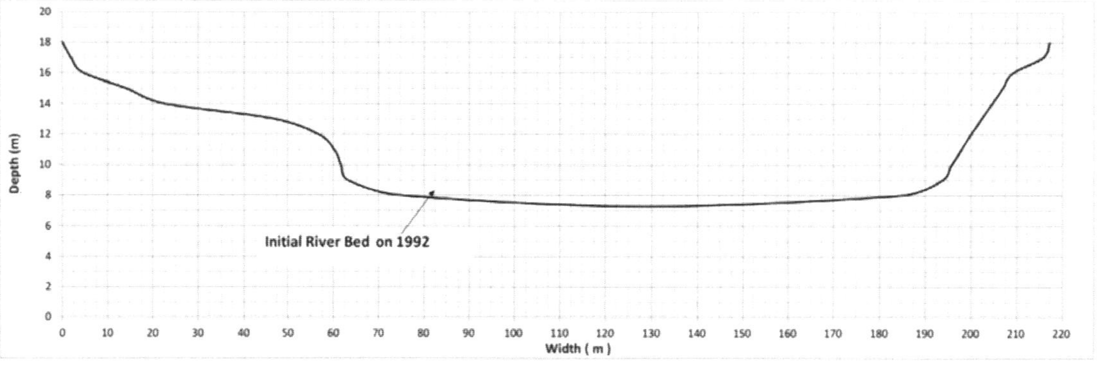

Figure 6. Cross Section of the river in Research Point 2.

4.2. *River Meander parameter*

River meander parameter on research point 1 and point 2 were explained in Table 1.

Table 1. River meander parameter

Research Station	River Meander Parameter					
	$r_c (m)$	λ (m)	a (m)	θ (°)	ϕ	W (m)
1	490	1900	395	41	120	292
2	1462	1809	261	8	61	179

4.3. *Discharge Fluctuation of Brantas River*

Discharge fluctuation in Brantas River from automatic water level recoded nearby research station 1 is arranged in Figure 7 to Figure 10

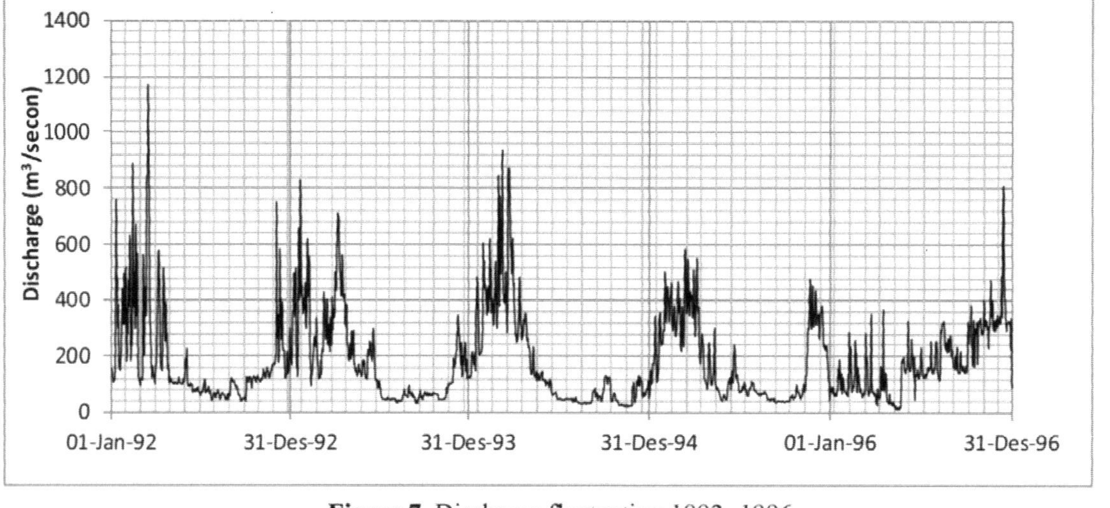

Figure 7. Discharge fluctuation 1992 -1996.

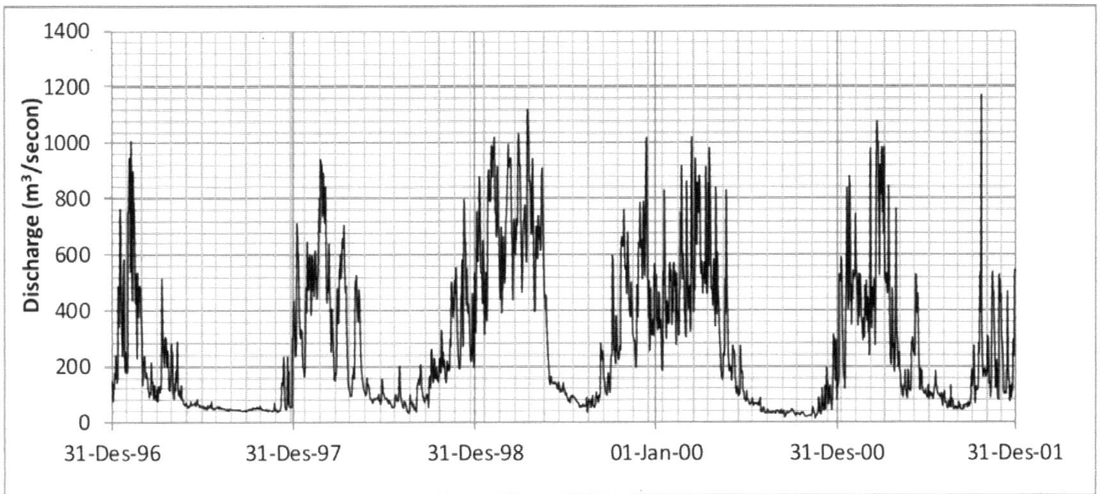

Figure 8. Discharge fluctuation 1997 -2001.

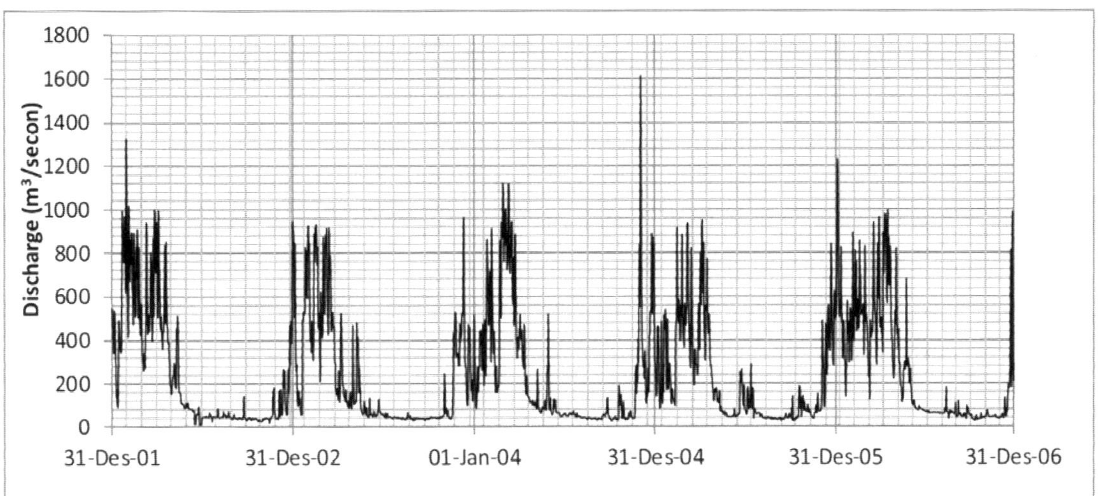

Figure 9. Discharge fluctuation 2002-2006.

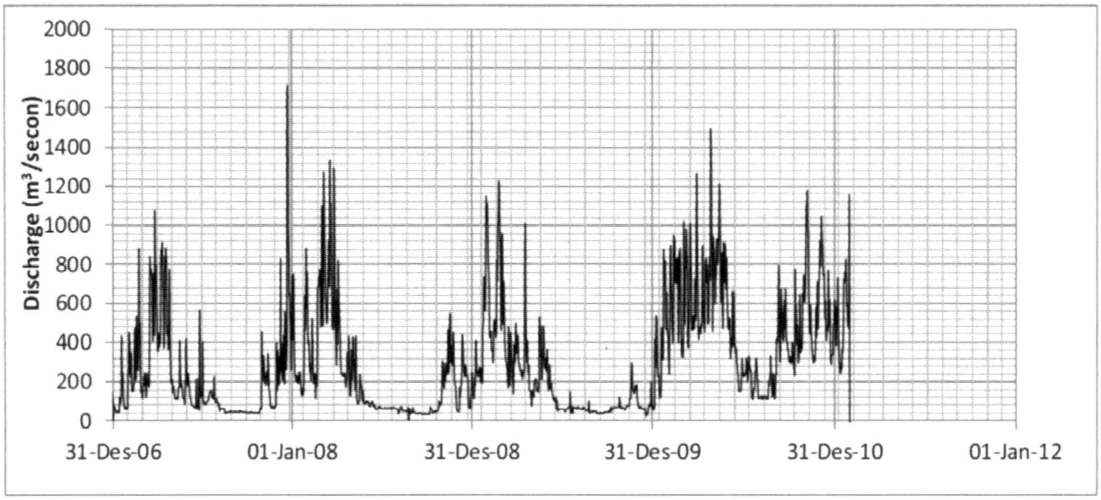

Figure 10. Discharge fluctuation 200-2010.

5. Result of river bed evolution from KUN-QArSHOV simulation methods

Result of river bed evolution expressed in the relationship curve between δQ with ΔH and ΔV from KUN-QARSHOV simulation methods with data input discharged in 1992-1996, 1997-2001, 2002-2006 and 2007-2010.[12],

5.1. The results of river bed evolution by KUN-QArSHOV simulation in research station 1

Output from KUN-QArSHOV simulation is the relationship chart between δQ with ΔH and ΔV. The relationship chart between ΔQ with ΔH and ΔV for research station 1 can be examined in Figure 11 to Figure 15.

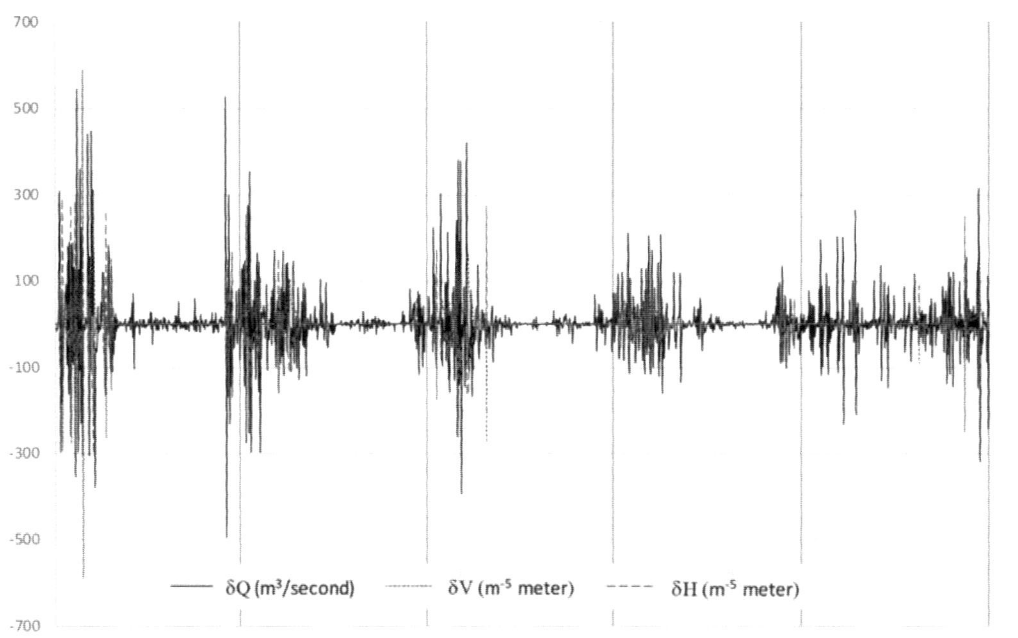

Figure 11. Discharge Fluctuation Effect on the Meandering River Bed Evolution in Research station 1 along 1992 to 1996.

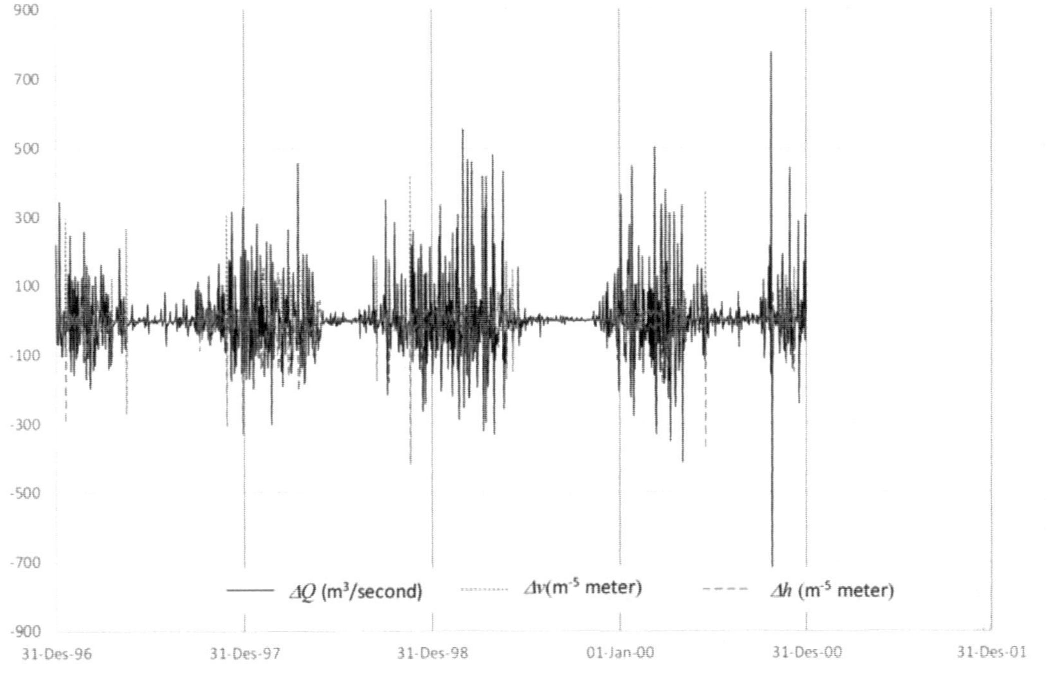

Figure 12. Discharge Fluctuation Effect on the Meandering River Bed Evolution in Research station 1 along 1997-2001.

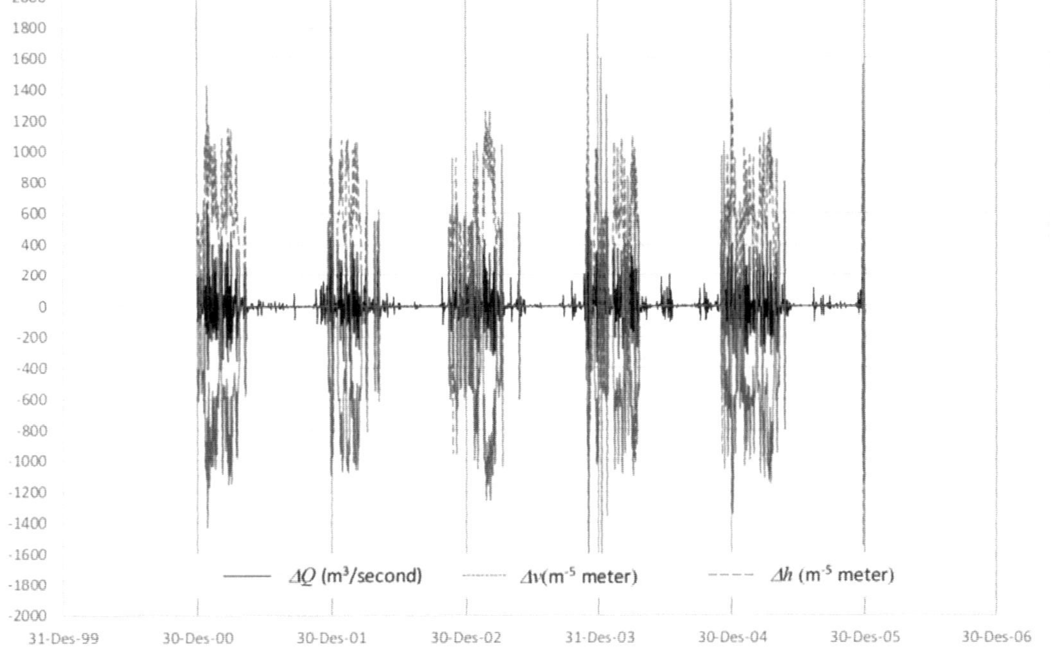

Figure 13. Discharge Fluctuation Effect on the Meandering River Bed Evolution in Research station 1 along 2002-2005.

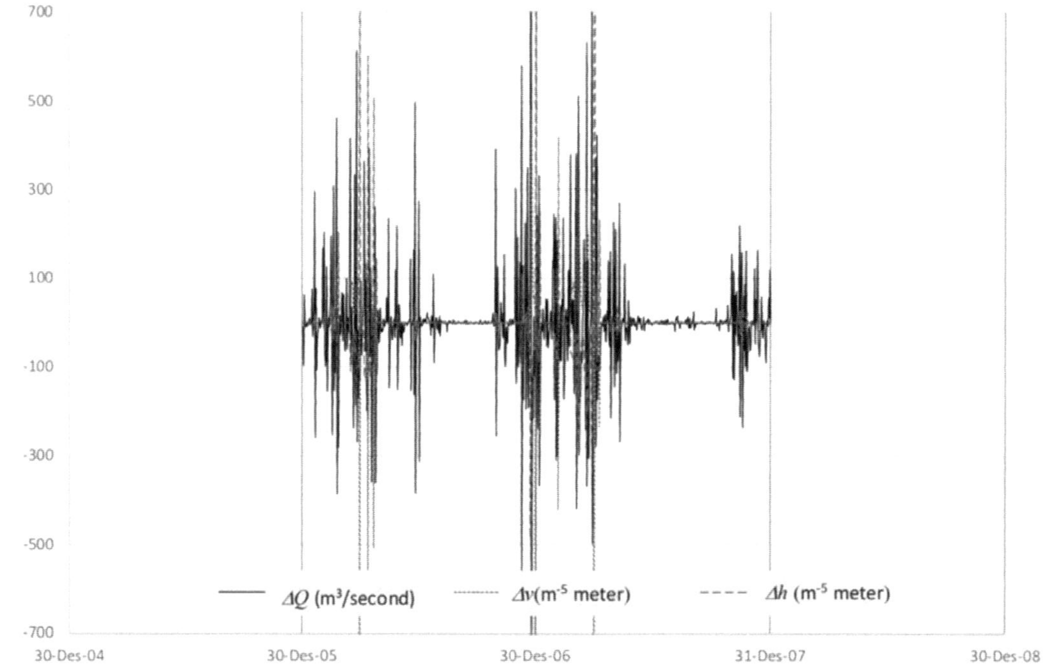

Figure 14. Discharge Fluctuation Effect on the Meandering River Bed Evolution in Research station 1 along 2006-2007.

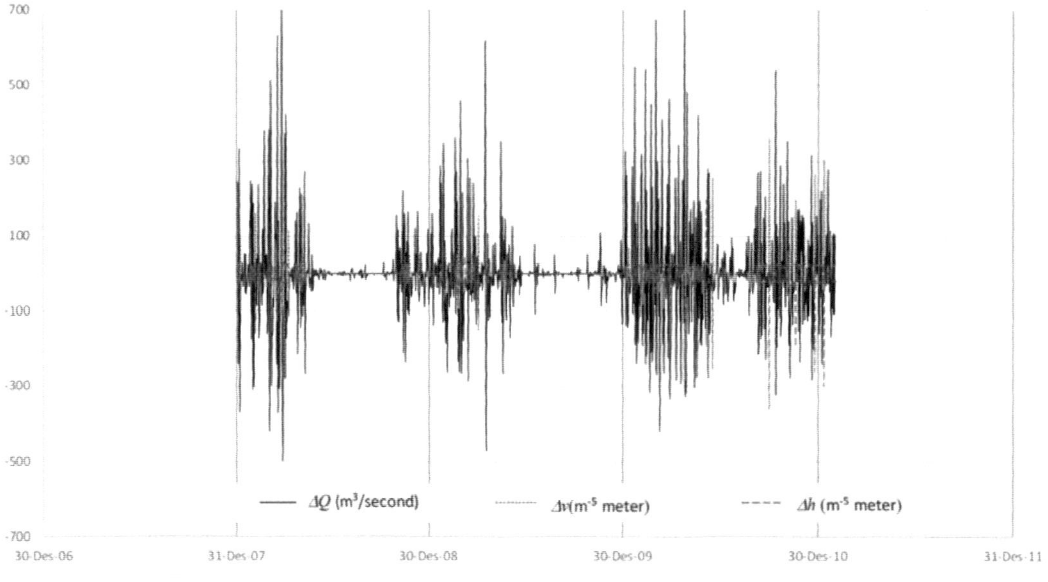

Figure 15. Discharge Fluctuation Effect on the Meandering River Bed Evolution in Research station 1 along 2008 to 2010.

5.2. The results of river bed evolution by KUN-QARSHOV simulation in research station 2

The relationship curve between ΔQ with ΔH and ΔV for research station 2 can be examined in Figure 16 to 19.

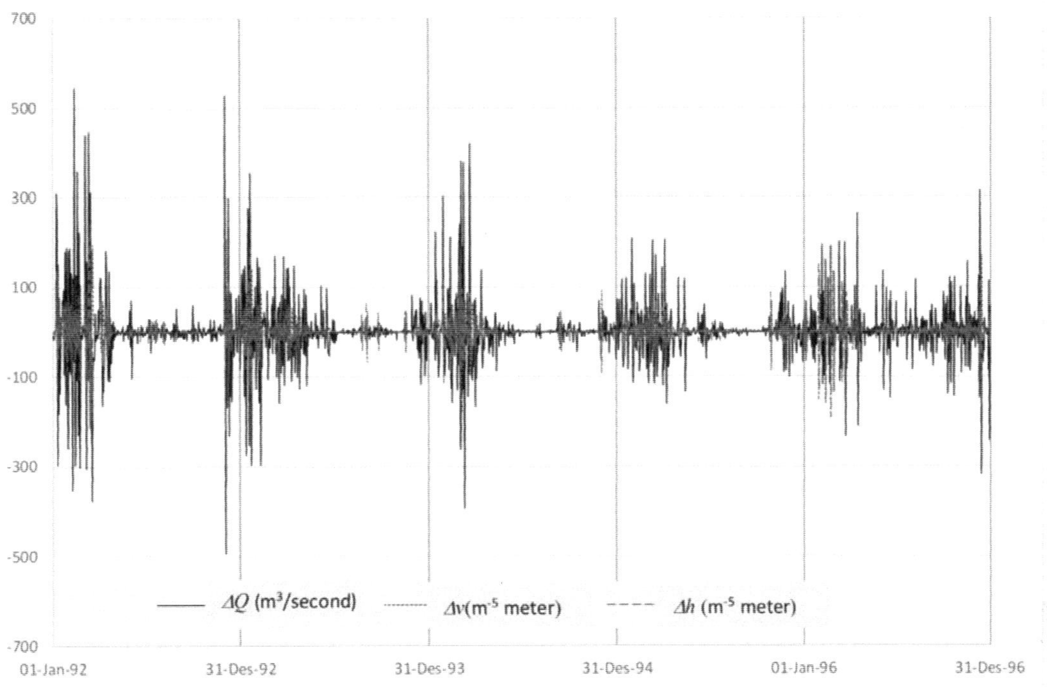

Figure 16. Discharge Fluctuation Effect on the Meandering River Bed Evolution in Research station 2 along 1992 to 1996.

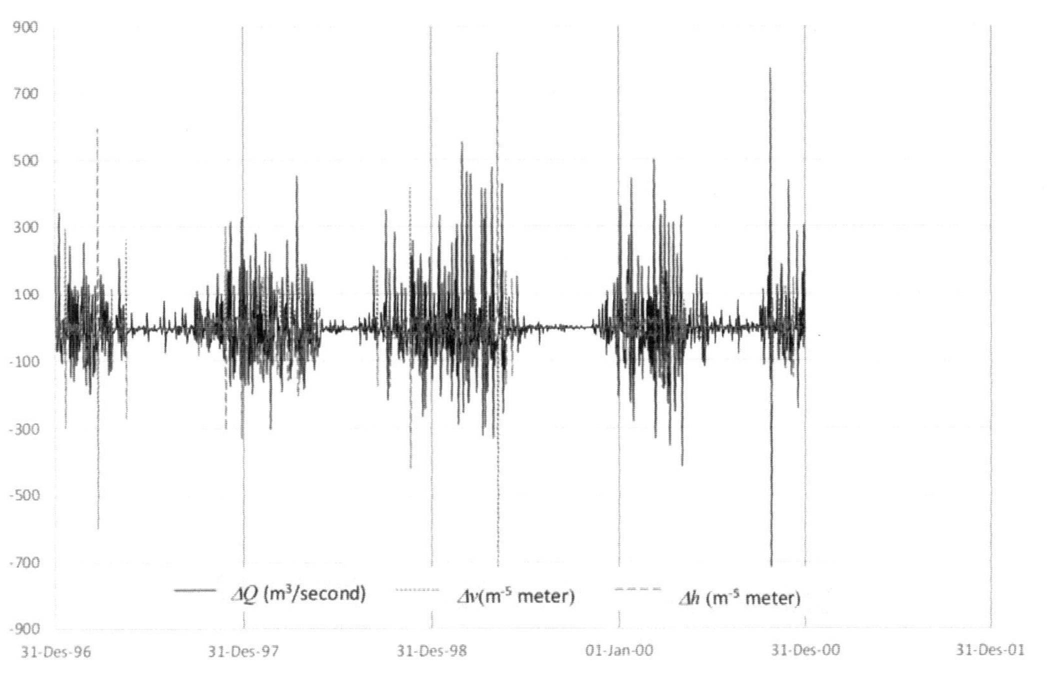

Figure 17. Discharge Fluctuation Effect on the Meandering River Bed Evolution in Research station 2 along 1997-2001.

IOP Conf. Series: Materials Science and Engineering **267** (2017) 012032 doi:10.1088/1757-899X/267/1/012032

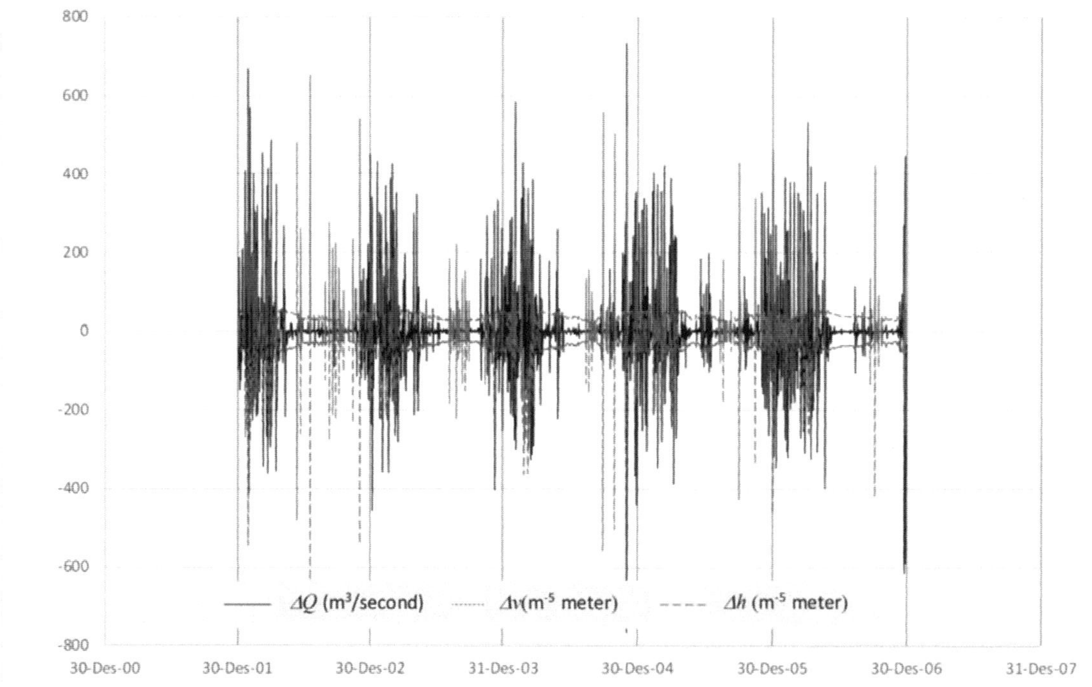

Figure 18. Discharge Fluctuation Effect on the Meandering River Bed Evolution in Research station 2 along 2002-2007.

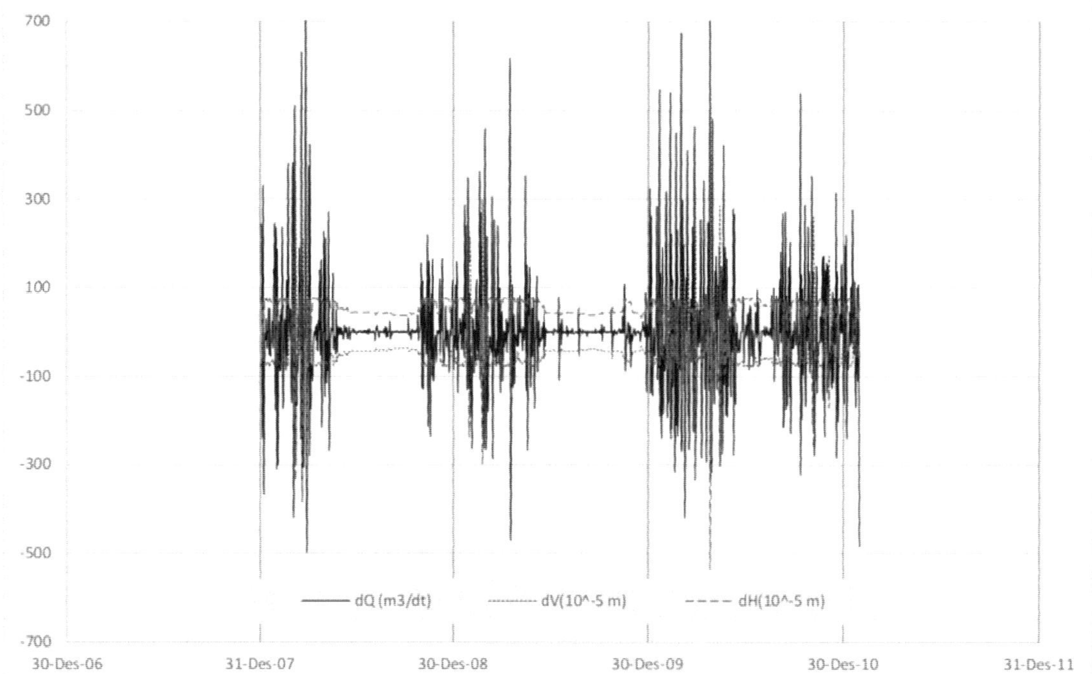

Figure 19. Discharge Fluctuation Effect on the Meandering River Bed Evolution in Research station 2 along 2008 to 2011.

6. Discharge fluctuation and meandering river evolution relationship

6.1 River bed sedimentation in river evolution

Evolution of the meandering river from sedimentation was analyzed on output of method KUN-QArSHOV modeling to get a relation between fluctuations discharge value ΔQ with the value of the sediment ΔS. The analysis results are shown in Figure 20.

Relationship between ΔQ with ΔS was satisfied to the erosion equation $\Delta S = 26.154e^{0,0033\,\Delta Q}$, where ΔS is sedimetation value and ΔQ discharge value. In constant discharge or $\Delta Q = 0$ occured sedimentation value was $26.154.10^{-5}$ meter.

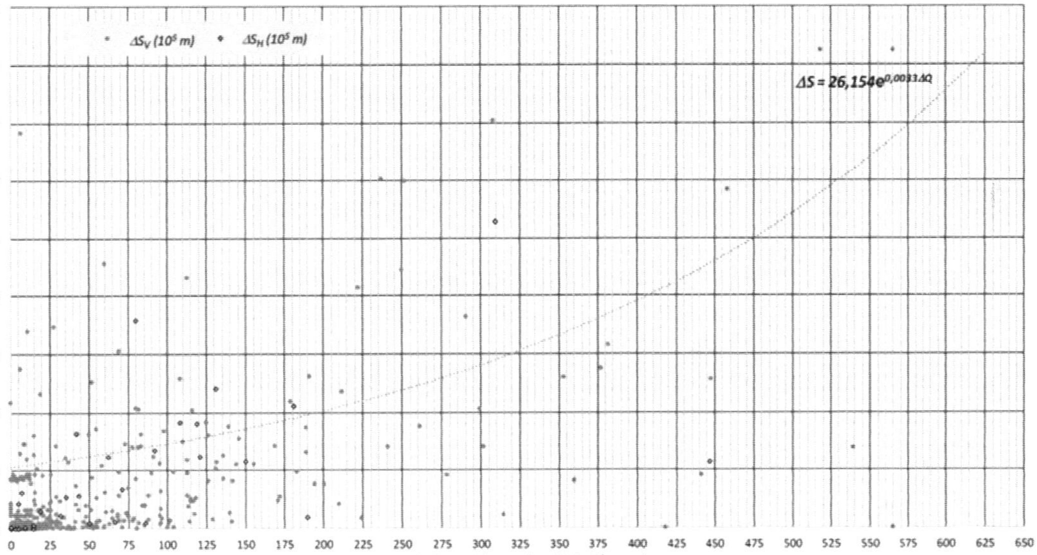

Figure 20. River bed sedimentation in river evolution ΔQ with ΔS.

6.2 River bed erosion in river evolution

Evolution of the meandering river from erosion is analyzed on output of method KUN-QArSHOV modeling to get a relation between fluctuations discharge value ΔQ with the value of the erosion ΔE. The analysis results are shown in Figure 21.

Relationship between ΔQ with ΔE was satisfied to the sedimentation equation $\Delta E = 8,3455e^{0,0075\,\Delta Q}$ where ΔE is erosion value and ΔQ discharge fluctuation value. On constant discharge, occured erosion value was $8.3455.10^{-5}$ meter

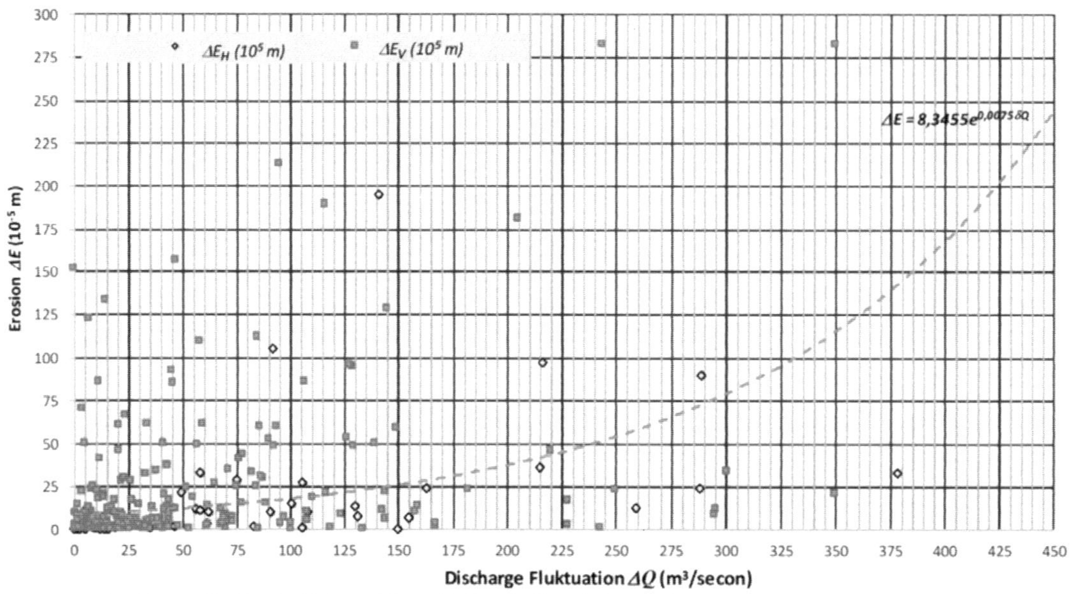

Figure 21. River bed erosion in river evolution ΔQ with ΔE.

6.3 *Simultaneous Erosion and Sedimentation in River Meandering Evolution*

Evolution of the bottom of meandering river is the difference between sedimentation value and erosion value in the same discharge and the right time.

Charts of the relationship between ΔQ with ΔE, and ΔQ with ΔS simultaneously was shown in Figure 22. This chart is observable the condition of a critical point between sedimentation and erosion happened on discharge between 269 m³/second and 270 m³/second, shown in Table 2

Table 2. Critical point between sedimentation and erosion

ΔQ (m³/second)	Sedimentation (10^{-5} m)	Erosion (10^{-5} m)
267	62.39	61.82
268	62.60	62.29
269	62.81	62.75
270	63.03	63.23
271	63.24	63.70
272	63.46	64.18
273	63.67	64.67

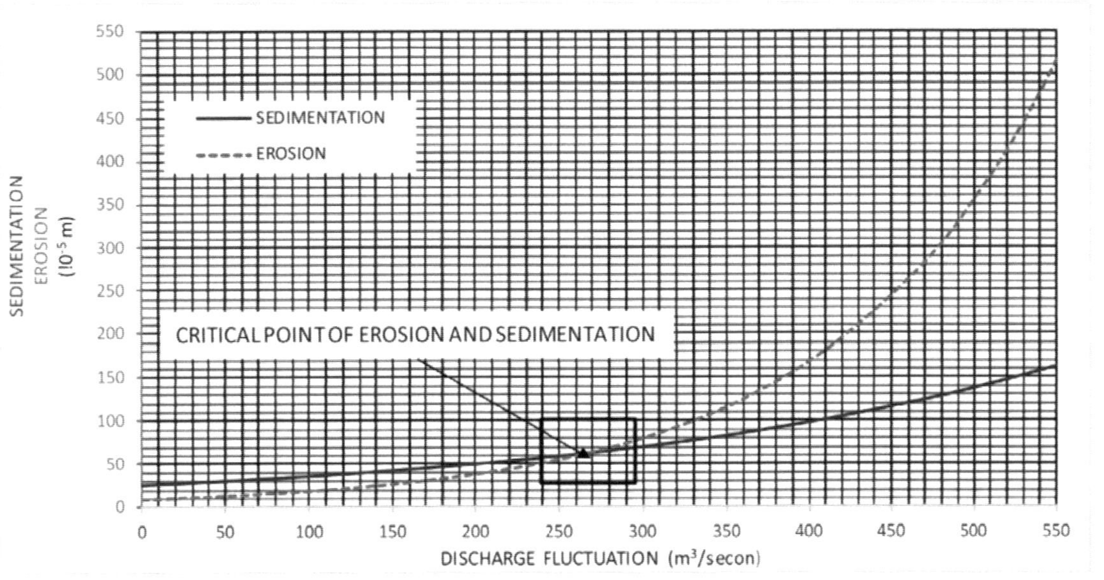

Figure 22. Charts of the relationship between ΔQ with ΔE, and ΔQ with ΔS simultaneously

6.4 Result of river bed evolution in research station 1

Result of river bed evolution in research station 1 is expressed in Figure 23

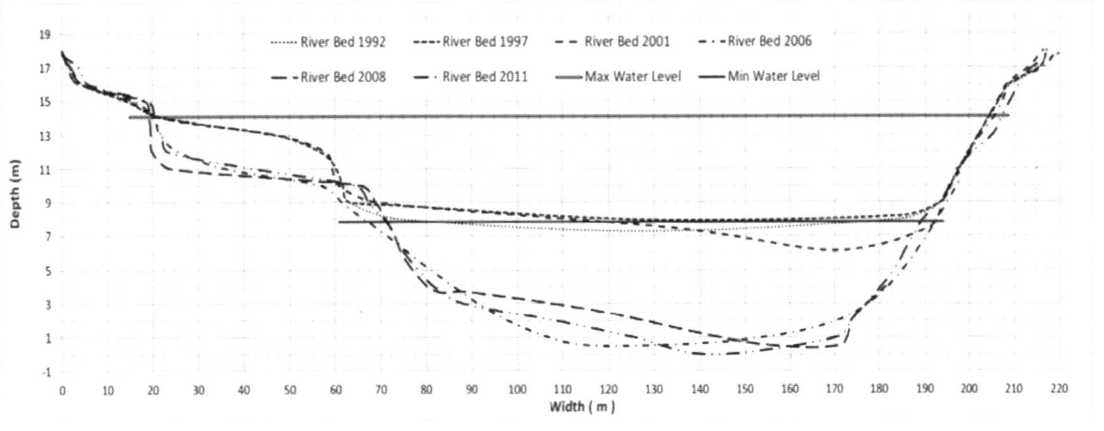

Figure 23. River bed evolution along 1992 to 2011 in research station 1.

6.5 Result of river bed evolution in research station 2

Result of river bed evolution in research station 2 is expressed in Figure 24

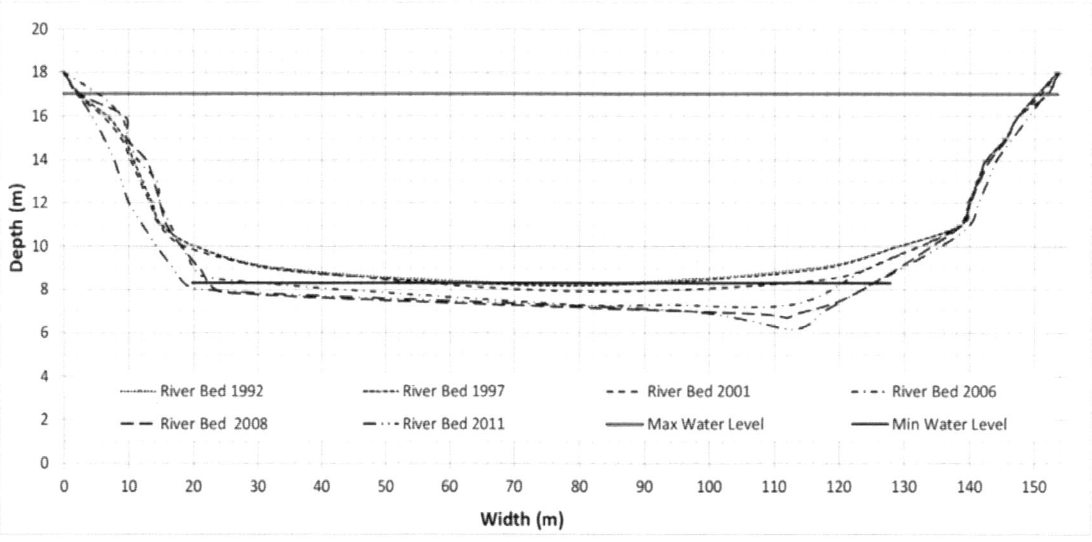

Figure 24. River bed evolution along 1992 to 2011 in research station 2.

7. Conclusions

- Erosion and sedimentation in the river happened simultaneously in one discharge value.
- Evolution of the bottom of the river is simultaneous between sedimentation and erosion, expressed in the relationship chart between ΔQ with ΔS, and ΔQ with ΔE, these chart are formulated to the equations as follows:
  - Erosion equation $\Delta S = 25,167e^{0,0034\,\Delta Q}$
  - On constant discharge, sedimentation value was $25.167.10^{-5}$ meter
  - Sedimentation equation $\Delta E = 8,3455e^{0,0075\,\Delta Q}$
  - On constant discharge, erosion value was $8.3455.10^{-5}$ meter
- Critical point between sedimentation and erosion happened on discharge between 269 m^3/second and 270 m^3/second

8. Advice

It needed annual cross section measurement in similar points for synchronizing data from Perum Jasa Tirta I measurements to measurement data that will be input in next research.

9. Acknowledgement

to officials and staff Perum Jasa Tirta I in Malang and the Balai Besar Wilayah Sungai Brantas in Surabaya over facilities, generosity, services and contributions data which is particularly beneficial for composing this research.

10. References

[1] Abed, J., and Garcia, M. H. 2006 *"RVR Meander:* A Toolbox for Re-meandering of Channelized Streams." Computers & Geosciences, 32, 92-101.

[2] A. J. Odgaard, 1987 *Streambank Erosion along Two Rivers in Iowa, in Water Resources Research,* 23(7), pp. 1225-1236.

[3] Brice, J.C. (1977). "*Lateral Migration of the Middle Sacramento River, California.*" *U.S. Geological Survey,* Water-Resources Investigations, 77-43.

[4] Chow, V. T.1959, *Open Channel Hdraulics*, McGraw-Hill Kogakusha, LTD

[5] D. M. Keady and M. S. Priest, 1977 *The Downstream Migration Rate of River Meandering Patterns,*" In Proceedings, Mississippi Water Resources Conference, Meeting 12th Mississippi Water Resources Conference, Jackson, MS, pp. 29-34.

[6] Duan, J. G. 2001 *Simulation of Streambank Erosion Processes with a Two-Dimensional Numerical Model. In Landscape Erosion and Evolution Modelling*, Harmon RS, Doe WW III (eds). Kluwer Academic/Plenum Publishers: New York; 389–427.

[7] Duan, J. G. and Pierre Y. Julien, 2005. *Numerical Simulation of the Inception of Channel Meandering, Earth Surf. Process.* Landforms 30, 1093–1110 (2005), Published online in Wiley Inter Science (www.interscience.wiley.com). DOI: 10.1002/esp.1264

[8] D. S. Biedenharn, P. G.Combs, G. J.Hill, Jr.Pinkard, C. F., and C. B. Pinkston, 1989 "*Relationship between Channel Migration and Radius of Curvature on the Red River," in Proc., Int. Symp. on Sediment Transport Modeling*, New Orleans, LO, pp.536-541.

[9] J. L. Briaud, F. Ting, H. C. Chen, Y. Cao, S. W. Han, and K. W. Kwak, 2001 *Erosion Function Apparatus for Scour Rate Predictions,* IEEE Trans.Journal of Geotechnical and Geoenvironmental Engineering, 127(2), pp. 105-113.

[10] J. M.Hooke, 1980 *Magnitude and Distribution of Rates of River Bank Erosion*, in Earth Surface Processes, 5(2), pp. 143-157.

[11] P. Ph. Jansen, L. van Bendegom, J van de Berg, M de Vries, and A. Zanen, 1979 "*Principles of Rover Engineering*", in Pitman Publishing Limited ISBN 0-273-01139-1, First Published.

[12] Kuntjoro, M. Bisri, A. Masrevaniah, A. Suharyanto, 2012.a,, "*Empirical Model Of River Meandering Geometry Changes Due To Discharge Fluntuation*" Journal of Basic and Applied Scientific Research, TextRoad Publication , J. Basic. Appl. Sci. Res., 2(2)1027-1033, 2012

[13] Kuntjoro, M. Bisri, A. Masrevaniah, A. Suharyanto, 2012.b., "*Modeling of discharge fluctuation influence on river meandering geometry change*". International Journal of Academic Research Part A; 2012; 4(6), 189-196. DOI: 10.7813/2075-4124.2012/4-6/A.26. Vol. 4. No. 6. November, 2012

[14] Kuntjoro, M. Bisri, A. Masrevaniah, A. Suharyanto, 2012.c., "*Empirical Model of Meandering River Geometry Change Under Discharge Fluctuation*", Doctor Dissertation, Civil Engineering Deartment Brawijaya University, Indonesia.

[15] Kuntjoro, M. Bisri, A. Masrevaniah, A. Suharyanto, 2013.a, *River Bed Erosion and Sedimentation Behavior in Discharge Fluctuation Condition*, Middle-East Journal of Scientific Research 18 (12): 1851-1858, 2013 ISSN 1990-9233

[16] Kuntjoro, Choirul Anwar, Pudiastuti, Didik Harijanto, Sungkono, 2013.b, *Inisiasi Perkiraan Arah Pergerakan Alur Sungai*, Jurnal APLIKASI ISSN.1907-753X, Volume 11, Nomor 2, Agustus 2013

[17] Kuntjoro, M. Bisri, A. Suharyanto, A. Masrevaniah, 2014, "*Impacts of River Groove Propagation on Irrigation Infrastructural Failures*", IPTEK, Journal of Proceeding Series, Vol. 1, 2014 (eISSN: 2354-6026)

[18] P. F. Hudson and R. H. Kesel, , 2000 "*Channel Migration and Meander-bend Curvature in the Lower Mississippi River Prior to Major Human Modification*," in Geology, 28(6), pp. 531-534.

[19] S. Ikeda, G. Parker, and K. Sawi, 1981 "*Bend Theory of River Meanders. I: Linear Development*," IEEE Trans Journal of Fluid Mechanics, 112, pp. 363-377.

IOP Conf. Series: Materials Science and Engineering **267** (2017) 012033 doi:10.1088/1757-899X/267/1/012033

An Evaluation of the Success Rate of Sermo Dam Management in Daerah Istimewa Yogyakarta

A Andriawan [1], Sobriyah [2], C Ikhsan [2]

[1] Student of Civil Engineering Master Program, Sebelas Maret University,

2 Lecturer, Civil Engineering Master Program, Sebelas Maret University

Email: aan.andriawan@gmail.com

Abstract. In dam operating and maintaining activities, there are some activities becoming the main function: the assessment of dam condition to keep monitoring and safeguarding the condition of dam as the main building. To achieve the maximum service, the maximal dam management is required as well and it should be followed with management evaluation. This case study was taken place in Sermo Dam of Daerah Istimewa Yogyakarta during 2015-2017. The method applied in this study was descriptive quantitative one, conducting a research using primary and secondary data. In this research, the assessment of dam condition was viewed from 1 (one) component, dam body, so that the component weight was 100%. The value of dam body condition was obtained from data of Sermo Dam monitoring in 2015-2016 and from the result of field survey in 2017. The result of research showed that the condition values of Sermo Dam with dam body component were 92.66% in 2015, 92.99% in 2016, and 93.99% in 2017. The result also showed that the value of dam body condition tended to increase during 2015-2017. To maintain the condition, the maximal operation and maintenance of dam was recommended.

1. Introduction

The news of Situ Gintung (Gintung Lake) collapsed on March 27, 2009 at 04.30 Local Time with more than 100 life tolls has stricken all of us. Because of its low dam/embankment, in addition to have large benefit, it stored a great destruction power. The incidence of Situ Gintung breakage was a failure in Dam Operating and Maintaining activities. In Dam operating and Maintaining activities, there are some activities becoming the main function: the assessment of dam condition to keep monitoring and safeguarding the condition of dam as the main building. Sermo dam (Waduk Sermo) has been beginning to operate since October 1996, with (gross) load of 25 millions m^3 at +136.6 m elevation (maximum normal water surface elevation) consisting of 21.9 millions m^3 as water load volume, 3.1 millions m^3 as dead storage volume, and dam puddle width of 1.57 km^2. The reservoir's catchment area width is 21.47 km^2. A big inspection has been conducted in 1996, 2004, and 2015, and Dam Monitoring Report is made annually. Dam management should be evaluated by considering the condition value of dam over years. This research is intended to find out whether or not the dam management tends to provide increasing, common or decreasing condition value in 2015-2017.

2. Methodology

The objective of dam management is to prevent the risk resulting from dam damage (Herlen, 2009). The management of PB. SOEDIRMAN, according to Public Work Minister's Regulation No. 72/PRT/1997, is to conduct geotechnic monitoring, dam instrumentation monitoring, survey monitoring, hydrology and climatology monitoring, dam sedimentation monitoring, and environment and water quality monitoring. . Marzulian Ramli and Bambang E. Yuwono, (2015) in their study also suggested that to minimize the failure of dam operation, there should be a maximal management. Yulius Heryant, et al (2014) stated that in operating and maintaining dam, there should be an assessment on dam as included in the Guidelines of Physical Dam Condition Assessment referring to the Government Regulation Number 37 of 2010 about Dam.

2.1 The Assessment of Dam Condition

The Guidelines of Physical Dam and Reservoir Condition Assessment issued or published by *Balai Bendungan* (Dam House) in 2010 governs the procedure of assessing Dam and its Reservoir. In maintaining dam correctly and well, there should be an assessment on dam as included in the Guidelines of Physical Dam Condition Assessment referring to the Government Regulation Number 37 of 2010 about Dam. The assessment on the physical condition of dam and its reservoir is conducted in some components: 1) Reservoir including: a) sedimentation, b) Reservoir's watershed area, c) Green Belt Area, and d) Reservoir's inflow; 2) Dam body, including: a) downstream slope, b) dam peak, c) upstream slope, d) supporting area, and e) leakage; 3) Complementary Building, including: a) withdrawal building, d) release building, and c) transferring building, and 4) complementary building, including: a) withdrawal building, b) release building, and c) transferring building. The weighing of each component and sub component in the assessment of dam's physical condition becomes the reference in estimating the damage index of dam's each component and sub component. The component and weight of assessment in every dam as well as its reservoir according to Standard Guidelines of Dam's Physical Condition Assessment of 2010 as follows:

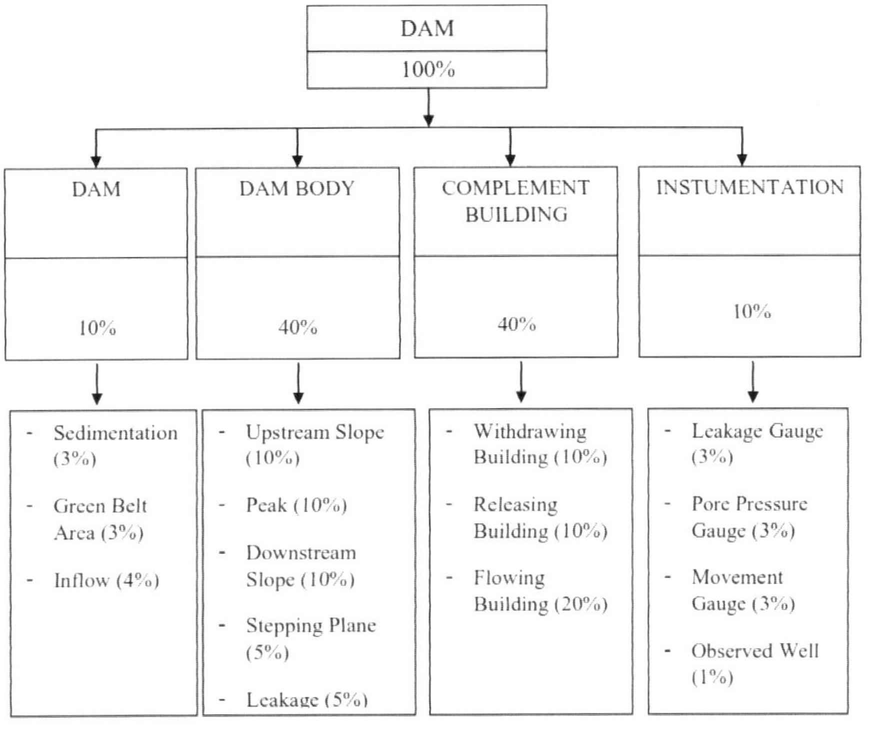

IOP Conf. Series: Materials Science and Engineering **267** (2017) 012033 doi:10.1088/1757-899X/267/1/012033

Figure 1. The scheme of component and weighing for the Assessment of Dam and Reservoir Physical Condition (Anonym, 2010).

The classification of condition assessment is made based on problem and damage level (severity level) of dam and reservoir. The classification of condition assessment can be seen in Table 1.

Table 1. Value of Dam Component Condition.

No.	Condition	Notes
	Good	If the condition value >90% and severity level <10% of the prior condition of building.
	Mildly damaged	If the condition value 80-90% and severity level 10-20% of the prior condition of building
	Moderately damaged	If the condition value 60-70% and severity level 21-40% of the prior condition of building.
	Severely damaged	If the condition value <60% and severity level >40% of the prior condition of building.

Source: Balai Bendungan (Anonym, 2010)

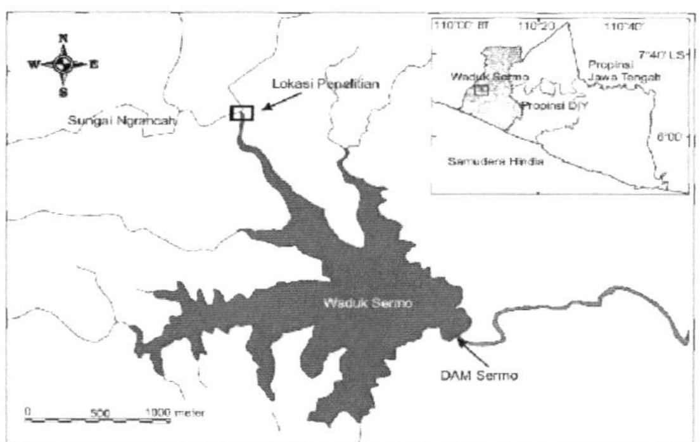

Figure 2. Research Location in Sermo Dam.

The first measure taken in this research was to collect secondary data in the form of data of Sermo Dam monitoring during 2015-2016 and data of big inspection on Sermo Dam in 2015 obtained from BWS Serayu Opak and primary data through field survey to obtain the condition value of Sermo Dam in 2017. Referred to the Guidelines of Physical Dam Condition Assessment referring to the Government Regulation Number 37 of 2010 about Dam, this research assessed the condition value of Dam viewed from 1 (one) component, dam body, so that the component weight become 100%. Subcomponents of dam body include downstream slope, dam peak, upstream slope, supporting area, and leakage. So that it can be illustrated in the scheme below.

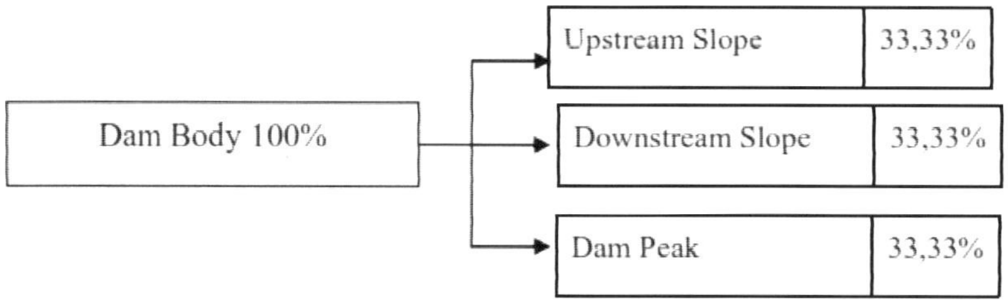

Figure 3. Scheme of component and weighing for the Assessment of Dam and Reservoir Physical Condition in Dam Body (Anonym, 2010).

3. Analysis and Discussion

3.1. The assessment on Dam Body in Sermo Dam in 2015

The assessment on Dam Body in Sermo Dam in 2015 used monitoring data of Sermo Dam in 2015 and big inspection data of Waduk Sermo in 2015.

3.1.1. Upstream Slope.

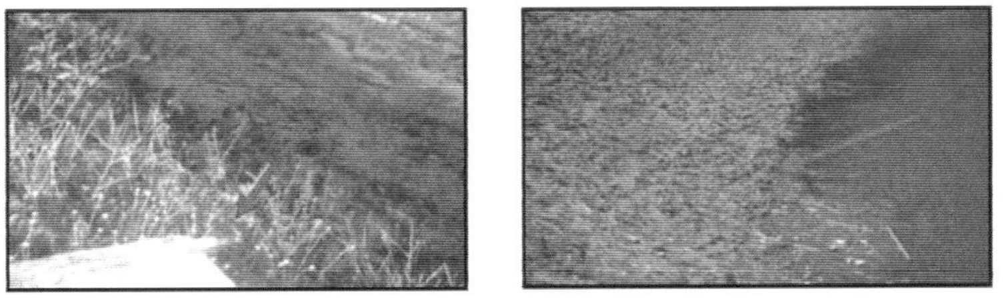

Figure 4. Condition of Sermo Dam Upstream Slope in 2015.

The slope protector consists of rip-rap (empty stone), apparently regular/stable slope inclination, no sign of movement found, no erosion and ruin flow found, weathering began to appear, wild plant including short grasses only, no subsidence found, and no animal hole or nest found. Watershed is not straight as the organization of replacing new rip-rap stone with old stone, the stone detached due to the fishermen's action. No water rotation existed along the watershed. This condition value of upstream slope obtains score of 94 (good).

3.1.2. Dam Peak.

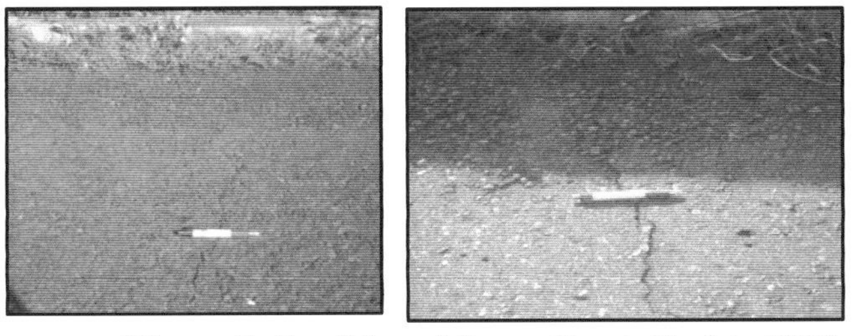

Figure 5. Condition of Sermo Dam's Peak in 2015.

The surface width of 8.00 m, asphalt hardening (hotmix), surface crack, (hair crack) on the right, the decrease (vertical movement) can be seen on the right side starting from Sta.15 to the right, the good/smooth straightness (horizontal movement), there is a safeguard fence (good condition) but there were some bolts vanishing in some parts. This condition value of dam peak obtains score of 91 (good).

3.1.3. Downstream Slope.

Figure 6. The condition of Sermo Dam Downstream Slope in 4.

Generally, its condition is as same as that of downstream slope; the protector of slope consists rip-rap (empty stone), apparently regular/stable slope inclination, no sign of movement found, no erosion and ruin flow found, weathering began to appear, wild plant including short grasses only, no subsidence found, no animal hole or nest found, no new leakage and wet area. The value of downstream slope condition obtains score of 93 (good).

3.2. The assessment of Dam Body Condition in Sermo Dam in 2016

3.2.1. Downstream Slope.

Figure 7. The condition of Sermo Dam Upstream Slope in 2016.

The condition of upstream slope was good, with the planned ratio of 1:1.8. The protector of slope includes rip-rap, there was a sign of uneven slope inclination condition due to weathering, there was no erosion, the plant was weeded once in a month, the decrease on the left and the right, no ruins, and no animal nest were found. There is a decrease of protecting layer stone in upstream part of dam peak in Sta.15 with the decrease of ± 30 cm. The value of upstream slope condition obtains score of 92 (good).

3.2.2 Dam Peak

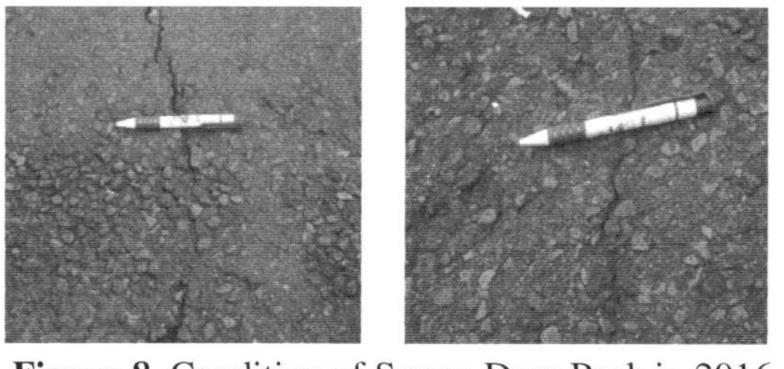

Figure 8. Condition of Sermo Dam Peak in 2016.

The surface width is 8.00 m, length is190 m, asphalt hardening (hotmix); there was no surface crack, there was no deflection, the straightness consistent with the plan. There is usually no exterior condition. The value of this dam peak condition obtains score of 94 (good).

3.2.3 Downstream Slope.

Figure 9. The condition of Sermo Dam Downstream Slope in 5.

IOP Conf. Series: Materials Science and Engineering **267** (2017) 012033 doi:10.1088/1757-899X/267/1/012033

The planned inclination condition is 1: 1.7; generally the condition is as same as that of upstream slope, the protector of slope consists of rip-rap (empty stone), apparently irregular/unstable slope inclination condition, no sign of movement, erosion flow and ruins found, no weathering apparent, wild plant including short grasses only, no subsidence found, no animal hole or nest found, no new leakage and wet area found. The value of upstream slope condition obtains the score of 93 (good).

3.3. The assessment of Dam Body in Sermo Dam in 2017

The assessment of Dam Body in Sermo Dam in 2017 was conducted using primary data, the data obtained from field survey in 2017.

3.3.1 Upstream Slope

Figure 10. Condition of Sermo Dam Upstream Slope in 2017.

Considering the result of survey on dam body, it could be found that the condition was good, the slope protector consists of rip-rap (empty stone), regular or unstable inclination condition. There is usually no exterior condition. The value of upstream slope condition obtains the score of 93 (good).

3.3.2 Dam Peak.

Figure 11. Condition of Sermo Dam Peak in 2017.

The condition of dam peak seems to be good, no surface crack was found, no deflection was found, no decrease was found, and the straightness was as planned. There is usually no exterior condition. The value of upstream slope condition obtained the score of 94 (good).

3.3.3 Upstream Slope.

Figure 10. Condition of Sermo Dam Upstream Slope in 2017.

IOP Conf. Series: Materials Science and Engineering **267** (2017) 012033 doi:10.1088/1757-899X/267/1/012033

Generally, the condition is still good, the protector of slope consists of rip-rap (empty stone), apparently irregular/unstable slope inclination condition, no sign of movement, erosion flow and ruins found, weathering beginning to appear, wild plant including short grasses only, no subsidence found, no animal hole or nest found, no new leakage and wet area found. The score of embankment slope is 95 (good).

Having obtained the assessment on each sub component of Sermo Dam, the assessment is accumulated in each sub component of Sermo Dam, and the condition of Sermo Dam in 2015, 2016, and 2017 is obtained, as shown in table 2.

Table 2. Condition Value of Dam Body in 2015-2017.

Subcomponent of Dam	Condition Value of Dam Body in 2015-2017		
	2015	2016	2017
Upstream Level	31.33	30.66	31.00
Dam Peak	30.33	31.33	31.33
Downstream Slope	31.00	31.00	31.66
Total	92.66	92.99	93.99

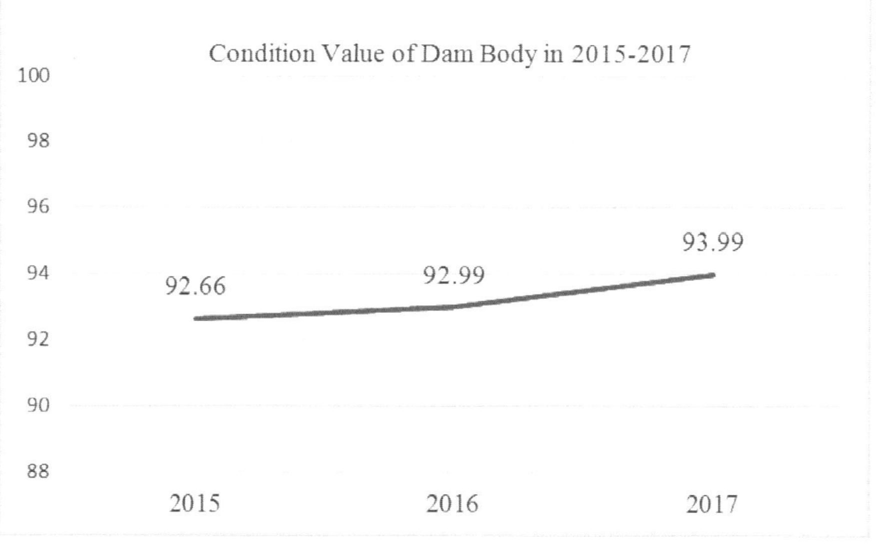

Figure 13. Chart of Sermo Dam Downstream Slope Condition in 2017.

The result of assessment on Sermo Dam condition with the dam body component obtain scores of 92.66% in 2015, 92.99% in 2016 and 93.99% in 2017, indicating that Sermo Dam condition is good (condition value >90%-100% and severity level <10%), viewed from the component of dam body in 2015-2017. From the value, it can be seen the trend of Sermo Dam condition value with dam body component. The trend increases over times despite insignificance.

4. Conclusion

From the discussion above, the following conclusion could be drawn:

1. The result of research showed that the condition value of Sermo Dam with dam body component were 91.33% in 2015, 91.66% in 2016, and 93.99% in 2017.
2. The result also showed that the value of dam body condition tended to increase during 2015-2017. To maintain the condition, the maximal operation and maintenance of dam was recommended.

IOP Conf. Series: Materials Science and Engineering **267** (2017) 012033 doi:10.1088/1757-899X/267/1/012033

5. References

[1] About Guidelines of Dam Operation, *Maintenance and Observation*, Water Resource Directorate General of Settlement and Infrastructure Department, Jakarta.

[2] Anonym, 2010, Government Regulation Number 37 of 2010, about Dam, Public Work Ministry, Jakarta.

[3] Herlen, 2009, The Management of Water Electricity Power Generator Dam of PB. SOEDIRMAN, Indonesian National Committee Seminar for Large Dam (KNI-BB), Yogyakarta June 10-11 2009

[4] M. Marzulian Ramli and Bambang E. Yuwono, 2015, Identifying the Dominant Factor of Dam Management Affecting the Performance of Dam, the 5th National Seminar of Civil Engineering in 2015 –UMS

[5] Yulius Heryant, et al, 2014, The Development of Dam's Physical Condition Assessment to Determine the Management of Dam Maintenance, A Case Study on Lodan Dam, Journal of Master of Civil Engineering of Sebelas Maret University Vol. II No.1 March 2014 ISSN: 2339-0271

IOP Conf. Series: Materials Science and Engineering **267** (2017) 012034 doi:10.1088/1757-899X/267/1/012034

Investigation of Counter-Rotating Wind Turbine Performance using Computational Fluid Dynamics Simulation

V A Koehuan[1, 2*], **Sugiyono**[1] **and S Kamal**[1]

[1]Department of Mechanical and Industrial Engineering, Faculty of Engineering, Gadjah Mada University, Yogyakarta, Indonesia.
[2]Department of Mechanical Engineering, Faculty of Science and Engineering, Nusa Cendana University, Kupang 85001, Indonesia.

[*)] Corresponding author: verdy_koehuan@yahoo.com

Abstract. Investigation of the dual rotor counter-rotating wind turbine (CRWT) performance using non-dimensional parameters of the rotor diameter ratio and the rotor axial distance ratio against the characteristics of power coefficient with tip speed ratio (TSR) as input parameters have been successfully carried through CFD simulation. CFD simulation used *k-e* turbulence realizable with hexahedral meshing to predict the CRWT performance to the rotor diameter ratio of $D_1/D_2 < 1$, $D_1/D_2 = 1$ and $D_1/D_2 > 1$ and rotor axial distance ratio with the s826 airfoil that has been applied to the single rotor wind turbine. The best CRWT performance obtained on the rotor diameter ratio of $D_1/D_2 = 1.0$ with the peak power coefficient of 0.5219 or increased to $\Delta C_{P,\,max} = 16.49\%$ from the single rotor. CRWT performance through the addition of rotor axial distance ratio showed the power coefficient of the front rotor continued to rise closely to the single rotor performance while the rear rotor will continue to decline. However, the overall CRWT performance were relatively stable after the ratio of the distance $Z/D_1 = 0.5$ with the peak power coefficient of 0.5348 or increased to $\Delta C_{P,\,max} = 19.37\%$.

1. Introduction

Smart model of the three blades of horizontal axis wind turbine (HAWT) is known as counter-rotating wind turbine (CRWT) with the double rotor, rotating on the same axis in the opposite direction which has successfully presented the findings are interesting in parentheses recent years. The CRWT which have two rotors with the same diameter or different diameter at the same time, drive an alternator whose aerodynamic performance is better than the single rotor. Rotor with large size is placed in front and small one is placed behind it or otherwise with the rotational speed of the rotor is the synchronous moving alternator [1]. Energy conversion efficiency of turbines called wind turbine power coefficient for the dual rotors are theoretically better than a single rotor wind turbine (SRWT), where peak power coefficient of the single rotor from the theory of momentum (Betz limit), is 59.3%, Eggleston et al. [2]. Meanwhile, according to Newman [3] based on actuator disc theory, dual rotor wind turbine without any losses has maximum power coefficient 16/25 (64%).

[1] To whom any correspondence should be addressed.

Ushiyama et al. [4], through experimental studies on dual rotor wind turbine with a diameter ratio of $D_1/D_2 = 0.6$ (D_1front rotor diameter and rear rotor diameter D_2) indicated that the power coefficient of counter-rotating turbine was better than co-axial and single rotor, while the co-axial turbines best performance of the rotor with two blades. Lee et al. [5] through free wake vortex lattice method also showed the counter-rotating turbine with a diameter ratio of $D_1/D_2= 1$ performance was better than the single rotor and co-axial. Yuan et al. [6] and Ozbay et al. [7] showed the same results by studying the effect of rotation direction of dual rotor wind turbine through laboratory-scale experiments with a diameter ratio of $D_1/D_2 = 1$ and the axial separation ratio $Z/D_1 \geq 0.7$ (Z is axial distance of the two rotors) on the 2MW atmospheric wind turbine conditions of boundary layer.

Experiments and numerical analysis (quasi-steady strip method) of the counter-rotating wind turbine with power output of 30 kW have a front rotor diameter of 5.5 m (D_1) and the rear rotor of 11 m (D_2) or ($D_1/D_2 = 0.5$) with the ratio of the distance $Z/D_1 = 0.5$ indicated an increase in power output of 9% [8]. Furthermore, this study followed by Kumar et al. [9] through variation within the front and the rear rotor (diameter ratio of 0.5) for converting power optimization gained a 9.67% increase in the maximum power at a distance ratio of 0.65. Lee et al. [10] through numerical analysis with a modified blade element momentum theory in CRWT best performance on the diameter ratio of 0.4 and TSR = 8. Rosenberg, et al. [11] and followed by Moghadassian et al. [12] studied the design parameters of the counter-rotating wind turbine with the use of the similar airfoil to the single rotor and kept fixed the rear rotor TSR = 7.55. This study shows that the aerodynamic performance CRWT 10.08% is better than a single rotor with a diameter ratio of 0.25 and axial distance ratio 0.8. Shen et al. [13] using CFD, EllipSys3D Solver combined with the actuator line method, for predicting the CRWT performance to diameter ratio $D_1/D_2 = 1$ at TSR = 5.81, were compared with the experiment of Nordtank 500 kW. The results of this study show an increased annual energy production to 43.5% and the peak power coefficient of 0.5. Lee et al. [14] through a combination of methods blade element momentum theory and momentum, improvements were made to the design parameters of a single rotor and counter-rotating wind turbine as the pitch angle, and rotor rotation obtained the CRWT performance with $D_1/D_2 = 1$ is 12% higher than the single rotor. Usui et al [15] with a tandem rotor use diameter ratio of 1.19 and a ratio of the distance of 0.08 through optimization of the blade profile and blade angle has obtained an increase in turbine efficiency.

This study aims to investigate the performance of double rotor with counter-rotating wind turbines by using non-dimensional parameters of rotor diameter ratio and rotor axial distance ratio against the characteristics of power coefficient with tip speed ratio (TSR) as input parameters using CFD simulations. Various types of the blade (airfoil) has been used for the design and analysis of the aerodynamic performance of CRWT, where each of these studies was conducted separately to the diameter ratio, distance ratio, and tip speed ratio (TSR). CRWT performance of the various studies conducted on the rotor diameter ratio $D_1/D_2< 1$, $D_1/D_2 = 1$ or $D_1/D_2> 1$ with the axial distance and a certain blade to calculate the best CRWT performance at maximum conditions. This phenomenon is very interesting for further investigation to diameter ratio of $D_1/D_2< 1$, $D_1/D_2 = 1$ and $D_1/D_2> 1$ and the rotor axial distance ratio, the CRWT is in the best performance with the use of a particular type of blade in a fairly wide range of tip speed ratio. Blade models used in the CFD simulation for investigating the performance of CRWT is a single rotor wind turbine from the experimental of the Blind test program [16] [17].

2. Numerical Method

2.1. Blade geometry and parameters study

Blade model used in this study is a single rotor wind turbine blade developed by Krogstad et al. [16] and Bartl et al [17] as well as several other researchers on the Blind test program. The turbine rotor blade with a rotor diameter of 0.944 m used a series airfoil S826 issued by NREL (National Renewable Energy Laboratory). Figure 1 shows the blade used in this study and the non-dimensional parameters of rotor diameter ratio and the ratio of the axial distance of two rotors presented in Table 1

IOP Conf. Series: Materials Science and Engineering **267** (2017) 012034 doi:10.1088/1757-899X/267/1/012034

and Table 2. The front rotor diameter and operating conditions in this study maintained the same as the single rotor, while the blade geometry such as the number of blades, pitch angle, the front rotor and rear rotor solidity set as a control variable where the value remains constant.

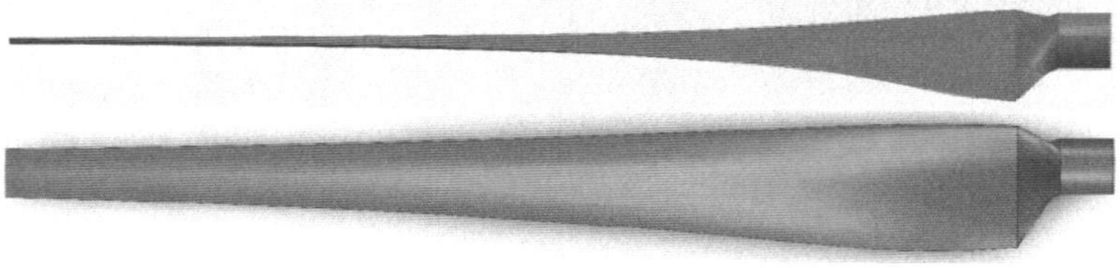

Figure 1. Blade model for counter-rotating wind turbine (CRWT).

Table 1. Parameters of rotor diameter ratio to axial distance ratio $Z/D_1 = 0.25$.

Parameters	Font (SRWT)	Rear				
		DR_1	DR_2	DR_3	DR_4	DR_5
rotor diameter, m	0.944	1.888	1.259	0.944	0.755	0.629
diameter ratio, D_1/D_2		0.5	0.75	1.0	1.25	1.5

Table 2. Parameters of rotor axial distance ratio.

Parameter	DZ_1	DZ_2	DZ_3	DZ_4	DZ_5	DZ_6	DZ_7	DZ_8
axial distance, m	0.1888	0.236	0.2832	0.3776	0.472	0.708	0.944	1.416
axial distance ratio, Z/D_1	0.2	0.25	0.3	0.4	0.5	0.75	1.0	1.5

2.2. Boundary conditions and meshing

Design and operational conditions of the single rotor turbine with a rotor diameter of 0.944 m that has peak power coefficient $C_{P,max} = 0.468$ at TSR 6 were applied to CRWT. The Numerical model used is the model with 1/3 of the domain (Figure 2) by assuming periodicity at the interface of cross-defined interfaces. While the mesh configuration is done through the process of volume, decomposition around the blade easily forms with hexahedral mesh element type using Gambit software. The single rotor turbine has 1.32 million of mesh nodes while CRWT follows the pattern of the single rotor with keeping the mesh density around the surface of the blade as shown in Figure 3.

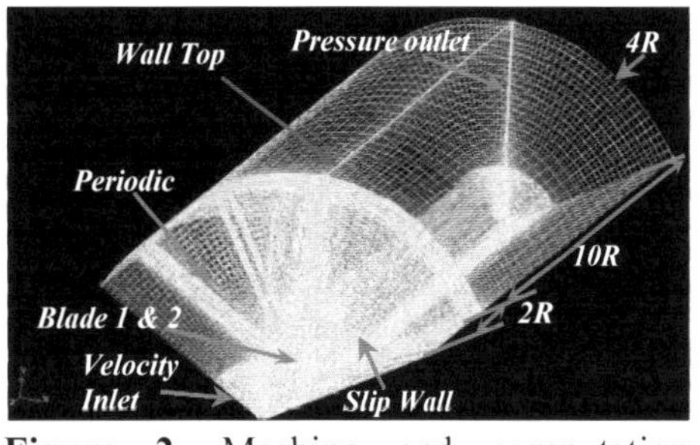

Figure 2. Meshing and computation domain, for rotor radius R = 0,472 m.

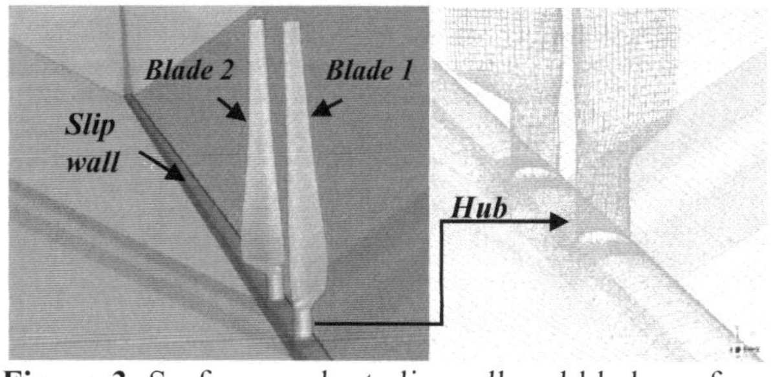

Figure 3. Surface mesh at slip wall and blade surface of the counter-rotating wind turbine.

2.3. Turbulent model

Determination of turbulence models depends on several considerations such as the physical characteristics of the flow, the experimental results for certain problems, the accurate level, the performance of existing computer and the time which are available for simulation. Turbulent model RANS (Reynolds-averaged Navier-Stokes) for incompressible and steady flow condition used by Taha et al. [18] is a model of turbulence *k-e* realizable, non-equilibrium pressure gradient was applied in this study.

Turbulent kinetic energy k and the dissipation rate of ε obtained from the general equation of fluid flow are the Reynolds-averaged Navier-Stokes (RANS) equations, where $\left(-\rho\overline{u_i^{'}u_j^{'}}\right)$ Reynolds stresses with components of speed, $u_i = \overline{u}_i + u_i^{'}$,

$$\frac{\partial}{\partial t}(\rho k)+\frac{\partial}{\partial x_i}(\rho k u_i)=\frac{\partial}{\partial x_j}\left[\left(\mu+\frac{\mu_t}{\sigma_k}\right)\frac{\partial k}{\partial x_j}\right]+G_k+G_b-\rho\varepsilon-Y_M+S_k \tag{1}$$

$$\frac{\partial}{\partial t}(\rho \varepsilon)+\frac{\partial}{\partial x_i}(\rho \varepsilon u_i)=\frac{\partial}{\partial x_j}\left[\left(\mu+\frac{\mu_t}{\sigma_\varepsilon}\right)\frac{\partial \varepsilon}{\partial x_j}\right]+C_{1\varepsilon}\frac{\varepsilon}{k}(G_k+C_{3\varepsilon}G_b)-C_{2\varepsilon}\rho\frac{\varepsilon^2}{k}+S_\varepsilon \tag{2}$$

Where G_k is the generation of turbulent kinetic energy due to the average speed gradient, G_b is turbulent kinetic generation due to buoyancy, Y_M represents the contribution due to compressibility and $k = 0.5\overline{(u_i^{'}u_j^{'})}$ is the turbulent kinetic energy. $C_{1\varepsilon}$, $C_{1\varepsilon}$, and $C_{1\varepsilon}$ are constants; σ_k and σ_ε are Prandtl turbulent numbers for k and ε. The turbulent viscosity μ_t are calculated from combining k and ε with $\mu_t = \rho C_\mu \frac{k^2}{\varepsilon}$, Where C_μ is a constant. The constants used in the above formula taken based on the default [19].

2.4. *Numerical solution method*

The numerical solution is done by the finite volume technique for the nonlinear differential equations Navier-Stokes into a set of algebraic equations using pressure-based solver. Type of pressure-based solver is used with absolute velocity formulations, while the coupled algorithm solution scheme is used for pressure-velocity coupling. In the iterative procedure, the method of solution for the completion of the pressure equation used second order, while for momentum and turbulent equations used QUICK scheme, which, according to the hexahedral mesh type was used in order to enhance the accuracy. The present work was done using a CFD code called Fluent that run in parallel on a personal computer with an Intel Core i3 4130 CPU, 16GB memory, VGA graphics card NVIDIA GeForce GT 730 2GB.

3. Results and discussion

3.1. *Validation of the numerical model*

The numerical model validation and the composition of the mesh is used in the simulation model of the single rotor wind turbine by RMSE (root mean square error) to compute errors that occur at power coefficient and thrust coefficient measurement reported by Bartl and Sætran [12]. The turbine power coefficient (C_P) which is the result of dividing the mechanical power output (P_{out}) and the total power in the wind flow (P_{in}) for the rotor swept area (A) was given. This analysis is formulated as follows,

$$C_P = \frac{P_{out}}{P_{in}} = \frac{Q.W}{\frac{1}{2}\rho V_0^3 A} \tag{3}$$

$$C_T = \frac{T_{out}}{T_{in}} = \frac{T_{out}}{\frac{1}{2}\rho V_0^2 A} \tag{4}$$

Torque output, Q and thrust output, T_{out} on the rotor was obtained from the CFD simulation results. Figure 4 and Figure 5 show comparisons of power coefficient and thrust coefficient of the single rotor wind turbine between CFD simulation and measurement data respectively. CFD simulation performed with turbulence model k-ε realizable has a peak power coefficient at TSR = 6 with $C_{P,\ max}$ = 0.448 or

RMSE value of 2.0% which was lower than the measurement data reported by Bartl and Sætran [12]. CFD simulation results shown by the power coefficient are close enough to experiment on low TSR (RMSE value of 0.85%), but at high tend to widen (RMSE value of 25.72%). The results shown by the thrust coefficient were very close to the data measured by the average value of RMSE were 2.53 %.

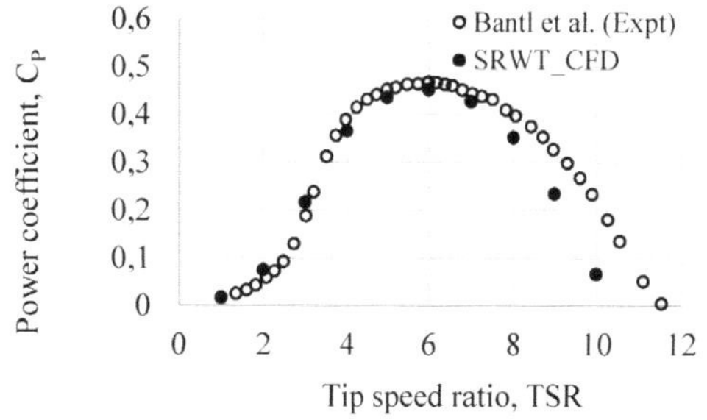

Figure 4. Comparison of SRWT power coefficient between CFD simulation and the experimental results.

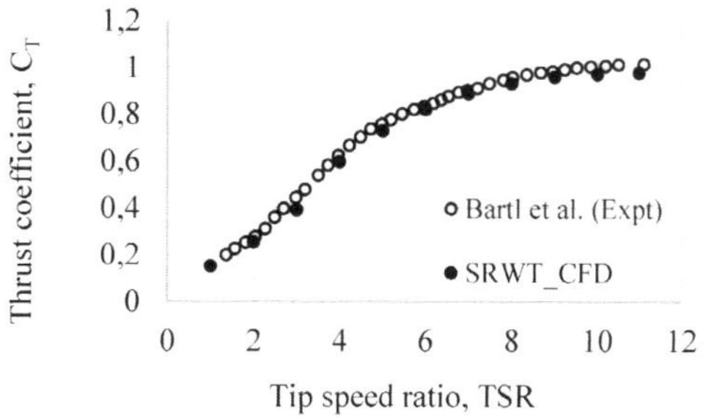

Figure 5. Comparison of SRWT Thrust coefficient between CFD simulation and the experimental results.

3.2. CRWT performance to rotor diameter ratio

Investigation of double rotor counter-rotating wind turbine performance through power coefficient characteristics with input parameters according to the single-rotor operating conditions were on tip speed ratio (TSR) between 1 and 11. The CFD simulations were performed using a constant speed of rotor rotation, which means that the front rotor and rear rotor rotation is kept constant with the ratio n_2/n_1 = 0.6 (n_1 and n_2 front rotor and rear rotor rotation respectively). Simulation parameters such as boundary conditions, the solution methods, and control solution, as well as the density of mesh nodes around the blade surface, made it equal to each variable in the rotor diameter and rotor axial distance of CRWT. While the convergence criteria for residual monitoring was set to 10^{-6} for all simulated parameters.

Comparison of the CRWT performance against SRWT from CFD simulations with variations of rotor diameter ratios performed at a rotor axial distance ratio of Z/D_1 = 0.25 for the range of tip speed ratios will be presented in two categories. Firstly, the increase in turbine power coefficient at the peak operating condition normalized by the SRWT expressed as $C_{P,\ max\ (CRWT)}/C_{P,\ max\ (SRWT)}$. The second is the increase in overall CRWT power coefficient that could be improved against SRWT at the range of TSR. From articles, review, the performance of CRWT tended to be partially related to the rotor diameter ratio was presented just for the peak performance with tip speed ratio (TSR) on its design condition. As we know that wind turbines are operated under atmospheric conditions, where natural freestream and design conditions will change with varied TSR ranges. Therefore, the performance of wind turbines must also be analysed for a wider TSR range so that turbine operations become more representative. The overall CRWT power coefficient would be obtained from the difference in the area that is formed by the curve of power coefficient and range of relative TSR. Figure 6 shows the power coefficient curve depicts a trend line by the second order polynomial agree well with the CFD result. Where the extent of this area is calculated by integrating the positive value of the power coefficient of relative tip speed ratio given in the input parameters. The area for representing the CRWT total power coefficient was normalized by SRWT which is presented as $A\_C_{P,\ CRWT}/A\_C_{P,\ SRWT}$. While the relative tip speed ratio (λ_R) is calculated based on the relative tip speed between the front rotor and rear rotor to the free stream flow, Usui et al. [15].

Comparison of the power coefficient between CRWT and SRWT with variations of rotor diameter ratio of CFD simulation results in Figure 6 shows the overall CRWT performance was better than SRWT. The peak power coefficient of the front rotor on D_1/D_2 = 0.5 and D_1/D_2 = 0.75 moved away from its operating conditions (TSR = 6) to TSR = 7 and TSR = 5 respectively, while for D_1/D_2 = 1.0 and D_1/D_2> 1.0 stayed at TSR = 6. The maximum CRWT performance obtained on diameter ratio of

D_1/D_2 = 1.0 (DR_3) with a maximum power coefficient of 0.5219 or increase of normalization of 1.1649 times SRWT ($\Delta C_{P,\ max}$ = 16.49%) at relative TSR = 9.6, while based on relative TSR the normalized total power coefficient is 1.8732 times higher than SRWT. From the CFD simulation, the higher CRWT performance in the range of diameter ratio D_1/D_2< 1.0 is D_1/D_2 = 0.5 (DR_1) and for diameter ratio D_1/D_2> 1.0 is D_1/D_2 = 1.25 (DR_4). Figure 6a and 6b show the CRWT performance on DR_1 and DR_4 with a maximum power coefficient of 12.45% and 8.54% at relative TSR = 10.99 and 8.88 respectively, while based on relative TSR the normalized overall power coefficient is 2.2067 for DR_1 and 1.7402 for DR_4 which is higher than SRWT. It appears that both front rotor power coefficient of DR_1 and DR_4 tended to rise and move closer to the single rotor (SRWT) to provide a positive contribution to overall CRWT performance.

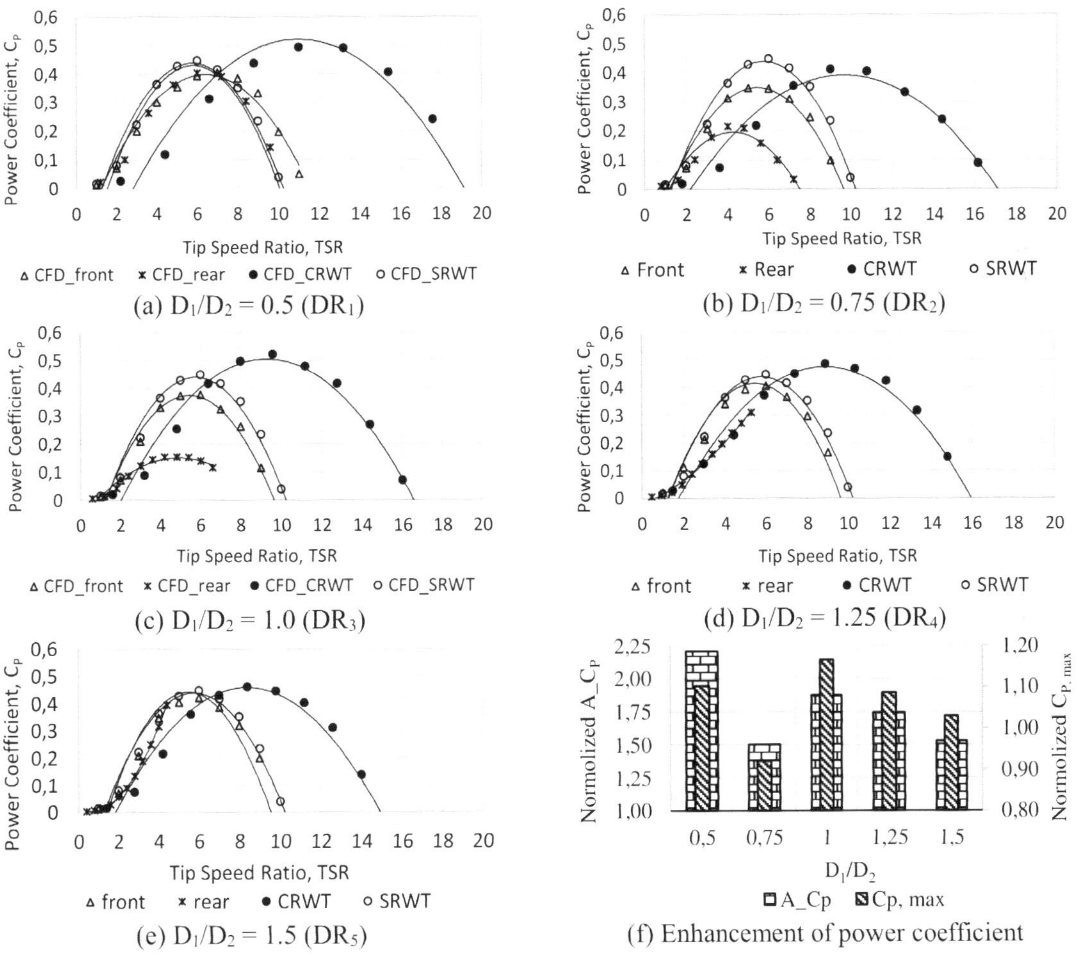

Figure 6. Comparison of power coefficient between CRWT and SRWT to tip speed ratios with a variation of rotor diameter ratio parameters at a rotor separation ratio of Z/D1 = 0.25.

The effects of the rear rotor diameter variation on the CRWT performance are when the rear rotor diameter is larger than the front rotors (D_1/D_2<1.0), then the CRWT performance increase with the addition of the rear rotor diameter is visible from both rising power coefficient for the front rotor and rear rotors. When the rear rotor diameter is smaller than the front rotor (D_1/D_2> 1.0), then the CRWT performance tends to drown if the rear rotor diameter was continuously scaled and characterized by the decrease of overall CRWT power coefficient (front and rear), although there was an increase in

IOP Conf. Series: Materials Science and Engineering **267** (2017) 012034 doi:10.1088/1757-899X/267/1/012034

front of rotor power coefficient. While CRWT achieves the maximum performance at $D_1/D_2 = 1.0$ which is geometrical when the front rotor diameter fixed and rear rotor diameter or the sweep area becomes larger ($D_1/D_2<1.0$), because the sweep area were proportional to power so increasing it the input power become higher. Conversely, when the rear rotor diameter becomes smaller ($D_1/D_2> 1.0$), the output power becomes low so that it effectively reduced turbine power coefficient. Because the rear rotor is operated in the wake of the front so that the rear rotor performance tends to be lower than the front rotors, although there is an increase in overall CRWT performance compared to SRWT. However, the diameter ratio of $D_1/D_2 = 0.5$, there is an increase on the rear rotor power coefficient than front rotor itself. The rear rotor diameter was two times greater than the front rotor or half of its blade length is not hindered by the front rotor and contacted directly to free stream so the loss due to the interference of flow from the front rotor is reduced. CRWT performance decline by the decreasing of the rear rotor diameter due to the its overall length on the interference of the front rotor. Flow disturbance resulted in a change of the magnitude and direction of inflow to the blade from design conditions, especially on the blade tip that produced vortex and separation on the blade surface.

3.3. CRWT performance to rotor axial distance ratio

CRWT performance against rotor axial distance ratio is shown in Figure 7 and Figure 8 with $D_1/D_2 = 1.0$, indicates that the overall power coefficient increases along with the increasing distance of the two rotors of $Z/D_1 = 0.2$ (DZ_1) to $Z/D_1 = 0.5$ (DZ_5). Performance of CRWT on $Z/D_1 = 0.2$ (DZ_1) is $\Delta C_{P,\,max} = 11.77\%$ and $Z/D_1 = 0.5$ (DZ_5) with $\Delta C_{P,\,max} = 19.37\%$ was better than single rotor. Furthermore, CRWT performance on the rotor axial separation ratio between $Z/D_1 = 0.5$ (DZ_5) and $Z/D_1 = 1.5$ (DZ_8) shows the front rotor power coefficient continued to increase with the addition of rotor axial distance to the single rotor performance. However, the opposite of the rear rotor power coefficient tends to fall but increased in CRWT performance and it was relatively stable after rotor axial distance ratio of Z/D1 = 0.5. By increasing the axial distance between two rotors, interference on the rear rotor was reduced and the velocity of the front rotor is recovered that increases the front rotor power coefficient, Lee et al. [14]. CFD prediction shown in Figure 8 indicated that interference of both front and rear rotors are quite stronger below the axial distance ratio of $Z/D_1 = 0.5$ as well as the lower front rotor power coefficient. However, after the axial distance ratio of $Z/D_1 = 0.5$, the power coefficient the front rotor was increased slightly and the rear rotor continues to decrease to $Z/D_1 = 1.5$, because of velocity recovery behind the front rotor and velocity deficit of the rear rotor decrease slowly.

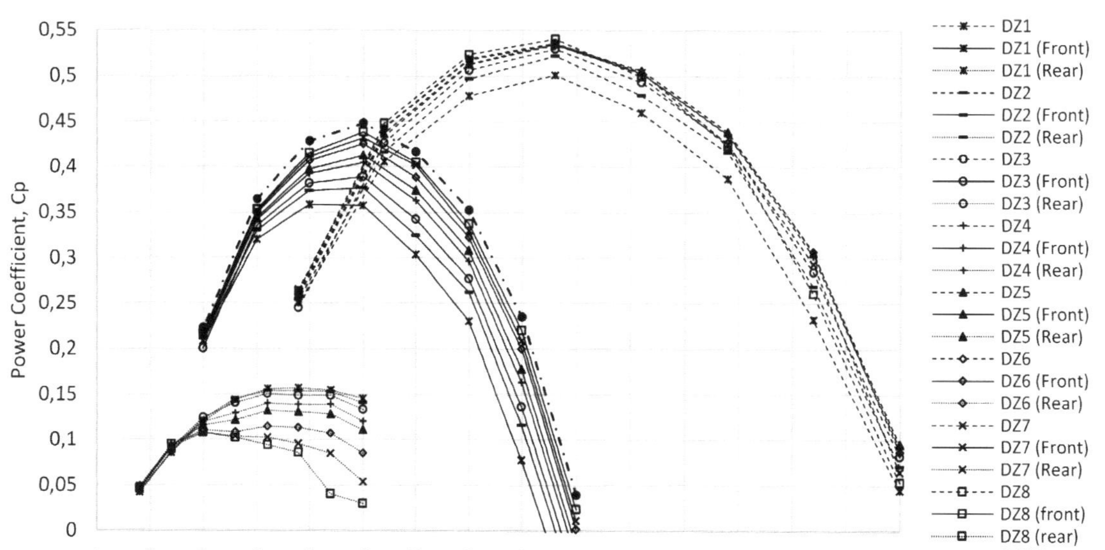

Figure 7. Power coefficient variation for CRWT, front and rear rotors against SRWT according to rotor axial distance ratio with diameter ratio $D_1/D_2 = 1.0$.

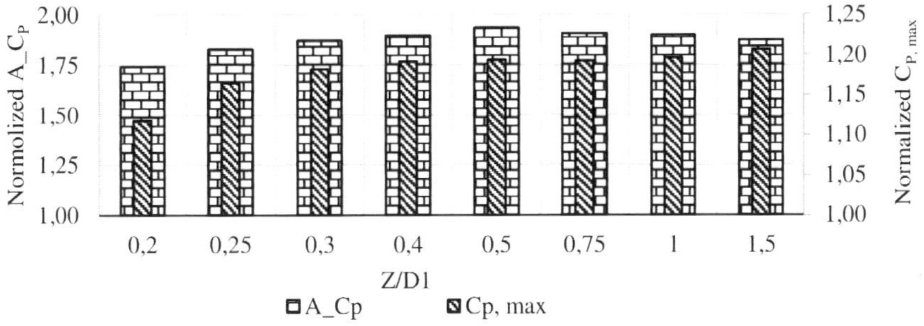

Figure 8. Enhancement of normalized maximum power coefficient and the normalization of overall power coefficient for CRWT against SRWT according to rotor axial distance ratio with diameter ratio $D_1/D_2 = 1.0$.

4. Conclusion

Investigation of counter-rotating wind turbines (CRWT) performance through power coefficient characteristics with a wide range input parameter of tip speed ratio, rotor diameter ratio and the ratio of the axial distance of the rotor has been successfully carried through CFD simulations. On the peak condition, CRWT performance obtained on diameter ratio of $D_1/D_2 = 1.0$ with a maximum power coefficient of 0.5219 or increased $\Delta C_{P,\,max} = 16.49\%$, while based on relative TSR the normalized total of power coefficient is 1.8732 times higher than SRWT. Evidence from CFD simulation results show the highest total CRWT power coefficient is in the diameter ratio $D_1/D_2 = 0.5$ (DR_1) and not in $D_1/D_2 = 1.0$ (DR_3) which has the highest peak power coefficient. CRWT performance to distance ratio shows the front rotor power coefficient continues to increase with the addition of rotor axial distance to near single rotor performance, the opposite of the rear rotor power coefficient tends to fall resulting in the increase of CRWT performance which was relatively stable after rotor axial separation ratio $Z/D_1 = 0.5$.

5. References

[1]. Kanemoto T and Galal A M 2006 Development of Intelligent Wind Turbine Generator with Tandem Wind Rotors and Double Rotational Armatures (1st Report, Superior Operation of Tandem Wind Rotors) *JSME Inter. Journal* **Series B 49(2)** 450–457

[2]. Eggleston D M and Stoddard F 1987 *Wind Turbine Engineering Design* (New York: Van Nostrand Reinhold)

[3]. Newman B G 1983 Actuator-disc Theory for Vertical-axis Wind Turbines *Journal of Wind Engineering and Industrial Aerodynamics* **15(1-3)**, pp.347–355

[4]. Ushiyama I, Shimota T, Miura Y 1996 An experimental study of the two-staged wind turbines, *Renewable Energy* **9(1)** p 909-912.

[5]. Lee S, Kim H, and Lee S S, 2010 Analysis of aerodynamic characteristics on a counter-rotating wind turbine *Current Applied Physics* **10(2)**pp. S339–S342.

[6]. Yuan W, Ozbay A, Tian W and Hu H 2013 An Experimental Investigation on the Effects of Turbine Rotation Directions on the Wake Interference of Wind Turbines *51st AIAA Aerospace Sci. Meeting including the New Horizons Forum and Aerospace Exposition*(Grapevine Texas: AIAA) p 607

[7]. Ozbay A, Tian W, and Hu H 2014 An Experimental Investigation on the Aeromechanics and Near Wake Characteristics of Dual-Rotor Wind Turbines (DRWTs) *32nd ASME Wind Energy Sym.*, 13-17 January 2014 (National Harbor Maryland: AIAA 2014-1085)

[8]. Jung S N, No T S, and Ryu K 2005 Aerodynamic performance prediction of a 30 kW counter-rotating wind turbine system *Renewable Energy* **30(5)** 631–644

[9]. Kumar P S, Abraham A, Bensingh R J, and Ilangovan S 2013 Computational and Experimental analysis of a Counter-Rotating Wind Turbine System *Journal of Scientific and Industrial Research* **72** May pp. 300–306
[10]. Lee S, Kim H, Son E, and Lee S S 2012 Effects of design parameters on aerodynamic performance of a counter-rotating wind turbine *Renewable Energy* **42**pp. 140–144.
[11]. Rosenberg A, Selvaraj S, and Sharma A 2014 A Novel Dual-Rotor Turbine for Increased Wind Energy Capture *Journal of Physics*: Conference Series **524** (012078) p 11(IOP Publishing).
[12]. Moghadassian B, Rosenberg A, Hu H, and Sharma A 2015 Numerical Investigation of Aerodynamic Performance and Loads of a Novel Dual Rotor Wind Turbine *33rd Wind Energy Symposium 5-9 January 2015* (Kissimmee Florida: AIAA SciTech) p 14
[13]. Shen W Z, Zakkam V A K, Sørensen J N, and Appa K 2007 Analysis of Counter-Rotating Wind Turbines *Journal of Physics*: Conference Series **75** (012003) p 10
[14]. Lee S, Son E, and Lee S S 2013 Velocity interference in the rear rotor of a counter-rotating wind turbine *Renewable Energy* **54**pp. 235–240.
[15]. Usui Y, Kubo K, Kanemoto T 2012 Intelligent Wind Power Unit with Tandem Wind Rotors and Armatures (Optimization of Front Blade Profile) *Journal of Energy and Power Eng.***6**pp. 1791-1799.
[16]. Krogstad P Å and Eriksen P E 2013 Blind test calculations of the performance and wake development for a model wind turbine, *Renewable Energy* **50** pp. 325-333.
[17]. Bartl J and Sætran L 2017 Blind test comparison of the performance and wake flow between two in-line wind turbines exposed to different turbulent inflow conditions *Wind Energy Science* **2(1)**, pp.55-76.
[18]. Taha Z, Sugiyono, Sawada T 2010 A comparison of computational and experimental results of Wells turbine performance for wave energy conversion *Applied Ocean Research* **32** pp. 83-90
[19]. Ansys Fluent 12.0 Documentation.

Acknowledgement
The authors thank the Ministry of Research, Tech., and Higher Education of the Republic of Indonesia and LPPM University of Nusa Cendana for supporting this research and paper through The Grant 2016.

IOP Conf. Series: Materials Science and Engineering **267** (2017) 012035 doi:10.1088/1757-899X/267/1/012035

A Novel of Buton Asphalt and Methylene Blue as Dye-Sensitized Solar Cell using TiO_2/Ti Nanotubes Electrode

Nurhidayani[1], M Z Muzakkar[1], Maulidiyah[1], D Wibowo[1] and M Nurdin[1,2]
[1]Department of Chemistry, Faculty of Mathematics and Natural Sciences, Universitas Halu Oleo, Jl. H.E.A. Mokodompit Kampus Baru Anduonohu, Kendari 93232 – Southeast Sulawesi, Indonesia

[2]E-mail: mnurdin06@yahoo.com

Abstract. A study of TiO_2/Ti nanotubes arrays (NTAs) based on Dye-Sensitized Solar Cell (DSSC) used Asphalt Buton (Asbuton) extract and methylene blue (MB) as a photosensitizer dye has been conducted. The aim of this research is that the Asbuton extract and Methylene Blue (MB) performance as a dye on DSSC solar cells is able to obtain the voltage-currents produced by visible light irradiation. Electrode TiO_2/Ti NTAs have been successfully synthesized by anodizing methods, then characterized by using XRD showed that the anatase crystals formed. Subsequently, the morphology showed that the nanotubes formed which has coated by Asbuton extract. The DSSC system was formed by a sandwich structure and tested by using Multimeter Digital with Potentiostat instrument. The characteristics of current (I) and potential (V) versus time indicated that the Asbuton was obtained in a high-performance in 30s of 14,000μV; 0.844μA, meanwhile MB dyes were 8,000μV;0.573μA. Based on this research, the Asbuton extract from Buton Island-Southeast Sulawesi-Indonesia was potential for natural dyes in DSSC system.

1. Introduction

The rapid development of technology and science in the few decades raised many problems as well as the energy problem [1,2]. Today, dependence on fossil energy causing the natural source supply to become thinner, are needed an alternative source of electrical energy as a substitute for fossil energy [3,4]. The renewable energy sources like solar energy is a more potential energy to serve as an alternative energy source [5,6]. The existence of solar energy from sunlight is very abundant on earth and it is the most potential as a power plant applied to solar cells with photovoltaic technology [7,8]. The principle of solar energy by converting sunlight into electrical energy, one of them is Dye-Sensitized Solar Cell (DSSC) [9]. DSSC is a solar cell based on the semiconductor sensitivity using dye as a sensitizer [10].

In practice, DSSC requires a semiconductor to be able to transfer electrons to generate potentials [11]. The semiconductors commonly used in DSSC are the metal oxide materials such as ZnO, TiO_2, Fe_2O_3, etc. [12,13]. Titanium dioxide (TiO_2) is commonly used as a DSSC material because of good optical properties, inert, harmless, and high-ability of dye absorption [14,15]. In addition, it also used for high-performance for electron transfer as a light-absorbing molecule [16,17].

Muhammad Nurdin, : +6281316551674

The usual dyes used as a sensitizer in DSSC are organometallic synthetic dyes such as ruthenium metal complexes [18]. Although it has achieved an efficiency more than 10%, these dyestuffs in natural are very limited because they are expensive and not easily synthesized [19]. Meanwhile, the natural dyes derived from plant extract such as leaves, flowers, and fruits, but the resulting conversion efficiency is lower than the ruthenium complex [20]. Various dyes have been widely used so it is needed the other alternatives of natural source like button asphalt (Asbuton) extract as a natural dye that has never been reported before compared to methylene blue (MB). This research, the Asbuton was used as a dye and a renewable alternative because it has a yellowish black color that predicted to absorb the visible light [21,22], while MB as the standard dye that previously used many researchers because of strong adsorption power. The photon absorption of visible light allows it to be carried out by the dyestuff as electric current and electron transfer to the TiO_2 material [23,24].

The surface area in a semiconductor is very influential on cell efficiency so the TiO_2 was made in nanotubes arrays (NTAs) in order to produce a wider particle surface. The high ordered TiO_2 NTAs was grow on a titanium (Ti) plate can be done on spatial condition, low-energy, and environmental friendly with an electrochemical system called an anodizing method [25,26]. This research, TiO_2/Ti NTAs was synthesized by anodizing method followed by a coating of Asbuton extract and MB dye for photon absorb into DSSC system. The Asbuton extract is expected to be sensitized on visible light irradiation in order to be able to convert the sunlight to electrical energy.

2. Experimental Method

2.1. Extraction of Asbuton
Asbuton bitumen preparation was performed by using mortal to minimize size, then it sieved at 100 mesh. The extraction of Asbuton was carried out by using two separation step using a soxhlet and an evaporator. The first step in the preparation of Asbuton extract was weighed 50 grams, then it inserted into filter paper and input into thimble tube. The 300 mL n-hexane as a solvent agent which was extracted for 3 hours with 30 cycles at a temperature of 70°C. The second step used an evaporator to separate the solvent with Asbuton extract. The separation process is stopped after the solvent has completely separated.

2.2. Preparation of Ti plate
Ti plate preparation was done by cutting Ti plate with 99% purity and 0.5 mm thickness which cut into size of 1 cm × 0.75 cm then it sanded using fine sandpaper size 1200CC until the surface is clean and shiny then washed by using a detergent, water, and distilled water (D.I. H_2O). Subsequently, the Ti plate was etched using a mixture of HF, HNO_3 and D.I. H_2O (1: 3: 6) solutions for 2 minutes. The final step was rinsed Ti plate and then dried [27].

2.3. Fabrication of TiO_2/Ti Nanotubes Arrays (NTAs)
The Titanium plate was inserted in the probe which had been filled with an electrolyte solution of NH_4F 0.27 M and D.I. H_2O in 98% glycerol. The anodizing process was carried out by placing the Ti plate as an anode and Cu plate as a cathode, as well as providing a bias potential of 25 Volts connected to the power supply. The anodizing process was done in 4 hours, then it annealed for 1.5 hours to a temperature of 500°C [28,29].

2.4. Design of DSSC
ITO glass was cut with size of 2.0 cm × 1.5 cm then it was cleaned with ethanol solution. A conductor substrate of ITO was dropped with TiO_2/Ti NTAs and Asbuton dyes. Meanwhile, the counter electrode was coated with a formed carbon layer. The solar cells were prepared with sandwich structures were the working electrode and counter electrode affixed face to face then it was added with electrolyte solution. The determination of current-potential used multimeter digital and potentiostat DY2100B to obtain data when irradiation used 500 Watts Halogen visible light.

3. Results and Discussion

3.1. Extraction of Asbuton

Asbuton extract has been prepared from large black rocks at Buton Island, Southeast Sulawesi. The microscale structure of Asbuton has been obtained with 200 mesh sieve to high extraction and uniform shape of asphalt rock to high contact with the solution agent [30]. The extraction process was done by using Soxhlet based on the high-temperature extraction in 30 cycles [31]. Then evaporated with n-hexane solution. The n-hexane solvent has good extraction ability to obtain the non-polar component [32,33]. The main purpose of Asbuton extraction process is to separate bitumen with other substances, either quartz sand or limestone. Asbuton bitumen extraction resulted through extraction method with soxhlet and evaporator obtained in this study from 50 grams of Asbuton bitumen by 14.22%. It shows that the inside of Asbuton consist about 80% impurities.

3.2. Fabrication of TiO_2/Ti Nanotubes Arrays (NTAs)

The fabrication of TiO_2 NTAs was carried out on Ti plate using anodizing method by giving a bias potential of 25.0 Volts for 4 hours in NH_4F and glycerol as electrolyte solution. The anodizing process based on an electrolysis process, in which Ti metal was placed at anode and Cu metal placed at cathode which was connected to the Power Supply [21,24]. The anodizing process, bias potential will cause Ti^{4+} ions to migrate from Ti metal to electrolyte and join the F^- ion forming the titanium hexafluoride complex $[TiF_6]^{2-}$. F^- ion in the electrolyte solution will move to the anode and will destroy the oxide layer formed to produce a tubular shaft based on the following reaction:

$$TiO_2 + 6F^- + 4H^+ \rightarrow [TiF_6]^{2-} + 2H_2O \quad (1)$$

The anodized Ti plate was calcined at 500°C for 3 hours to evaporate an organic solvent trapped in the Ti plate and to oxidize the chained and attached ligand. In addition, the calcination process was performed to obtain an anatase type of TiO_2 crystal structure.

Figure 1. The result of TiO_2/Ti NTAs by Anodizing method.

According to Nurdin et al. anatase structure is the highest structure to be applied in solar cells because it has a larger surface area compared to other crystal structures [34]. The larger surface area will be absorbing dyes so the photons will be converted into electric current. TiO_2/Ti formed on the surface of the Ti electrode (Figure 1) shows the difference of Ti surface between TiO_2 resulting from the anodizing method with TiO_2 which hasn't overgrown yet. The grayish color indicated that the anodizing results obtained a smooth surface with a fairly good layer thickness. TiO_2 NTAs has a tube-shaped structure and a more orderly shape compared to other mesoporous structures. It has nano size scale, wider surface, then it was resulted in the amount of dyestuff adsorbed on the TiO_2/Ti surface will become more numerous [17]. It is proportional to the absorption of photons will also increase to produce a more efficient light absorption process [25].

3.3. Coated Asbuton-TiO_2/Ti NTAs

TiO_2/Ti electrode has been formed by the anodizing method, then coated with Asbuton extract using a dip-coating technique. The dip-coating method has been widely used to form thin layers on the TiO_2/Ti surface because the good transparency and adhesion [21]. Asbuton-TiO_2/Ti NTAs was heated in an oven at 80°C temperature for 5 minutes to obtain the high binding between Asbuton extract and TiO_2/Ti NTAs.

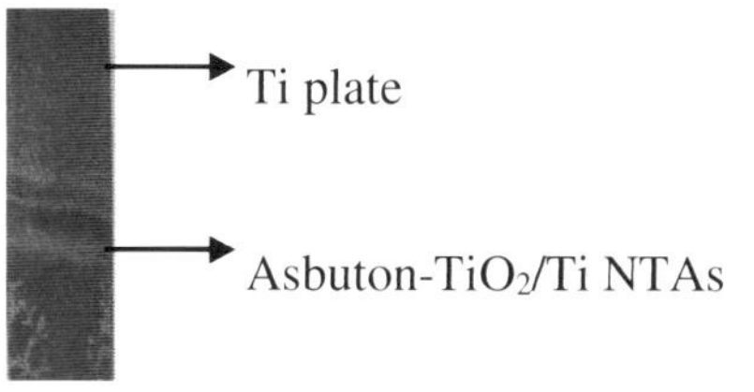

Figure 2. The result of coated Asbuton-TiO_2/Ti NTAs.

3.4. Characterization Materials

3.4.1. X-ray diffraction (XRD).

The XRD characterization was performed to obtain the crystal structure information of a synthesized TiO_2/Ti semiconductor. The crystal structure obtained is anatase crystals as an evidence by comparing the sample peaks against the standard peaks of JCPDS No.21-1272. Figure 3 shows the results of XRD analysis with specific peaks which is the form identity of TiO_2 crystals.

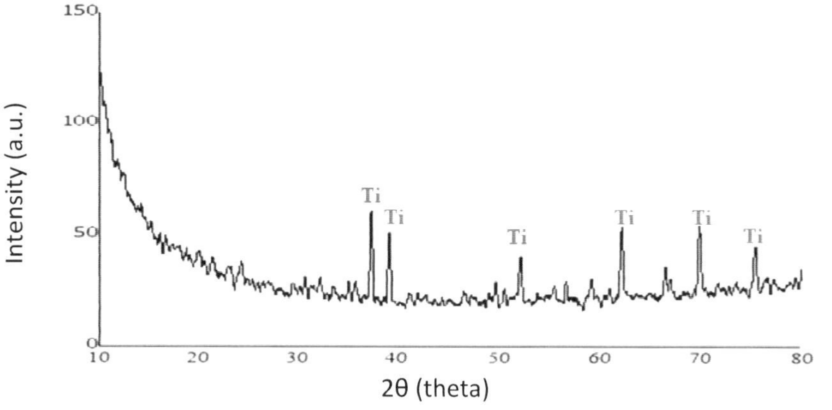

Figure 3. XRD pattern of TiO_2/Ti NTAs.

3.4.2. Scanning Electron Microscopy-Energy Dispersive X-ray (SEM-EDX).

Characterization using SEM-EDX was performed to provide information about the form of Asbuton-TiO_2/Ti surface morphology, and the doped substance of Asbuton on TiO_2.

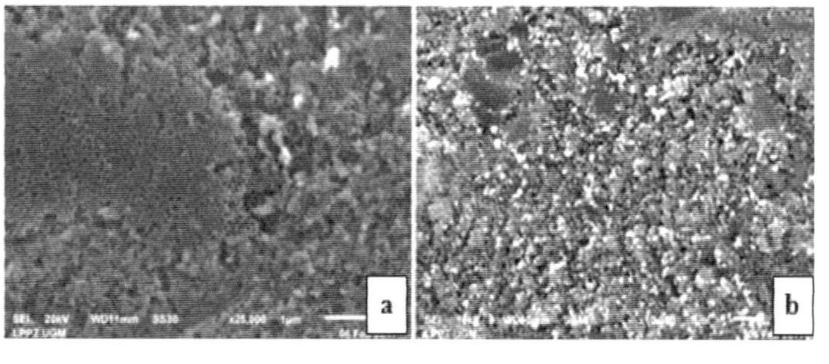

Figure 4. Morphology of Asbuton-TiO_2/Ti NTAs (a) Magnification 1 μm, (b) Magnification 10 μm.

Figure 4 shows nanotubes formed on Asbuton-TiO_2/Ti NTAs, the presence of Asbuton dopant didn't damage or mask NTAs all pores in TiO_2/Ti surface. The coating process results of Asbuton extract on TiO_2/Ti NTAs showed the uniform formation. It is a good condition for sensitizer using Asbuton as a dye in DSSC. In addition, to provide the success of dip-coating process coated on a substrate. Supported by EDX data, showed that the present Carbon (C) element of Asbuton coated on TiO_2/Ti NTAs. According to Odebunmi et al. Asbuton extract contain C element such as saturated hydrocarbons, hydrogenated polycyclic aromatic compounds, and high molecular weight phenols and carboxylic acids [35].

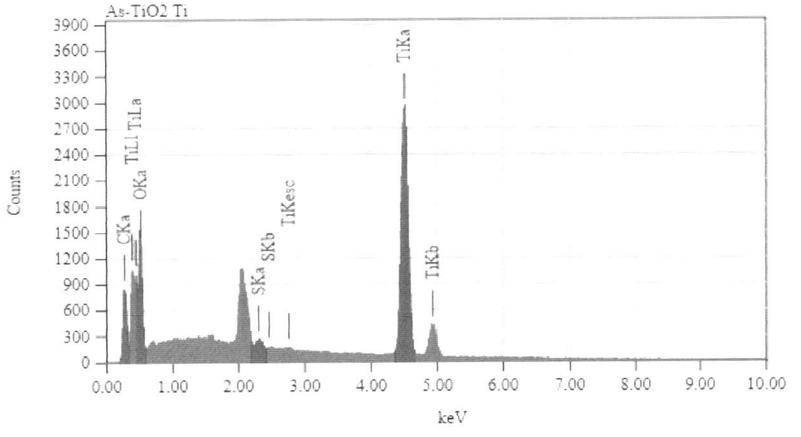

Figure 5. EDX Spectrum of Asbuton-TiO_2/Ti NTAs.

EDX characterization (Figure 5) shows the results of Asbuton-TiO_2/Ti NTAs that were the presence of element content on the electrode surface. The intensity of S and C peaks as an indication of sulfur (S) and carbon (C) elements on the TiO_2/Ti electrode ($Ti_1O_2C_1S_{0.025}$). According to Chilingar and Yen have reported that Asbuton contains an organic component of hydrocarbons, asphaltenes, and other organic compounds with sulfur, oxygen and nitrogen content [36]. The composition data of Asbuton-TiO_2/Ti electrode elements can be seen in Table 1.

Table 1. Element composition on Asbuton-TiO_2/Ti NTAs by EDX

Element	KeV	Composition (Wt%)
C	0.277	15.54
O	0.525	36.95
S	2.307	0.83
Ti	4.508	46.67

3.5. *Performance Test of DSSC*

The performance quality of solar cell can be determined by efficiency because it is a measure in determining the characteristics of voltage-current (I-V) using poteniostat, and digital multimeter. The solar cell testing was performed using the illumination of 500 Watts halogen visible light lamp as a light. The light source of the halogen lamp was directed perpendicularly on the solar cell surface.

The DSSC solar cells have been fabricated by sandwich structure then current and potential were characterized by digital multimeter. Determination of potentiometer based on resistance (R), a potential (V), and current (I). In this research, DSSC has been tested by visible light irradiation showed the high-performance of Asbuton-TiO_2/Ti NTAs compared to MB-TiO_2/Ti NTAs in 30s was obtained that the potential and current were 14,000 μV; 0.844 μA and 8,000 μV; 0.573 μA, respectively. The high time irradiation has decreased the data because of exposure of visible light which made damage of dyes and the electrolyte will be evaporated. The high resistance which

IOP Conf. Series: Materials Science and Engineering **267** (2017) 012035 doi:10.1088/1757-899X/267/1/012035

reducedthe current flowing in the DSSC was not detected on the multimeter. We can the determine of current with the Ohm Law equation,

$$I = \frac{V}{R} \tag{2}$$

Where, I is the current through the conductor in units of amperes, V is the voltage/potential measured across the conductor in units of volts, and R is the resistance of the conductor in units of ohms. Based on the data obtained in measurement results are presented in Table 2.

Table 2. Results of DSSC performance test using MB and Asbuton extract.

Time (seconds)	DSSC MB		DSSC Asbuton extract	
	V (μV)	I (μA)	V (μV)	I (μA)
30	8000	0.573	14000	0.844
60	6000	0.508	12000	0.802
90	6000	0.504	11000	0.771
120	4000	0.430	9000	0.758
150	2000	0.393	9000	0.643
180	800	0.383	6000	0.614

The performance quality of each solar cell is presented by voltage and current graph changes in both DSSC cells against to time presented in Figure 6.

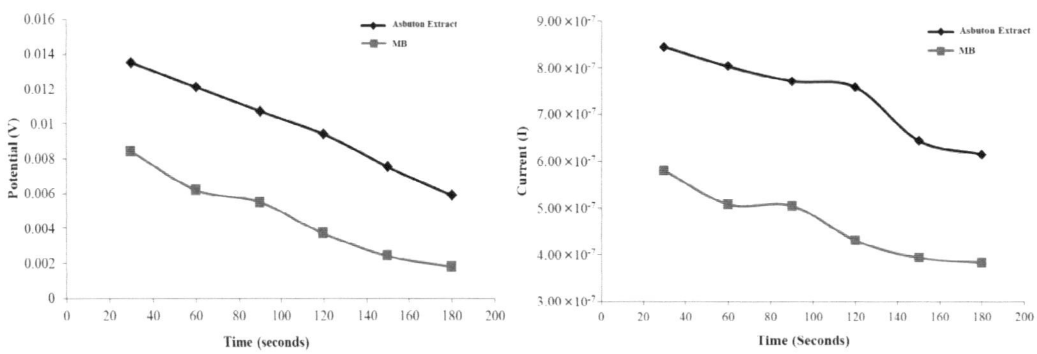

Figure 6. Current and potential changes; (A)The potential-time (V-t), and (B) The current-time (A-t).

This condition has saturated on TiO_2/Ti NTAs surface and the coating process of organic dyes were not equal. In other hands, the electrolyte effect has been decreased caused by visible light irradiation with high intensity that produced thermal condition. Based on the electrolyte function as the electron transfer medium, if the electrolyte is discharged or even absent, the electron transfer activity will be reduced or virtually absent, this affects the output current and voltage generated in the DSSC system.

DSSC cell performance which continued its measurement using Potentiostat is an electrochemical instrument consisting of electrical circuits that are useful for maintaining potential and regulating a fixed potential at a certain value. This instrument is intended to provide a voltage scan of the working electrode while measuring the magnitude of the voltage and current passed through the working electrode. The capabilities of each solar cell are presented with graphs of the voltage-to-current relation can be seen in Figure 7.

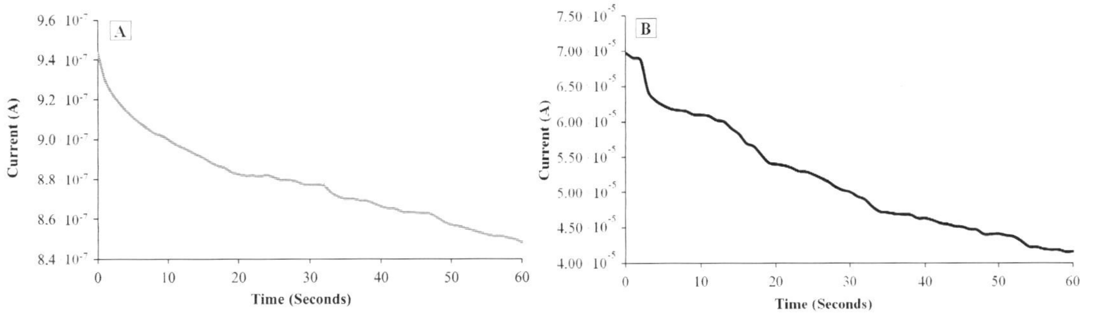

Figure 7. The performance test of DSSC current (A) against time (s), (A) MB-TiO_2/NTAs (B) Asbuton-TiO_2/NTAs.

The performance test based on I-t (Figure 7) shows that the two tested DSSC have produced different current outputs and the the longer of irradiation time that produced the low-current. The results show that Asbuton extract was in high-performance than MB dye due to the black characteristic which is able to absorb energy in visible light.

The overall results obtained in this study ranging from current and voltage measurements using digital multimeters and portable potentiostats showed that Asbuton extract emited greater current and voltage than MB dye, so Asbuton extract can be increased as a dye in the DSSC system, and can be used as a renewable alternative, since the utilization of Asbuton extract on DSSC manufacture has not been previously reported. The ability of each solar cell is presented with a graph of the time-relation voltage in Figure 8.

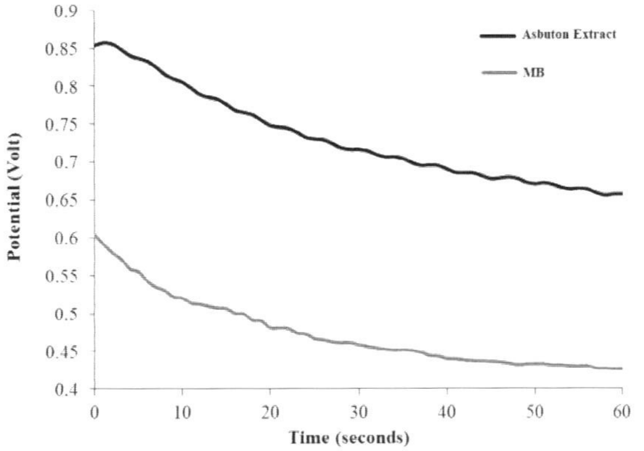

Figure 8. Potential-time (V-t) graph of Asbuton-TiO_2/Ti NTAs and MB-TiO_2/Ti NTAs.

4. Conclusion

In this paper, we have employed an Asbuton extract and MB as dyes in DSSC system using TiO_2/Ti NTAs and have evaluated its current-potential as a sensitizer. The characteristics of current (I) and potential (V) against time indicated that the Asbuton was obtained in high-performance in 30 s of 14,000μV; 0.844μA, meanwhile MB dyes were 8000μV;0.573μA, respectively.

5. References

[1] Maulidiyah, Nurdin M, Fatma F, Natsir M, Wibowo D 2017 *Analytical Chemistry Research* **12** 1.
[2] Sari A A, Kurniawan H H, Nurdin M, Abimanyu H, 2015 *Energy Procedia* **68** 254.
[3] Dahnum D, Tasum S O, Triwahyuni E, Nurdin M, Abimanyu H, 2015, *Energy Procedia* **68** 107.
[4] Triwahyuni E, Hariyanti S, Dahnum D, Nurdin M, Abimanyu H 2015 *Procedia Chemistry* **16** 141.

[5] Lee J-H, Ahn K, Kim S H, Kim J M, Jeong S-Y, Jin J-S, Jeong E D, Cho C-R 2014 *Current Applied Physics,* **14** 856.
[6] Umar A A, Nafisah S, Saad S K M, Tan S T, Balouch A, Salleh M M, Oyama M 2014 *Solar Energy Materials & Solar Cells,* **122** 174.
[7] Guai G H, Song Q L, Lu Z S, Ng C M, Li C M 2013 *Renewable Energy* **51** 29.
[8] Upadhyaya H H, Senthilarasu S, Hsu M-H, Kumar K D 2013 *Solar Energy Materials & Solar Cells* **119** 291.
[9] Chang H, Chen C-H, Kao M-J, Chien S-H, Chou C-Y 2013 *Applied Surface Science* **275** 252.
[10] Kim K-D, Lim D C, Seo H O, Lee J Y, Seo B Y, Lee D J, Song Y, Cho S, Lim J-H, Kim Y D 2013 *Applied Surface Science* **279** 380.
[11] Lee Y, Kang M 2010 *Materials Chemistry and Physics,* **122** 284.
[12] Nurdin M, Maulidiyah, Watoni A H, Abdillah N, Wibowo D 2016 *Int. J. ChemTech Res.* **9** 483.
[13] Nurdin M, Zaeni A, Maulidiyah, Natsir M, Bampe A, Wibowo D 2016 *Orient. J. Chem.* **32** 2713.
[14] Maulidiyah, Nurdin M, Erasmus, Wibowo D, Natsir M, Ritonga H, Watoni A H 2015 *Int. J. ChemTech Res.* 8 416.
[15] Maulidiyah, Nurdin M, Widianingsih E, Azis T, Wibowo D 2015 *ARPN J. Eng. Appl. Sci.* **10** 6250.
[16] Maulidiyah, Ritonga H, Faiqoh C E, Wibowo D, Nurdin M 2015 *Biosci. Biotechnol. Res. Asia* **12** 1985.
[17] Maulidiyah, Nurdin M, Wibowo D, Sani A 2015 *Int. J. Pharma. Pharmaceu. Sci.* **7** 141.
[18] Parisi A, Pernice R, Ando A, Cino A A C, Franzitta V, Busacca A C 2017 *Optik* **135** 227.
[19] Adeloye A O, Ajibade P A 2014 *Molecules* **19** 12421.
[20] Taya S A, El-Agez T M, Elrefi K S, Abdel-Latif M S, 2015 *Turkish J. Physics* **39** 24.
[21] Nurdin M, Muzakkar M Z, Maulidiyah M, Nurjannah M, Wibowo D 2016 *J. Mater. Environ. Sci.* **7** 3334.
[22] Ruslan, Mirzan M, Nurdin M, Wahab A W 2016 *Int. J. Applied Chem.* **12** 399.
[23] Arham Z, Nurdin M, Buchari B 2016 *Int. J. ChemTech Res.* **9** 113.
[24] Maulidiyah, Tribawono D S, Wibowo D, Nurdin M 2016 *Anal. Bioanal. Electrochem.* **8** 761.
[25] Maulidiyah, Ritonga H, Salamba R, Wibowo D, Nurdin M 2015 *Int. J. ChemTech Res.* **8** 645.
[26] Nurdin M 2014 *Int. J. Pharm. Bio. Sci.* **5** 360.
[27] Mursalim L O A M, Ruslan, Safitri R A, Azis T, Maulidiyah, Wibowo D, Nurdin M 2017 *IOP Conference Series: Materials Science and Engineering* xx xx.
[28] Wibowo D, Ruslan, Maulidiyah, Nurdin M 2017 *IOP Conference Series: Materials Science and Engineering* xx xx.
[29] Hikmawati, Watoni A H, Wibowo D, Maulidiyah, Nurdin M 2017 *IOP Conference Series: Materials Science and Engineering* xx xx.
[30] Maulidiyah, Wibowo D, Hikmawati, Salamba R, Nurdin M 2015 *Orient. J. Chem.* **31** 2337.
[31] Maulidiyah, Cahyana A H, Suwarso W P, Nurdin M 2015 *Int. J. ChemTech Res.* **8** 782.
[32] Maulidiyah, Sabarwati S H, Safutra E, Nurdin M 2016 *Int. J. Pharm. Bio Sci.* **7** 159.
[33] Maulidiyah, Imran, Muntu W, Nurdin M 2016 *Int. J. Applied Chem.* **12** 347.
[34] Nurdin M, Maulidiyah 2014 *Int. J. Sci. Technol. Res.* **3** 122.
[35] Odebunmi E O, Olaremu A G 2015 *Open J. Applied Sci.* **5** 485.
[36] Chilingar G V, Yen T F 1078. *Bitumens, asphalts, and tar sands*, New York, Elsevier Science.

Acknowledgement
We acknowledge for financial support of the Ministry of Research, Technology and Higher Education, the Republic of Indonesia.

IOP Conf. Series: Materials Science and Engineering **267** (2017) 012036 doi:10.1088/1757-899X/267/1/012036

Optimization of Energy Efficiency and Conservation in Green Building Design Using Duelist, Killer-Whale and Rain-Water Algorithms

T R Biyanto*[1], Matradji[1], M N Syamsi[1], H Y Fibrianto[2], N Afdanny[1], A H Rahman[1], K S Gunawan[1], J A D Pratama[1], A Malwindasari[1], A I Abdillah[1], T N Bethiana[1] and Y A Putra[1]

[1] Department of Engineering Physics, Institut Teknologi Sepuluh Nopember Surabaya, Surabaya 60111 East java, Indonesia

[2] Department of Industrial Engineering, Pusan National University, South Korea

*Email: trb@ep.its.ac.id; trbiyanto@gmail.com

Abstract. The development of green building has been growing in both design and quality. The development of green building was limited by the issue of expensive investment. Actually, green building can reduce the energy usage inside the building especially in utilization of cooling system. External load plays major role in reducing the usage of cooling system. External load is affected by type of wall sheathing, glass and roof. The proper selection of wall, type of glass and roof material are very important to reduce external load. Hence, the optimization of energy efficiency and conservation in green building design is required. Since this optimization consist of integer and non-linear equations, this problem falls into Mixed-Integer-Non-Linear-Programming (MINLP) that required global optimization technique such as stochastic optimization algorithms. In this paper the optimized variables i.e. type of glass and roof were chosen using Duelist, Killer-Whale and Rain-Water Algorithms to obtain the optimum energy and considering the minimal investment. The optimization results exhibited the single glass Planibel-G with the 3.2 mm thickness and glass wool insulation provided maximum ROI of 36.8486%, EUI reduction of 54 kWh/m^2·year, CO_2 emission reduction of 486.8971 tons/year and reduce investment of 4,078,905,465 IDR.

Keywords: Green Building, Mixed-Integer-Non-Linear-Programming, Duelist Algorithm, Killer-Whale Algorithm, Rain-Water Algorithm

1. Introduction

The development in the field of building construction grows faster in terms of both design and quality of the building. Currently, most buildings are designed based on green building requirement. In the United States, investors obtain the opportunities of green building for a long-term investment. It's due to green building operational costs are more efficient than conventional buildings [1].

In Indonesia, the development of green building has been endorsed by the Green Building Council Indonesia (GBCI). GBCI has assessment standards called Greenship rating [2]. The highest point in Greenship rating is Energy Efficiency and Conservation (EEC). According to Allen, et al. [3], improving energy efficiency in buildings can reduce the serious pollutions such as solid and liquid waste as well as CO_2 emission reduction that affects human health. The highest energy consumption is used to serve the electrical demand in office building. Hartungi and Jiang [4], this type of the building consume huge energy compared to other type of buildings. The amount of energy used in office buildings are used for the cooling system of the building. Uribe, et al. [5], the external load is a major disturbance in maintaining the temperature and humidity of the building. External load is influenced by the value of Overall Thermal Transfer Value (OTTV) and Roof Thermal Transfer Value (RTTV).

The value of OTTV is influenced by the type of glass and walls. Specifications of glass types that affect the value of the cooling load is the Shading Coefficient (SC) and Light Transmittance (LT) of the glass. Value of SC affects the heat radiation of sunlight in OTTV calculation. The value of SC is proportional with the value of OTTV. According to Yik, et al. [6], finally it will reduce the energy consumption in the cooling system. Another specification of glass is LT. If the value of LT increases, then the daylight transmit to inside the building will increase due to the decreasing of artificial lighting energy consumption [7]. In addition, the value of LT on glass specification is proportional to the value of SC. If the value of LT increases, then the value of SC will increase. Although the lighting load decreases, the increasing of SC will increase external heat load. It shows the external heat load and heat load due to lighting SC and LT values on glass are non-linear. In addition, the specification of glass and the price have wide range in the market. Thus, the amount of glass investment cost related to glass specification is complex.

As mentioned before, the amount of external load is also influenced by type of roof. Bojic, et al. [8], the utilization of insulation inside the roof can be reduce the external load through the roof. Consequently, the additional cost is required due to the effort to reduce OTTV and RTTV. Finally, energy efficiency and conservation that promise higher profit or saving required higher investment as capital cost. Hence, optimization of return on investment (ROI) that consider saving and investment cost is required. On the other hand, targeted EEC point to reach certain level of Greenship is required extra investment. Since these optimized variables i.e. the type of glass is non-linear and the present of insulation in the roof is an integer, hence, the optimization problem is classified as Mixed Integer Non-Linear Programming (MINLP). Venkataraman [9], the suitable algorithm to solve MINLP problems is stochastic optimization algorithm such as Duelist, Killer-Whale and Rain-Water Algorithms due to those capability to obtain the global optimum solution.

2. Literature Review

Eighty percent of modern human activities are performed inside the building. These activities require energy, water supplies, and also produce waste materials such as solid, liquid and gas wastes that affect to planet sustainability. This impact cannot be avoided but rather be mitigated to be minimized. Moreover, green building concern to minimize the utilization of material by minimizing the waste of construction material and maximizing the use of green material that can be recycled and available near the building under construction [2]. Green building has contributed to minimize global warming by utilization of renewable energy, optimization of energy consumption and air vent [10].

2.1 Greenship Rating

Greenship rating is a GBCI standard to assess the quality of green building for certification purpose. There are 6 categories namely Appropriate Site Development (17 points), Energy Efficiency and Conservation (26 points), Water Conservation (21 points), Material Resources and Cycle (14 points), Indoor Health and Comfort (10 points), and Building and Environment Management (13 points). Refer to those categories, the EEC has the biggest point in Greenship rating. The EEC category, contain some criteria as shown in table 1 [2].

Table 1. Detail points in EEC

Criteria	Point
Electrical Sub Metering	Prerequisite
OTTV Calculation	Prerequisite
Energy Efficiency Calculation	20
Natural Lighting	4
Ventilation	1
Climate Change Impact	1
On Site Renewable Energy (Bonus)	5

Source: GBCI, 2013

In the EEC category, there are three main calculations namely OTTV calculation, simulation or calculation of natural lighting and artificial lighting, and the calculation of energy consumption. Refer to SNI 03-6389-2011, the calculation of OTTV can be described in the following equation:

$$OTTV = \alpha[U_W \, x \, (1 - WWR)] \, x \, TDek + U_f \, x \, WWR \, x \, \Delta T + SC \, x \, WWR \, x \, SF \quad (1)$$

Where $OTTV$ is the building envelope in one direction (W/m^2), α is heat absorption, U_w is thermal transmittance of opaque wall (W/m$^2\cdot$K), WWR is window to wall ratio, $TDek$ is equivalent temperature difference (K), U_f is thermal transmittance of glass (W/m$^2\cdot$K), ΔT is temperature difference (K), SC is shading coefficient of glass and SF is solar factor for vertical surface (W/m^2). Natural and artificial lighting calculations are performed using DIALux software. Calculation of building energy efficiency using GBCI standards is performed by comparing the baseline and design data [9]. Baseline data is the reference data that available in ISO/ASHRAE/GBCI standards. Meanwhile, the design data is building target data that is used by architects in planning of the building.

2.2 Cost of Investment

One of the purposes of calculating the energy consumption in green building is to obtain the maximum points in Greenship rating by minimizing the cost of investment. Analysis of investment cost in green building is performed by calculating the cost of building baseline investment and cost of building design investment where cost of building baseline investment is cost of investment that is calculated using the reference data of the building with a default value ISO/ASHRAE/GBCI standards. Meanwhile, the cost of building design investment is obtained from the cost of the investment as consequent of targeted EEC point. In general, the calculation of baseline cost of EEC investment is shown in the following equation.

$$C_{EEC\,B} = C_{glass} + C_{AC\,System} + C_{lamp} \quad (2)$$

Where $C_{EEC\,B}$ is cost of EEC baseline investment (IDR), C_{glass} is cost of glass investment (IDR), $C_{AC\,System}$ is cost of cooling system investment (IDR), and C_{lamp} is cost of lamp investment (IDR). While the calculation of cost of EEC investment on green building design is shown in the following calculations.

$$C_{EEC\,D} = C_{glass} + C_{AC\,System} + C_{lamp} + C_{ESM} + C_{lux\,sensor} \quad (3)$$

Where $C_{EEC\,D}$ is cost of EEC design investment (IDR), C_{lamp} is cost of lamp investment (IDR), C_{ESM} is cost of electrical sub metering investment (IDR) and $C_{lux\,sensor}$ is cost of lux sensor investment (IDR). It's also necessary to determine the cost of extra investment that is calculated from cost of baseline investment minus cost of design investment as shown in the following equation.

$$Cost\ of\ Extra\ investment = C_{EEC\,D} - C_{EEC\,B} + C_{other} \tag{4}$$

Where C_{other} is cost of extra investment of other green building category (IDR). Cost of extra investment of other green building categories are calculated using the following equation.

$$C_{other} = C_{ASD} + C_{MRC} + C_{WAC} + C_{IHC} + C_{BEM} \tag{5}$$

Where C_{ASD} is ASD category cost of investment (IDR), C_{MRC} is MRC category cost of investment (IDR), C_{WAC} is WAC category cost of investment (IDR), C_{IHC} is IHC category cost of investment (IDR) and C_{BEM} is BEM category cost of investment (IDR). The implementation of green building concept will be reduced the energy consumption compare to the baseline building. Reduce of energy consumption will increase the saving over the time. The amount of saving will be equal with extra investment at the certain time. It called return of investment (ROI). ROI calculation is shown in the following equation.

$$\text{ROI} = \frac{\text{Saving}}{\text{Extra Investment Cost}}\ x\ 100\% \tag{6}$$

2.3 Duelist Algorithm (DA)
Duelist Algorithm (DA) is an evolutionary computation technique inspired by how duelist improve their capabilities in a duel that was developed by Biyanto, et al. in 2016 [11]. The algorithm is tested by using 11 different optimization problems and the result shows that the proposed algorithm is able to surpass the other recent stochastic algorithms and present robust results.

2.4 Killer-Whale Algorithm (KWA)
Killer-Whale Algorithm (KWA) is a new evolutionary computation algorithm that is inspired by the life of Killer Whale that was developed by Biyanto in 2016. KWA is tested by using 6 objective function containing Gaussian noise, Uniform noise and Seldom Cauchy noise, the results obtained from this test are compared with algorithm such as CMA-ES, GA, ICA and SA. The results show that the KWA is able to outperform the four algorithms used for comparison with a success rate as much as 4 of 6 objective function in 5 and 20 dimensions [12].

2.5 Rain-Water Algorithm (RWA)
Rain-Water Algorithm (RWA) is a new algorithm that is inspired by the pattern of physically rain water movements from air to the lowest place on the earth that was developed by Biyanto in 2016. The results show that the Rain-Water Algorithm is able to outperform the four algorithms used for comparison in term of speed of convergent [13].

3. Research Method
In this research, an office building was optimized using GBCI standard. The optimized variables were 154 glass types and 3 insulation types. The objective function was calculated using equation (7) as follows:

$$\text{Obj. Function} = \text{Max ROI} \tag{7}$$

Cost of extra investment for the building was calculated based on 2013 market prices and tabulated in Table 2 [14].

Table 2. Cost of water consumption and other cost of investment

Description	Cost
Cost of water consumption	154,053,075 IDR
ASD cost of investment	490,875,000 IDR
MRC cost of investment	898,475,000 IDR
WAC cost of investment	1,063,592,524 IDR
IHC cost of investment	76,100,000 IDR
BEM cost of investment	630,000,000 IDR

Source: Biyanto, et al., 2013

In general, the flow chart of research methodology is shown in Figure 1.

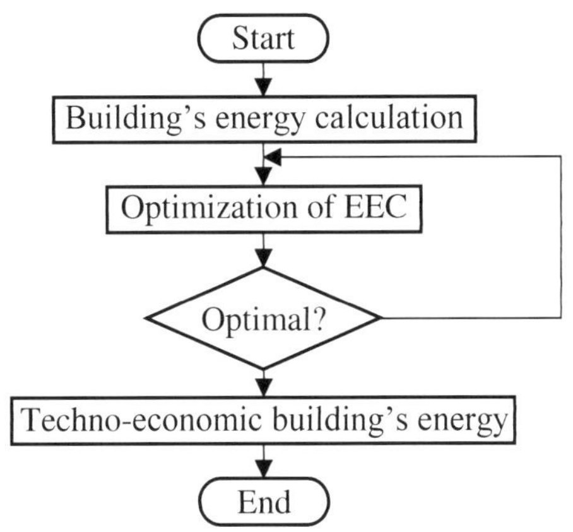

Figure 1. EEC optimization flow chart

4. Results and Discussions

In this research, Duelist, Killer-Whale and Rain-Water Optimization Algorithms were used to optimize ROI in green building design. These algorithms exhibited similar optimization results, as it may be due to the optimization problem which can be classified in simple problem. Hence, in the rest of discussions, the optimization results were provided from those optimization algorithms. The simulation of artificial lighting using LED lamps Philips Core-Line Batten Module 40 W type (BN120CL1200 1Xled38S/830) for each floors is shown in Figure 2. Figure 2 shows that most of artificial lighting area was above 350 lux and it was met with SNI 03-6575-2001 standard. The average of electrical consumptions due to lighting load of all floors was 5.80 W/m^2.

Table 3. Combination the chosen design of any EEC

EEC Point	Design Combination		ROI
	Type of Glass	Insulation	%
14	Panasap Dark Grey (3 mm)	Glasswool	30.2424
15	Panasap Green (5 mm)	No Insulation	31.9647
16	Planibel G (3.2 mm)	Glasswool	36.8486
17	Panasap Green (8 mm)	Glasswool	29.5671
18	Sunergy Green (6 mm)	Glasswool	20.6389

19	Stopsol Blue Green (6 mm) + Air + Clear (6 mm)	Glasswool	18.7667
20	Stopsol Green (8 mm) + Air + Planibel G (6 mm)	Glasswool	15.1297

The results of optimization show that the combination of the type of glass and insulation results different EEC points and ROI as shown in Table 3.

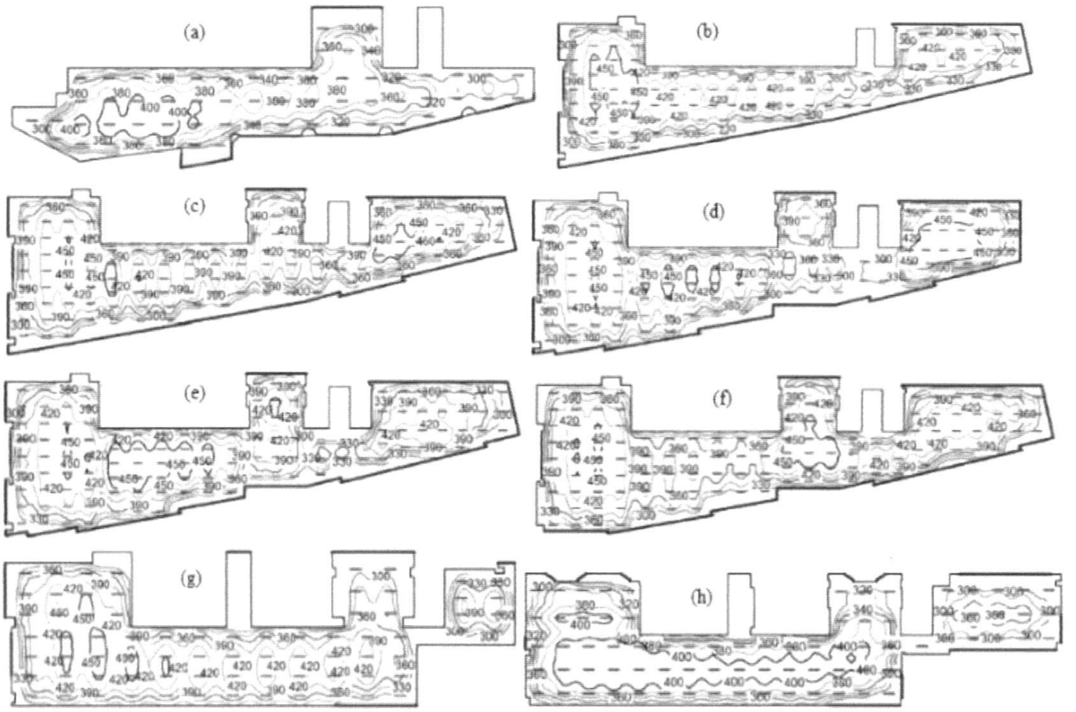

Figure 2. Artificial lighting at (a) ground floor and mezzanine floor, (b) 2nd floor, (c) 3rd floor, (d) 4th floor, (e) 5th floor, (f) 6th floor, (g) 7th floor, and (h) 8th until 16th floor

Table 3 shows that the EEC 16 point has maximum ROI value of 36.8486%. Meanwhile, the EEC 20 point has lower ROI than the EEC 16 point. The amount of Energy Use Intensity (EUI) also affected by EEC point as shown in Figure 3.

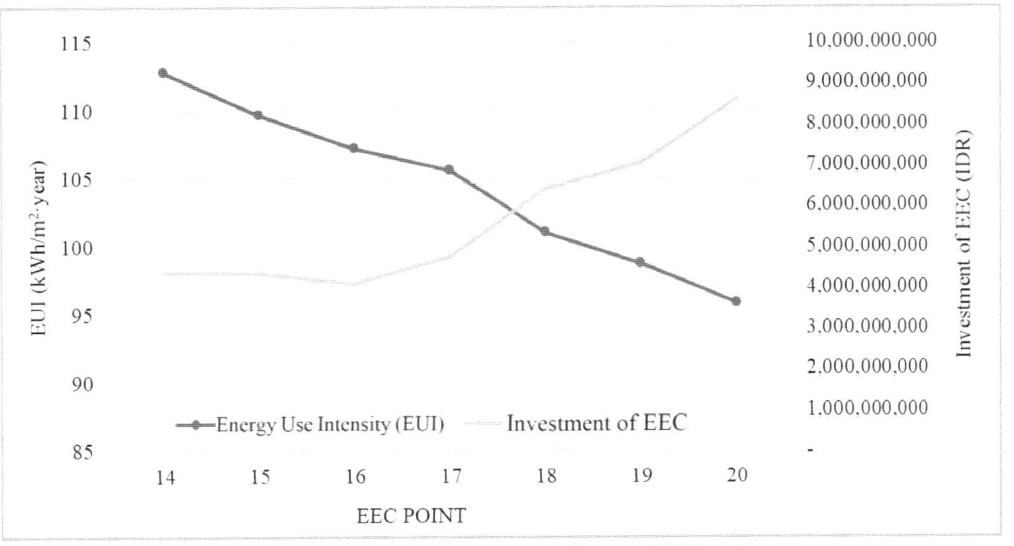

Figure 3. EUI and investment for any EEC points

Table 4 shows that the results of optimization has lower cooling load and EUI than the baseline building. The amount of EUI also affect the amount of CO_2 emission from the building. From an economic of view, the optimization results provide lower investment due to decreasing the price of the cooling system and glasses.

Table 4. Comparison between building baseline and optimization

Description	Unit	Comparison	
		Baseline	Result
Cooling System Capacity	TR	242.84	200.11
EUI	$kWh/m^2 \cdot year$	161	107
CO_2 emission	tons/year	1456.195	969.298
Cost of Investment Cooling System	IDR	1,457,069,420	1,200,675,063
Cost of Investment CHWP	IDR	118,110,930	66,088,490
Cost of Investment CWP	IDR	73,819,331	41305307
Cost of Investment AHU	IDR	188,443,896	93751561
Cost of Investment Glass	IDR	1,455,111,856	335,795,044
Cost of Investment Insulation	IDR	-	27,750,000
Cost of Investment Lamp	IDR	1,522,320,000	2,218,500,000
Cost of Investment ESM	IDR	-	24,000,000
Cost of Investment Lux Sensor	IDR	-	71,040,000

5. Conclusion

In this paper, Duelist, Killer-Whale and Rain-Water Optimization Algorithm were used to optimize EEC in green building design. The single glass Planibel G with 3.2 mm thickness and glasswool insulation were chosen as optimized variable that provided maximum ROI of 36.8486%, EUI reduction of 54 $kWh/m^2 \cdot year$, CO_2 emission reduction of 486.8971 tons/year and reduce investment of 4,078,905,465 IDR due to lower investment in the cooling system and glass.

6. References

[1] Bradshaw W B 2006 *Buying Green* (Cambridge, Massachusetts: Massachusetts Institute of Technology)

[2] Rating Development Departement 2013 *GREENSHIP New Building Version 1.2* (Jakarta: Green Building Council Indonesia)

[3] Allen J G, MacNaughton P, Laurent J G C, Flanigan S S, Eitland E S and Spengler J D 2015 Green Building and Health *Global Environmental Health and Sustainability* **2** 250-258

[4] Hartungi R and Jiang L 2007 Energy efficiency and conservation in an office building: a case study *International Journal of Energy Sector Management* **6** (2) 175-188

[5] Uribe O H, Martin J P S, Garcia-Alegre M C, Santos M, Guinea D and Passaro V M N 2015 Smart Building: Decision Making Architecture for Thermal Energy Management *Sensors (Basel)* **15** (11) 27543–27568

[6] Yik, F W H and Wan K S Y 2005 An evaluation of the appropriateness of using overall thermal transfer value (OTTV) to regulate envelope energy performance of air-conditioned buildings *Elsevier Ltd: Energy* **30** 41-71

[7] Mardaljevic J, Heschong L and Lee E 2009 Daylight Metrics and Energy Saving *Lighting Research Technol.* **41** 261-283

[8] Bojic M, Miletic M and Bojic L 2014 Optimization of thermal insulation to achieve energy savings in low energy house (refurbishment) *Elsevier Ltd: Energy Conversion and Management* **84** 681-690

[9] Venkataraman P 2002 *Applied Optimization with MATLAB Programming* (New York: John Wiley & Sons)

[10] Ng Ban H and Zainal A 2011 An overview of Malaysia green technology corporation office building: a showcase energy-efficient building project in Malaysia *Journal of Sustainable Development* **4** (5) 212-228

[11] Biyanto T R, Fibrianto H Y, Nugroho G, Listijorini E, Budiati T and Huda H 2016 Duelist algorithm: an algorithm inspired by how duelist improve their capabilities in a duel, *Int. Conf. in Swarm Intelligence: Advances in Swarm Intelligence* pp 39-47

[12] Biyanto T R 2016 *Killer Whale Algorithm: An Algorithm Inspired by the Life of Killer Whale* (Surabaya: Institut Teknologi Sepuluh Nopember)

[13] Biyanto T R 2016 *Rain-Water Optimization Algorithm* (Surabaya: Institut Teknologi Sepuluh Nopember)

[14] Biyanto T R 2013 *Project Group: Gedung Waskita Office Jakarta Timur* (Jakarta: Green Building Council Indonesia)

Enhancing Output Voltage of Piezoelectric-Based Nanogenerators Using Zinc Oxide Material Codoped with Aluminum and Cobalt.

D Subagiyo[1] and Suyitno[2]

1) Student of Department of Mechanical Engineering, Sebelas Maret University, Jl. Ir. Sutami 36 A, Surakarta, Indonesia, 57126

2) Lecture of Department of Mechanical Engineering, Sebelas Maret University, Jl. Ir. Sutami 36 A, Surakarta, Indonesia, 57126

E-mail: subagiyo.mesin10@gmail.com

Abstract. The cobalt and aluminum codoped into zinc oxide (ZnO) crystals are studied for enhancing the output voltage of piezoelectric-based nanogenerators (PENGs). The nanogenerators were simply fabricated from ZnO nanofibers (NFs) synthesized by electrospinning machine. The electrospinning processes were conducted at a flow rate of 4 μL/min and at various sintering temperature of 450, 500, 550, and 600°C. The results show that the highest output voltage of PENGs was obtained at a sintering temperature of 500°C, i.e. 127 mV. Therefore, PENGs fabricated from ZnO NFs codoped with aluminum and cobalt are challenging for the next generation of self-powered devices. The output voltage of PENGs at sintering temperatures of 450°C, 550°C, and 600°C are 75 mV, 105 mV, and 85 mV, respectively. The PENGs which fabricated from ZnO NFs was codoped with aluminum and cobalt will generate high output voltage, and are interesting candidates for exploiting in self-power devices.

Keywords: Fibers; Cobalt; Aluminum, Codoping; Zinc oxide; Electrospinning; Piezoelectric; Nano generator

1. Introduction

Piezoelectric-based nanogenerators (PENGs) are promising self-powered devices which convert mechanical energy into electrical energy [1-3]. The PENGs were commonly fabricated from piezoelectric materials having a non-centrosymmetric crystalline property; e.g. zinc oxide [2, 4], lead zirconate titanate (PZT) [5], bismuth sodium titanate $(Bi_{1/2}Na_{1/2})TiO_3$(BNT)–barium zirconate titanate $Ba(Ti,Zr)O_3$ (BZT) [6], polyvinylidene fluoride (PVDF), and so forth. The materials with a non-centrosymmetric crystal are polarized between atoms and excited charge carried when subjected to an external force [7]. Among them, ZnO is a good material exploiting for PENGs because of biocompatible, biodegradable, and biosafe [8]. However, utilizing ZnO materials for PENGs are proving to have low output power density.

Many efforts have been done to enhance the performance of PENGs with ZnO materials, mainly on physical morphology [9], electrical-mechanical properties [10], and heterojunction or hybrid structures [11]. The issues in physical morphology are to vertically well-aligned ZnO nanorods with a high aspect ratio [9]) and to synthesizing highly porous membrane [12]. Enhancing the electro-mechanical properties of ZnO material is commonly used by doping. Li-doped ZnO confirms that the NWs were p-type with a high carrier mobility of 14.59 cm^2/V s and an effective hole carrier concentration of 1.13×10^{17} cm^{-3}[10].

IOP Conf. Series: Materials Science and Engineering **267** (2017) 012037 doi:10.1088/1757-899X/267/1/012037

For self-charging devices, the critical factor of PENGs is in the output voltage. PENGs with low output voltage means that they cannot charge electrically. Therefore, enhancing the output voltage of PENGs has become essential. This study will present the sintering temperature effect for enhancing the output voltage of PENGs with ZnO NFs (nanofibers).

2. Methods

The materials which used in this study were zinc acetate dehydrate ($(CHOCOO)_2Zn.2H_2O$), ZnAc, Merck), aluminum chloride hexahydrate ($AlCl_3.6H_2O$, AlCl, Merck), cobalt acetate tetra hydrate ($(CoCH_3COO).4H_2O$, CoAc, Sigma Aldrich) Polyvinyl alcohol ($(-C_2H_2O)n$, PVA, MW 72.000, Merck), and aquades (H_2O) which used without further purification.

The fabricating of PENGs from ZnO NFs codoped with cobalt and aluminum consisted of three steps, namely (1) preparing precursor solutions, (2) producing green fibers with electrospinning machine, (3) sintering fibers, and (4) fabricating and testing PENGs. The synthesized NFs were then fabricated into PENGs and undergone for properties and performance test. The process of fabricating PENGs can be seen in Figure 1.

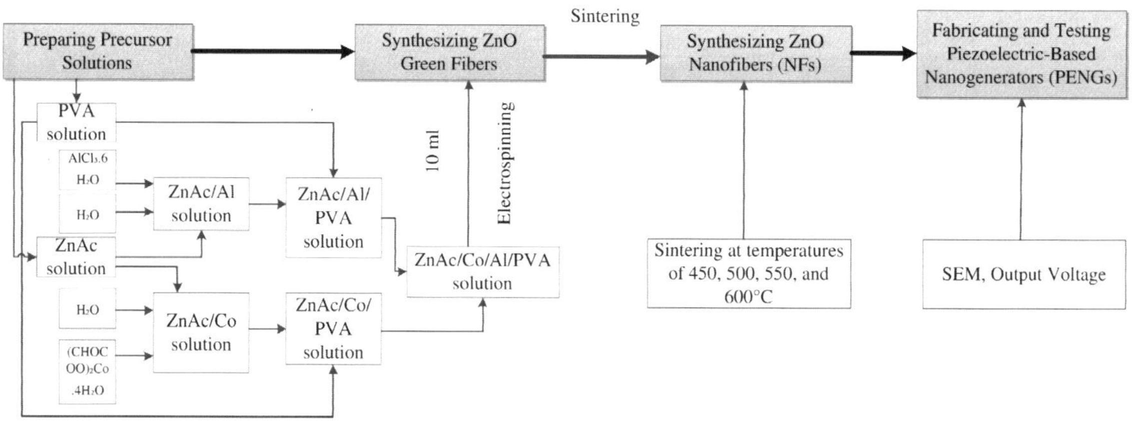

Figure 1. Scheme process of fabricating PENGs

To prepare PVA solution, the solid PVA was mixed with H_2O (1:10 wt%), stirred at 70°C for 4 hours, and followed by keeping it at room temperature for 8 hours. Subsequently, preparing the ZnAc solution was conducted by mixing 2 g ZnAc with 8 mL of water and stirring it for 1 hour at the temperature of 70°C.

ZnAc solution doped with 10 wt% of aluminum (ZnAc/Al solution) was prepared by mixing 4 g of ZnAc solution with 10 wt% $AlCl_2$, then poured it into 16 ml H_2O, and stirred at a temperature of 70°C for 1 hour. The same procedure was done when preparing ZnAc solution doped with 9 wt% of cobalt to obtain ZnAc/Co solution.

The Production of ZnAc/Al/PVA and ZnAc/Co/PVA solutions were done by mixing ZnAc/Al and ZnAc/Co solution with PVA solution at a mass ratio of 1:4 and stirred for 1 hour at a temperature of 70°C.

Furthermore, preparing a ZnAc solution codoped with Al and Co at a ratio of 25% Al and 75% Co to become ZnAc/Al/Co solution was conducted by mixing ZnAc/Al/PVA solution and ZnAc/Co/PVA solution at a ratio of 25%:75%, and so forth.

Second step was producing green fibers using electrospinning process. Each precursor solution of 1 ml was injected by syringe pump at a flow rate of 4 μl / minute where the distance between electrode is 8 cm. When the high voltage was turned on, the precursor solution started to be pushed out, attracted by the electrostatic field, and attached itself on the surface of the collector plate becoming green fibers.

Step of sintering process was conducted by heating green fibers in an electric furnace at temperatures of 450, 500, 550, and 600°C for 4 hours. The purpose of sintering is to remove organic matters and to grow ZnO crystallization. The ZnO NFs were then simply fabricated into PENGs.

The output voltage of PENGs was examined by applying cyclic load of 0.5 kg in every 3 seconds. The voltage was recorded by ADAM Advantech 4018 series of data acquisition. The magnitude load of 0.5 kg was performed to represent the applied conditions of weak finger pressure [2, 13]. The performance of PENGs was then calculated based on the amount of power (Watt) generated by providing various resistances (Ohm).

3. Results and Discussion

PENGs were simply fabricated from ZnO NFs doped with Al/Co. At sintering temperatures of 450 and 500°C, the output voltage increases with the increase of the electrical resistance used. At a sintering temperature of 550°C, a large output voltage was generated at a resistance of 1 MΩ and then decreased as the resistance went up.

This was more pronounce in PENGs with a small ratio of Al/Co in which Co core was easier to make in the crystallization process lead to the formation of small crystalline sizes. The resulting voltage of PENGs fabricated at a sintering temperature of 500°C can reach 127 mV and greater than those which made by the sintering temperatures of 450°C, 550°C, and 600°C, i.e. 75 mV, 105 mV, and 85 mV, respectively as shown in Figure 2.

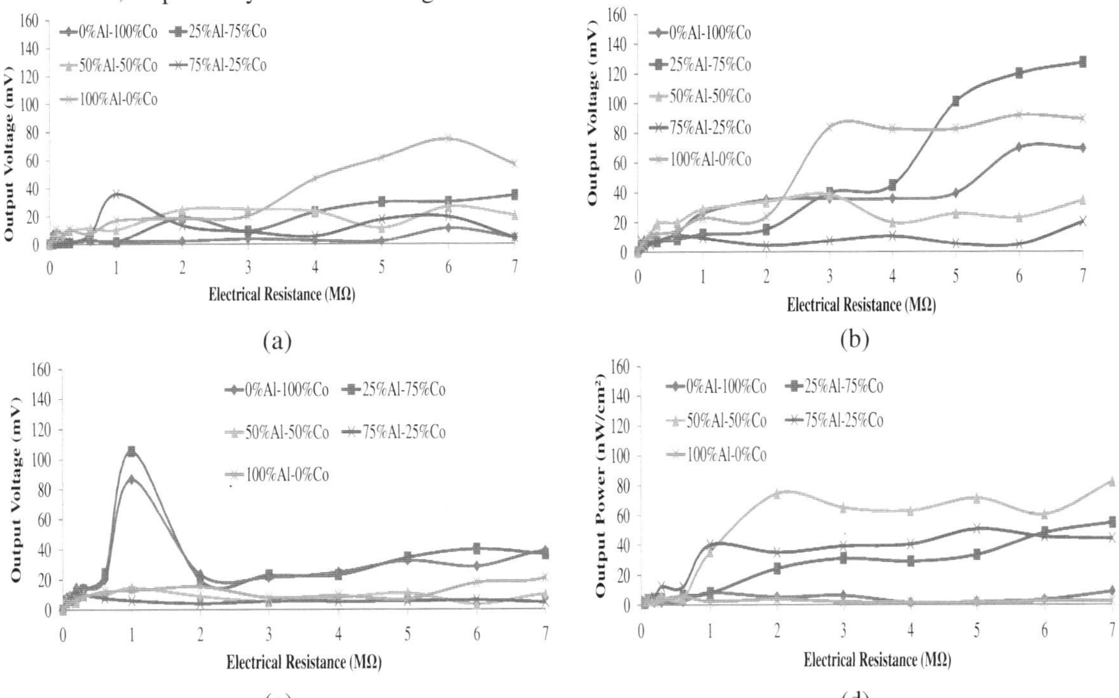

Figure 2. The output voltage of PENGs fabricated with ZnO NFs doped with Al/Co at sintering temperatures (a) 450°C, (b) 500°C, (c) 550°C, and (d) 600°C.

The output voltage was also low at low electrical resistance. The high output voltage was generated at a resistance load higher than 4 MΩ. At high electrical resistance, the current is difficult to flow causing the increase of output voltage. Meanwhile, the effect of the mass ratio of codoping between aluminum and cobalt on the output voltage of PENGs did not show a regular pattern. This is due to irregularity of the orientation from ZnO NFs as seen in Figure 3.

Figure 3 also showed that micro changes in diameter occur as the sintering temperature increases. The ZnO NFs have a large fiber size at a sintering temperature of 550°C compared to other sintering temperatures. The increasing sintering temperature make it more clear to see the form of ZnO NFs. However, at sintering temperatures greater than 550°C, some fibers appear to be joined to form larger fibers. This phenomenon caused the output voltage of PENGs made from ZnO NFs synthesized at sintering temperatures greater than 550 has decreased.

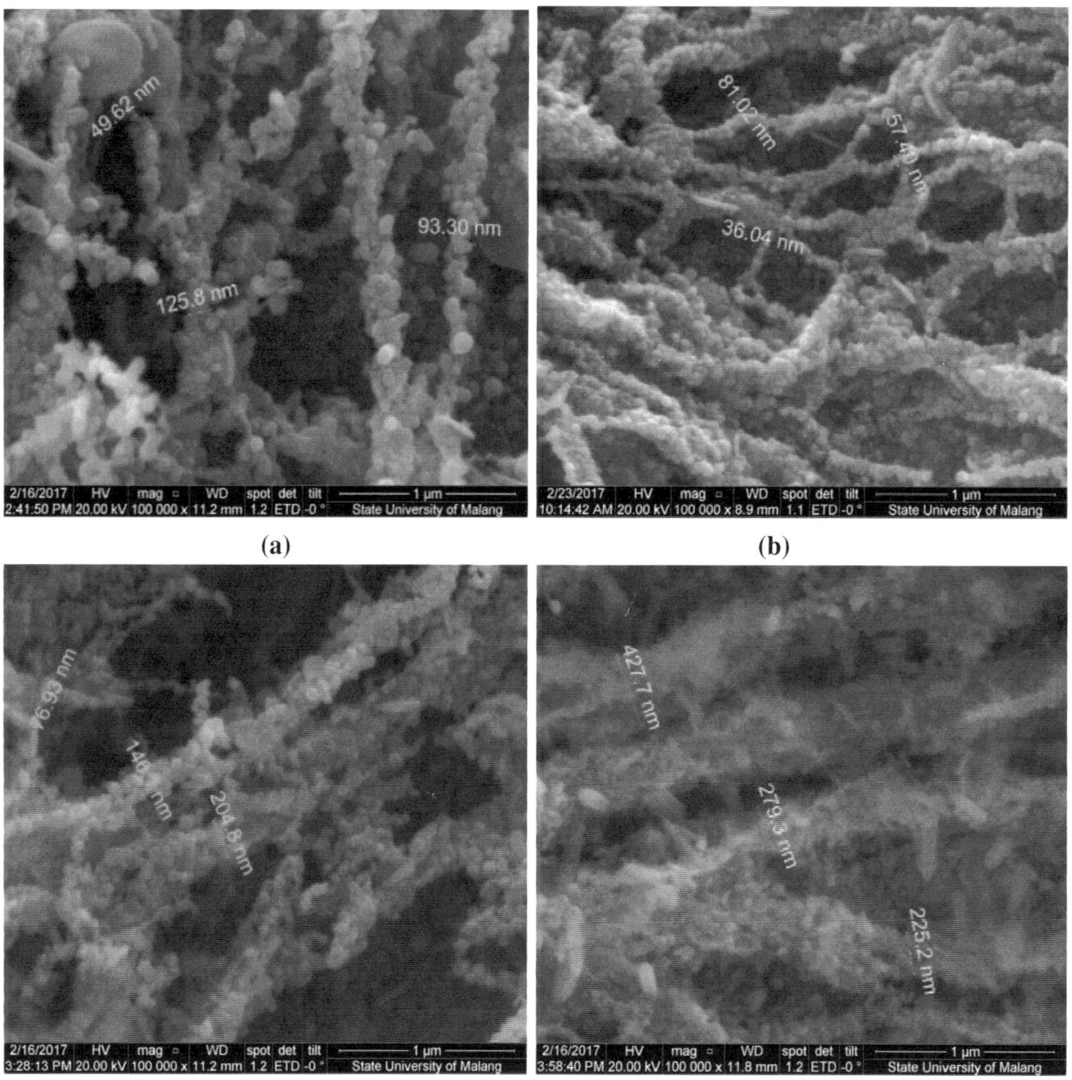

Figure 3. SEM images of ZnO NFs codoped with Al/Co and synthesized at sintering temperature (a) 450°C, (b) 500°C, (c) 550°C, and (d) 600°C.

4. Conclusions

PENGs were successfully and simply fabricated by electrospinning method from ZnO NFs doped with aluminum and cobalt. The highest output voltage of PENGs was obtained at a sintering temperature of 500°C, i.e. 127 mV. The output voltage of PENGs at sintering temperatures of 450°C, 550°C, and 600°C are 75 mV, 105 mV, and 85 mV, respectively. The PENGs fabricated from ZnO NFs codoped with aluminum and cobalt generated high output voltage, and are interesting candidates for exploiting in self-power devices.

5. Acknowledgements

The authors thank to the Ministry of Research, Technology, and Higher Education of the Republic of Indonesia for the research funding through Hibah Kompetensi FY 2015-2017. The authors would like to thank also for the research facilities supported by LPPM UNS.

6. References

[1] Suyitno S, Huda S, Arifin Z and Hadi S 2014 Repeatability, reproducibility, and durability of zinc oxide fibre-based nanogenerator synthesized by simple electrospinning machine *Advanced Science Letters* **20** 2299-303

[2] Suyitno, Purwanto A, Hidayat R L L G, Sholahudin I, Yusuf M, Huda S and Arifin Z 2014 Fabrication and Characterization of Zinc Oxide-Based Electrospun Nanofibers for Mechanical Energy Harvesting *Journal of Nanotechnology in Engineering and Medicine* **5** 011002

[3] Hananto, D.R. S and JUlius 2011 Application of Piezoelectric Material Film PVDF (polyvinylidene flouride) as Liquid Viscosity Sensor *Jurnal Neutrino* **3 (2)**

[4] Ubaidillah, Suyitno, Juwana W E, Prabandono B and Purwanto A 2013 Characteristics of zinc oxide nanorods synthesized by low power DC thermal plasma *Scientia Iranica* **20** 2348-55

[5] Xu Q and Qin Y 2017 Theoretical study of enhancing the piezoelectric nanogenerator's output power by optimizing the external force's shape *APL Materials* **5** 074101

[6] Peng C, Li J-F and Gong W 2005 Preparation and properties of (Bi1/2Na1/2)TiO3–Ba(Ti,Zr)O3 lead-free piezoelectric ceramics *Materials Letters* **59** 1576-80

[7] Momeni K, Odegard G M and Yassar S 2010 Nanocomposite electrical generator based on piezoelectric zinc oxide nanowires *Journal of Applied Physics* **108 (11)** 114303

[8] Zhang Y, Ram M K, Stefanakos E K and Goswami D Y 2012 Synthesis, Characterization, and Applications of ZnO Nanowires *Journal of Nanomaterials* **2012** 22

[9] Kim D Y, Lee S, Lin Z-H, Choi K H, Doo S G, Chang H, Leem J-Y, Wang Z L and Kim S-O 2014 High temperature processed ZnO nanorods using flexible and transparent mica substrates for dye-sensitized solar cells and piezoelectric nanogenerators *Nano Energy* **9** 101-11

[10] Chang Y-T, Chen J-Y, Yang T-P, Huang C-W, Chiu C-H, Yeh P-H and Wu W-W 2014 Excellent piezoelectric and electrical properties of lithium-doped ZnO nanowires for nanogenerator applications *Nano Energy* **8** 291-6

[11] Liu C, Peng M, Yu A, Liu J, Song M, Zhang Y and Zhai J 2016 Interface engineering on p-CuI/n-ZnO heterojunction for enhancing piezoelectric and piezo-phototronic performance *Nano Energy* **26** 417-24

[12] Kim Y-S, Xie Y, Wen X, Wang S, Kim S J, Song H-K and Wang Z L 2015 Highly porous piezoelectric PVDF membrane as effective lithium ion transfer channels for enhanced self-charging power cell *Nano Energy* **14** 77-86

[13] Choi D, Lee K Y, Lee K H, EokSuKim, Kim T S, Lee S Y, Kim S-W, Choi J-Y and Kim J M 2010 Piezoelectric Touch-Sensitive Flexible Hybrid Energy Harvesting Nanoarchitectures *Nanotechnology* **21**

Maximizing Drilling Performance through Enhanced Solid Control System

S Irawan [1], **B I Kinif** [1(*)], **R Bayuaji** [2]

[1]Department of Petroleum Engineering
Universiti Teknologi PETRONAS, Seri Iskandar, Perak, Malaysia
[2]Department of Civil Engineering
Institut Teknologi Sepuluh Nopember, Surabaya, East Java,

E-mail: imros_kinif@yahoo.com

Abstract. High solid content in drilling mud may affect its properties and result in uncertainties at downhole condition. It eventually contributes to poor rig operation performance and operating cost. This research focus on developing solid control system that is suit for drilling 12.25-inch hole. The first part discussed the performance of Rate of Penetration (ROP), Equivalent Circulating Density (ECD) and drill string drag while the second part of the research discussed about the effect of solid control system performance to mud properties Plastic Viscosity (PV), Yield Point (YP) and Low-Gravity Solid (LGS). The input parameters were gathered from two different set up of solid control systems that were used in Well A and Well B. The result is mainly based on the performance of original solid control system new design versus old design. Installation of distributor tank and channel the mud to respective shale shakers significantly enhanced the system and operational performance. The ROP at 12.25-inch drilling were improved by 20%. New design improved average the ECD margin by reducing additional pressure exerted using original mud from 4.9% to 2.9%. High ECD margin is not recommended because it can break the weak formation. Mud properties while drilling the 12.25-inch hole section; PV, YP and LGS values were improved by 14 %, 17 % and 25 % respectively. Proper mud flow control and routing system at new develop design of solid control system effectively removed the solid in the drilling fluid. This improvement minimizes the tendency of frequent mud flow, screen mesh plugging and tool wear issue. Mud properties such PV, YP and LGS were maintained with an acceptable mud design envelope.

1. Introduction

Effective performance of solid control equipment to remove drilled solids from drilling mud is a key factor to have good mud properties [1]. Solid and particles that are not removed during the circulation and remain in the system may reduce in size until it becomes difficult to remove using mechanical Solid Control Equipment (SCE). Mud properties must be maintained within the acceptable specification to allow efficient cuttings transport to surface [2]. This research focus on evaluating the performance of two different solid control systems that were used drilling the 8.5 inch and 12.25-inch wellbore sections. Preliminary finding revealed that most of Non-Production Time (NPT) on the rig was always associated with mud performance and solid control equipment. Mud data and drilling parameters were carefully studied to understand the root cause of poor drilling performance. The condition of unequal flow distribution over the shale shakers resulted in screen mesh plugging, shakers overflow, cones plugged, and exhibit rope frequently seen at rig site. These contributed to increased solid content that eventually affected the properties of PV, YP and LGS. Inability of the system to control these properties was the contributing factor for the poor ROP, ECD and drag. Therefore, the performance of drilling mud and role in this context was seen as vital for the whole drilling process.

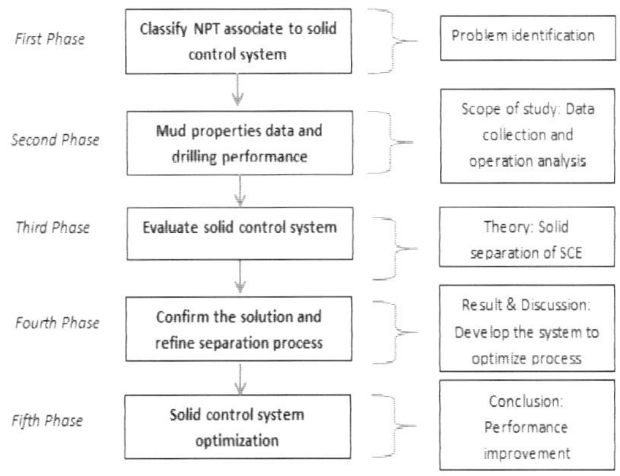

Figure 1: Solid Control Phases Diagram

Reduction of high drilled solid content in the mud system can significantly increase the ROP, minimize the string drag effect, prevent lost circulation and enhanced hole stability. Improvement in these parameters can increase the service life of the mechanical equipment, drill bits and more pumps. In operation perspective, the drilling mud must be capable to provide sufficient hydrostatic pressure and avoid formation influx. Therefore, the development of a solid control system capable to prevent severe problem drilling and attributed by drill solid contaminants is required. Mud properties within the acceptable specification ensure efficient hole cleaning and good cutting transport to surface.

2. Solid Content and Mechanical Solid Control System

The commercial solids (barite) and non-commercial solid (drilled cuttings) are two primary sources of solid particles that exist during drilling. Solid that are concentrated in the mud is classified either based on Specific Gravity (SG) or density and particle size. Solid SG higher than 4.2 (weight material) are

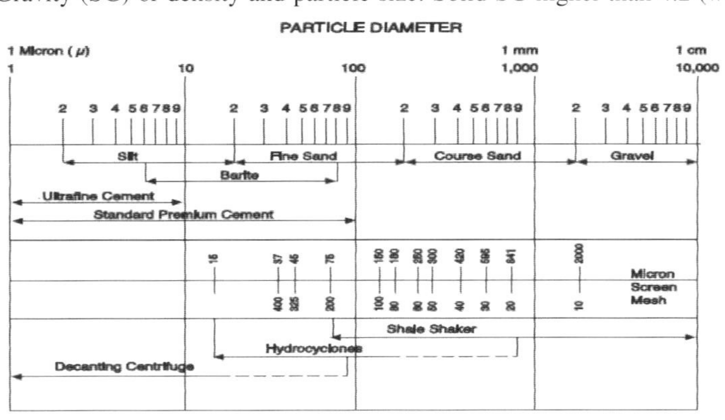

Figure 2: Particle Diameter

categorized as High-Gravity Solid (HGS) while solid with SG 1.6 to 2.9 (average 2.6) is categorized as Low-Gravity Solid (LGS). The solid control system sequence consists of shale shakers, hydrocyclones (desander and desilter) and centrifuge. Each of mechanical equipment is separate and work based on the solid particles sizes. The equipment is connected in series and each stage of processing performance is partly dependent upon the previous equipment. By-passing the shaker screen is not recommended because failure to discard the solid particles at the shale shakers unit would overload the downstream equipment. Drilled solids that are not removed during the first circulation through the surface equipment are always subject to mechanical degradation [3]. The smaller the particle, the greater the surface area may develop, and it is causing greater effect to the mud system.

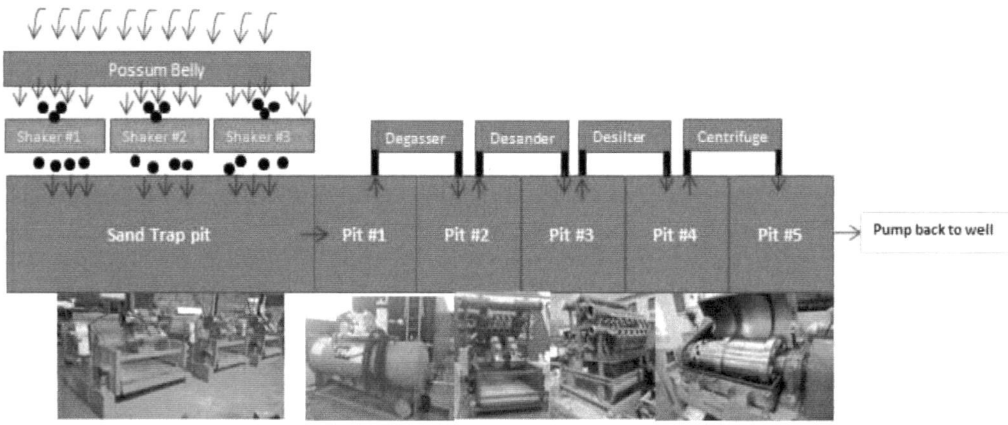

Figure 3: Old design solid control system

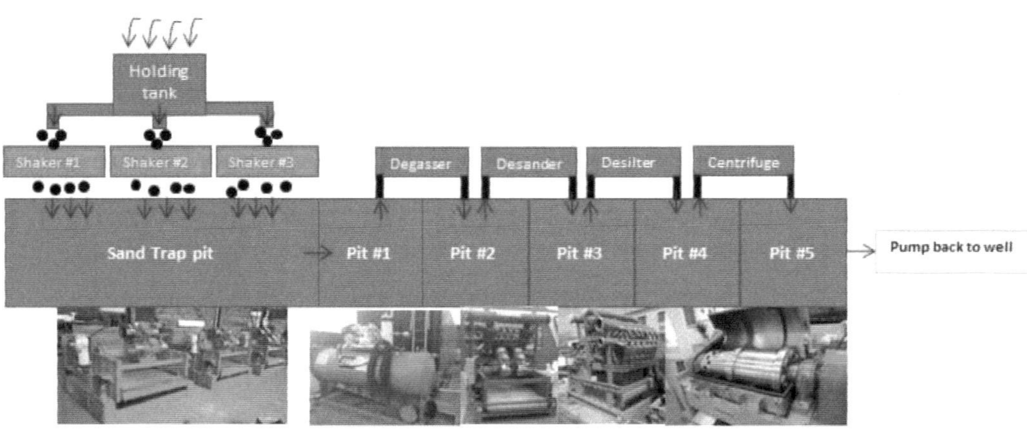

Figure 4: New design solid control system

3. Development of Solid Control Equipment

During drilling operation, the circulated mud carries drilled cuttings to surface. These solids is then treated at mechanical solid control system by separating them from the mud according to the particles size. In this paper, the research focus on maximizing the equipment performance and separation efficiency because the old design was inefficient to remove the solid from the mud. The flow line distributor was installed at the end of return line to the shale shakers and replaced the high-pressure hose to rigid piping system for reducing pressure loss due to vibration. This system allows the solids-containing drilling mud and drilling mud to be divided equally between the desire numbers of shale shakers. Mud circulation without any reduction in solid concentration may thickening the fluid viscosity and develops resistance to flow character of the mud. This brings to the changes in its density, rheology and other properties. Utmost advantage is to remove as much solids as practical to reduce down time resulting from plug flow lines, fluid end repair, drill string erosion and re-drilling of solids as drilling fluid recycles itself through the mud system loop [1].

3.1.1. Effect of Solid particles on MW

Density or MW is measured and expressed in pounds per gallons (lb/gal), pound cubic feet (lb/ft^3) and grams per cubic centimeter (g/cm^2). It is an important property to maintain the well hydrostatic and prevent gas influx migration into the wellbore. Some of weight material such as barite and hematite are added into the mud to increase the mud weight. Mud in the wellbore column must exert a greater pressure than the fluids in porous rocks that are penetrated by the bit. The pressure exerted by the drilling mud at any depth or gradient of pressure is related to its density. Too dense or viscous mud may exert excessive pressure to the wellbore and causing loss of circulation.

3.1.2. Effect of Solid particles on well Bottom Hole Pressure

High solids content ultimately can increase the density and viscosity of the mud. Hence, it exposes the well to high BHP and mud properties contamination. Moreover, high horse power is also required to break up the gel and pump the mud for circulation. This not only trigger to hydraulic fracturing effect but also induce the tendency of mud losses into the formation. Under the primary well control requirement, the mud density must be formulated to compensate the desire BHP either meet or exceed the pore pressure of the rock formation. Failure to control the build of solid and regrinding the same drilled cutting may increase the mud surface area. become very difficult to remove using mechanical solid control equipment. High solids are abrasive and can increase filter cake thickness. The higher of filter cake, the higher chances of drill string to get stuck due to differential sticking effect. Thin and impermeable filter cake important to reduce contact area across the drill string.

3.1.3 Effect of Solid particles on PV

PV is a function of solids concentration, size and shape of the solid particles and viscosity ofliquid phase [3]. It is regarded as a guide to solid control for field application [5]. PV increases if the volume percent of solid increases, or if the volume percent remains constant and the size of the particles decreases. Decreasing particle size may increase surface area that lead to fractional drag problem. This plastic viscosity is sensitive to the concentration of solid and depends largely on the bulk volume of solids in the mud [6]. A low PV implies lower ECD exerted at bottom while high PV trigger to an increase of ECD because high pumping pressure is needed to break the gel. YP/PV ratio is a significant indicator of drilling fluid condition, low ratio indicates smaller tendency for gas cutting, swabbing pressure and greater settling velocity of cuttings whereas high ratios indicate coagulation and flocculation [7]. Removal of drilled solids from a drilling fluid will decrease plastic viscosity and if this solid remain in the fluids, it will grind into smaller and more numerous particles which increases plastic viscosity and decreases drilling performance [8].

3.1.4 Effect of Solid on YP

YP is the initial resistance of the fluids to flow caused by the electrochemical forces between the particles. It is also expected to be a function of the solid concentration of the solids and those factors, such as surface charges and potential, which affect the inter-particle forces [9]. YP and gel strength should be low enough to allow sand and shale cuttings to settle out and entrained gas to escape, minimize swabbing effect during pulling the string out of hole and permit the circulation to be started at low pump pressure [10]. Efficient elimination of drilled solids right after the fluid leaves the annulus was the best solution to avoid drilling fluid-cutting interaction that subsequently can increase the fluid density [11]. A change in the PV of drilling mud can cause small changes in YP. Therefore, it is always important to keep the viscosity of a mud from getting too low. The mud should have minimum viscosity properties to lift the cuttings from bottom of the hole to surface. The mud must capable to keep the weighting material and drilled cuttings in suspension while circulating or stop pumping. Normal reaction in the event of poor cutting transport is to increase the YP of the mud. However, significant increase in YP may result poor performance of finest mesh at shaker screen. Changing the mesh screen to a coarser screen decreases the quantity of drilled solid that can be removed [12].

3.1.5 Effect of Solid on ROP

Rheological and filtration properties become difficult to control when the concentration of drilled solid become excessive [1]. High particulate solids in the mud reduce ROP because of increase in mud density and viscosity. The higher the mud density, the greater the differential pressure exerts. ROP decreases when differential pressure increases. Lower mud density may decrease the dynamic chip hold down and permitting faster RPM. Low viscosity mud promotes fast penetration because of good scavenging of drilled cuttings. Despite applying more WOB and RPM can comfortably achieve the desire ROP but drilling with contaminated mud properties decrease in ROP for a long run. Darley cited that low concentration of non-colloidal drilled solid below 4% capable to maintain ROP at high level [3]. Mud properties such as PV and Yield stress/gel strength showed that although these properties have effects on ROP, but not very significant, only annular pressure losses seemed to drastically affect the ROP which is directly related to ECD [5].

3.1.6 Effect of Solid on Drag and ECD

The fluid rheology plays important role for solid transport and optimize the hole cleaning [14]. The best way to pick solid is with a low viscosity fluid in turbulent flow. Hole cleaning can be optimized by the use drilling mud with low gel strength and with low viscosity within the shear rates exposed to the annular flow [13]. In situations where ECD is not a limiting factor, high – viscosity fluids with high YP/PV ratios are preferred. Under situation where ECD is a limiting factor, the use of thin fluids in turbulent flow should be considered. Driller must ensure the ECD as well as its static density is within the safe limit. ECD is the effective density of a moving fluid and slightly more than the static density because of the friction pressure drop in the annulus. ECD depends on the pump rates and fluid viscosity. Therefore, maintain ECD within limits means keeping viscosity low. The main cause of elevated viscosity is low gravity solid (LGS) increased. Close monitoring on solid control equipment must be performed to ensure LGS are kept to a minimum [14].

IOP Conf. Series: Materials Science and Engineering **267** (2017) 012038 doi:10.1088/1757-899X/267/1/012038

4. Research Methodology

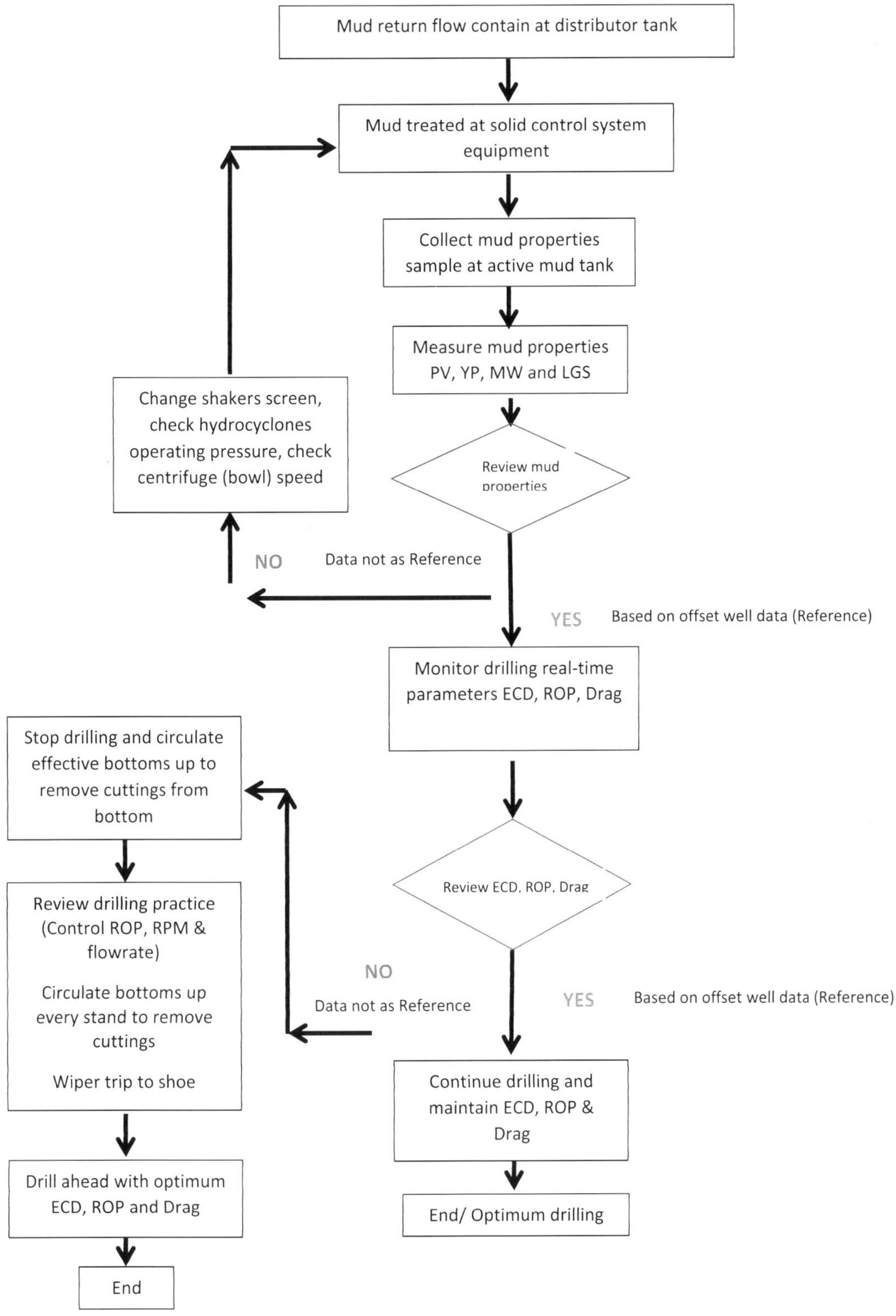

Figure 5: Flow Diagram of the Methodology

4.1.1. Setup of Solid Control System

The observation not limited but including drill cutting trends (shape and flow) and equipment working performance. The solids control comprises of three (3) shale shakers, hydrocyclones (Desander & Desilter) and centrifuge. The introduction of flow distributor tank at the end of the flowline and redistribute the mud through lines to respective shale shakers is designed to optimize the mud flow performance. This improvement was minimized the tendency of shale shakers overflow and reduce processing overloading.

4.1.2. Data Acquisition and Measurement

Data collection and evaluation of the mud properties including PV, YP and LGS. This measures to be used as a tool to evaluate the efficiency of the mechanical equipment. The drilling parameters including ECD and ROP are obtained from real-time downhole acquisition tool. Drill string drag is recorded after each drill pipe connection to monitor the hole condition. Mud parameters and drilling data are correlated to oversee the drilling performance. Gradual changes in mud properties, high ECD and poor ROP are significantly reflect to the ineffective of solid control system.

5. Result and Discussion

Field data are obtained from onshore drilling in Borneo Block. All data used are obtained from 2 different wells with similar lithology called Well A and Well B. The Well A was drilled using original solid control system while Well B was drilled with New developed design. The performance of both systems was compared while drilling the 12.25-inch section. A total of 40 mud samples were collected and measured to evaluate the PV, YP and LGS. These results act as a preliminary step to investigate the performance of the mud on ROP, ECD and drill string drag to justify the performance of the new design in this analysis.

5.1.1. Performance of solid control system on plastic viscosity.

Figure 6 shown the tabulated PV of New Design and Old Design solid control system. At the start of drilling operation, PV for both systems performed on the same trend. This reflects the solid separation and treatment was working effectively. As drilling deeper at Well A with Old system and more drilled cuttings excavated, the mud viscosity was getting thicker that resulted the PV to gradually increased. Increase of PV is subjected to drilling the Well B with New system design had improved the solid removal processing by 14% compared to Old Design. The performance of New system was economical and reliable as system capability justifies it to maintain the PV reading throughout the operation. The inability of the Old Design to eliminate rapid development of mud contamination significantly lead to overloading works at downstream equipment which increase solid contents in the drilling mud. Frequent mesh screen plugging, discharge rope from the Hydrocyclone and solid recirculation contributed to poor solid removal and PV increment. Spray discharge was not achieved because the Old Design utilized high pressure hose as a suction line to Desander and Desilter. Pressure generated to feed the mud into Desander and Desilter through suction hose caused vibration, pressure loss and inconsistent mud flow.

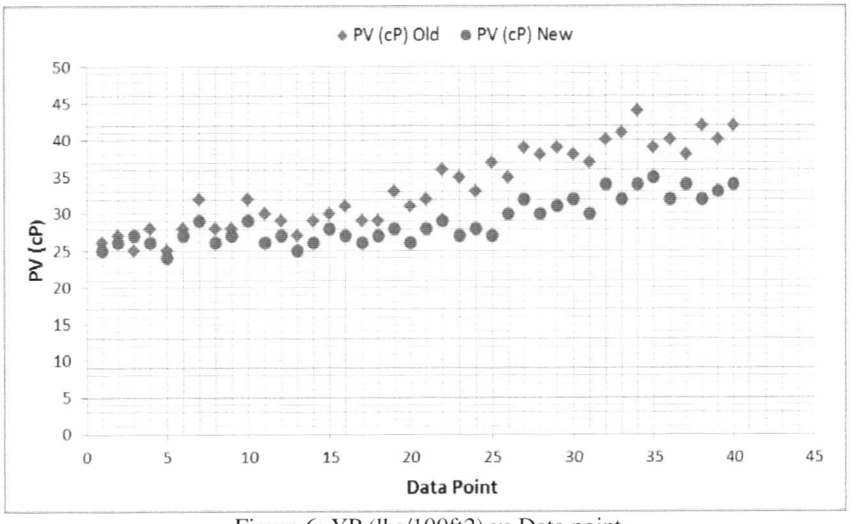

Figure 6: YP (lbs/100ft2) vs Data point

5.1.2. Performance of solid control system on yield point.

Figure 7 shown the tabulated YP for Old Design and New Design. The YP of Old Design gradually increased because solid in the drilling mud was not properly discarded while drilling Well A. Mud overflow on the shale shakers was frequently observed and occasionally the unit was bypassed to minimize the surface losses. By passing the unit significantly overload the downstream mechanical equipment and result the equipment incapable to remove the solid efficiently. Changes in low shear rate viscosity reflect to the mud YP. In this condition colloidal clay platelets link together (flocculate) with consequent by increase in their specific surface area. When mud is at static condition, the mud contain high solid and becomes attractive and repulsive. Stable and consistent mud flow distribution to shale shakers was helpful in controlling the YP build up. The ability of the New Design to maintain the YP showed that the solids were properly separates by the system and result of in low pressure loss while the drilling mud was circulated. Consistent value of YP at New Design while drilling the section typically provides good cutting carrying capacity (CCC) of the drilling fluid. Good control of YP reduces the chances of pressure spike that can break the formation which may result lost in circulation. Sufficient YP and gel strength were achieved at acceptable gel strength to help for cutting suspension while circulating and pump shut down. Moreover, the mud was capable to lift the cuttings from bottom of the hole to surface.

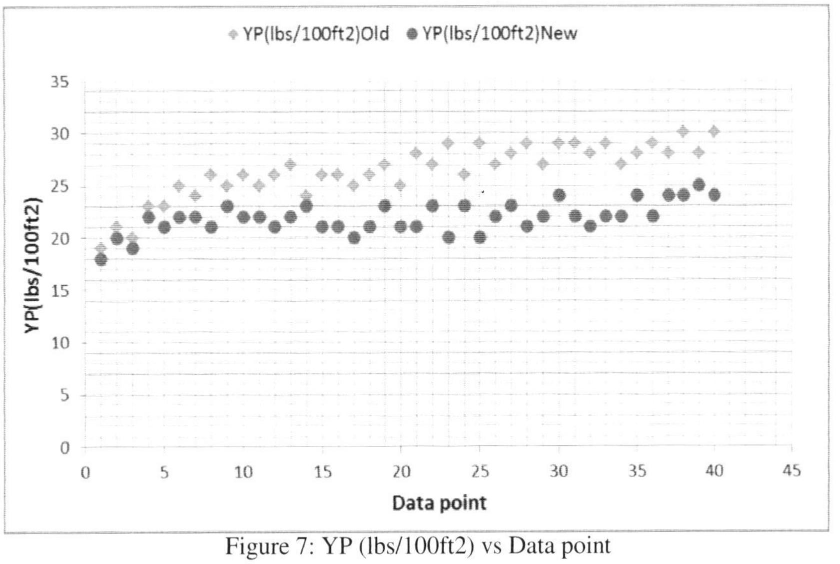

Figure 7: YP (lbs/100ft2) vs Data point

IOP Conf. Series: Materials Science and Engineering **267** (2017) 012038 doi:10.1088/1757-899X/267/1/012038

5.1.3. Performance of solid control system on low-gravity solid.

Figure 8 shows the tabulated LGS for Old Design and New Design. Rapid increment of LGS while drilling the Well A was obviously due to inability of the Old Design to remove the solid efficiently from drilling mud. At the start of drilling operation, both designs removed the solid effectively from the drilling mud as the LGS was tabulated at the range of 7% to 8%. As drilling the well, a deeper, the solid control equipment (SCE) of Old Design was observed getting poor in handling the mud return from the well. Frequent shale shakers overflow and bypassing the shakers in order to prevent massive surface loss of expensive fluids significantly created additional risk to the solid removal processing. The situation results of the LGS in the mud system to rapidly increased to 14%. High solid content in this mud was considerably abrasive and may degrade down the drilling equipment through silt size. The smaller the particles the more pronounced the effect on the mud properties because smaller particles are more difficult to remove or control its effect on the fluid. Re-circulating of mud that contained drilled solid may gradually deteriorate mud properties. The upper limit of the solid fraction should be in the range of 6% to 8% by volume. The New Design of solid control system as tabulated in figure 8 shows that the system LGS was improved with system removal by 25% while drilling the Well B. The average reading of LGS was 8.7%.

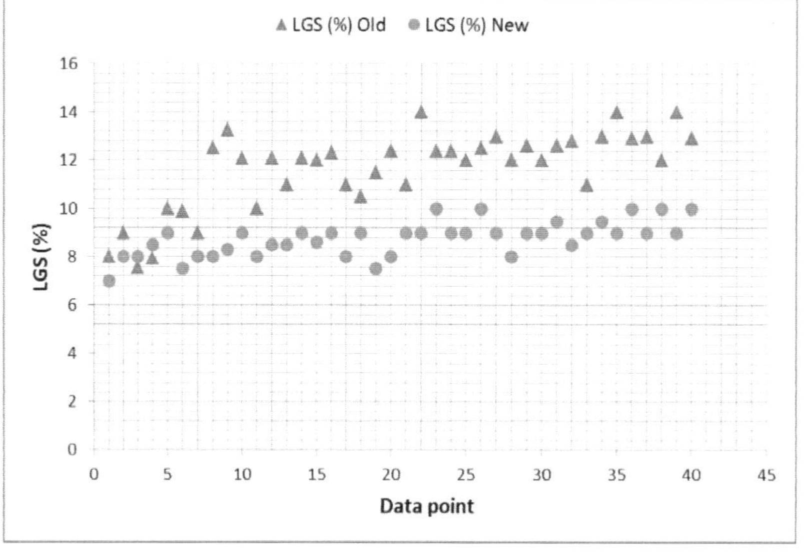

Figure 8: LGS (%) vs Data point

5.1.4. Performance of solid control system on mud weight and equivalent circulating density.

Well A was drilled using old design solid control system. The section was from 760m MDRT to 2163m MDRT. MW was gradually increased from 10ppg – 11.6ppg mud to maintain the hydrostatic pressure in the column. Early tabulated data on the top section demonstrated a good ECD trend with no significant pressure spike. As drilling reached to 1124m MDRT depth, ECD spike was increased up to 12.4ppg. Failures to transport the cutting effectively to surface exposed the formation to pressure spike. It occurred because drilled cutting that remain in the well created tight tolerances between hole and drill string geometries. Poor hole cleaning triggered high ECD that can break the formation and losses of mud. ECD trend progressively built up and pressure spike was observed at 1244m MDRT and 1474m MDRT. Well circulation was commenced until the hole clean, but the entire operation progress became slows. The flow rate was ramped up in controlled manner to minimize the risk of pressure spike and high ECD. The ROP was also controlled to reduce the impact of high ECD. The average ECD additive factor was 0.6 ppg that is equivalent to 4.9 % increment from original MW during mud than in dynamic condition.

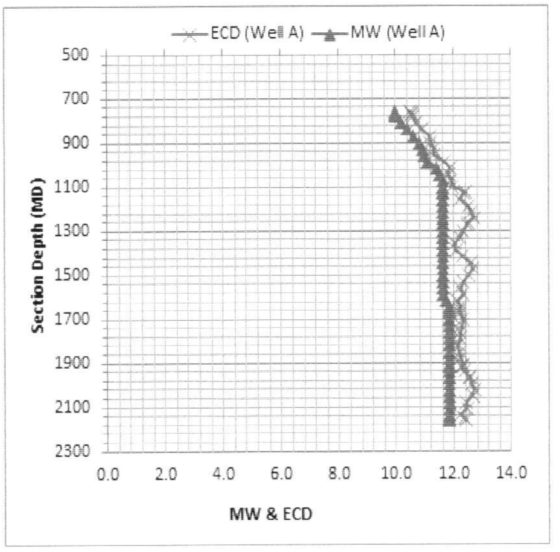

Figure 10: Section Depth vs MW & ECD (Old Design)

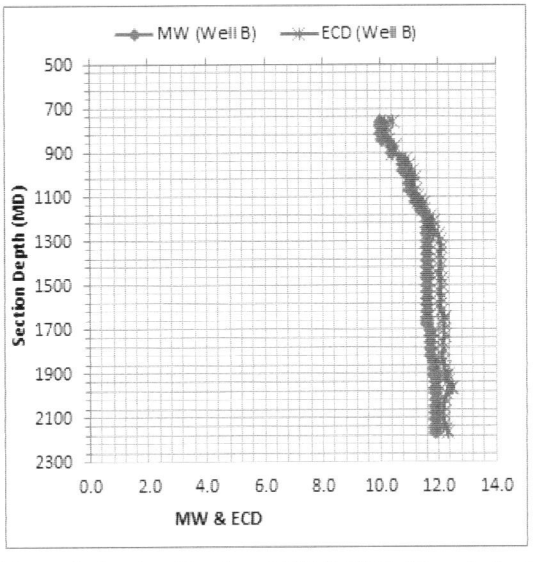

Figure 9: Section Depth vs MW & ECD (New design)

Well B was drilled using new design of solid control system. The section was drilled from 750m MDRT to 2200m MDRT. The MW used to drill this section was 10.0ppg to 11.9ppg and gradually increased to maintain the hydrostatic pressure in the column. The tabulated data shows that the ECD was consistent at safe drilling margin until section TD at 2150m MDRT. No significant ECD spike was observed throughout the operation. A slight increase was observed at 1935m MDRT with 12.32ppg to 12.5ppg but the impact is still below the fracture gradient. Increase in ECD occurred because more cutting bed accumulated on the low side of drill string. Due to gravity effect at deviated hole, heavier cuttings and mud along the lower side of the hole moved at lower rate than the clean mud on the upper side. Peak performance of solid separation equipment is essentially important to ensure the LGS, YP and PV are maintained within the acceptable envelope. These parameters provide good cutting transport and no excessive pumping pressure required to break the gel. The average ECD additive factor was 0.3 ppg that is equivalent to 2.9 % increment from original MW during mud flow in dynamic condition.

IOP Conf. Series: Materials Science and Engineering **267** (2017) 012038 doi:10.1088/1757-899X/267/1/012038

5.1.5. Performance of solid control system on rate penetration.

The ROP trend of old design was seen concentrated at range between 40m/hr to 60 m/hr. ROP was controlled in such manner due to ECD surge. Mud properties specification was not within the recommended limit and it significantly influenced the ROP consistency. Continuous of solid built up in the mud system altered the mud properties which eventually caused poor drilling performance. Drilled solids circulated up the annulus increased the pressure differential and lead to slower drilling. Instantaneous maximum ROP achieved using old design SCE was 62m/hr with 48.78m/hr on average. Drilling the same section for Well B using new design solid control system showed improvement on ROP trend. Maximum instantaneous ROP was registered at 77m/hr with 61m/hr in average. Efficient solid removal using new design improved the ROP performance by 25%.

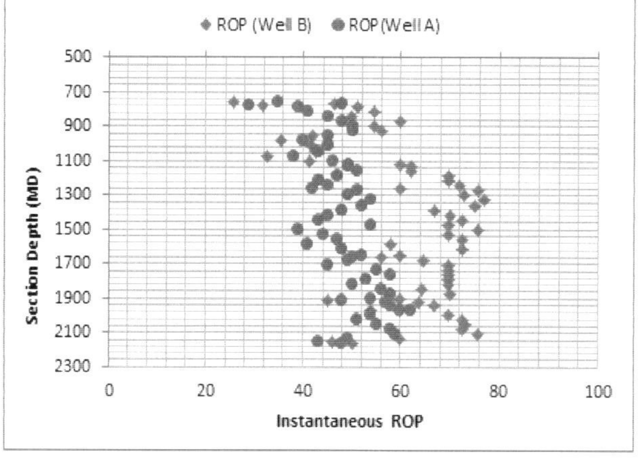

Figure 11: Section Depth vs ROP

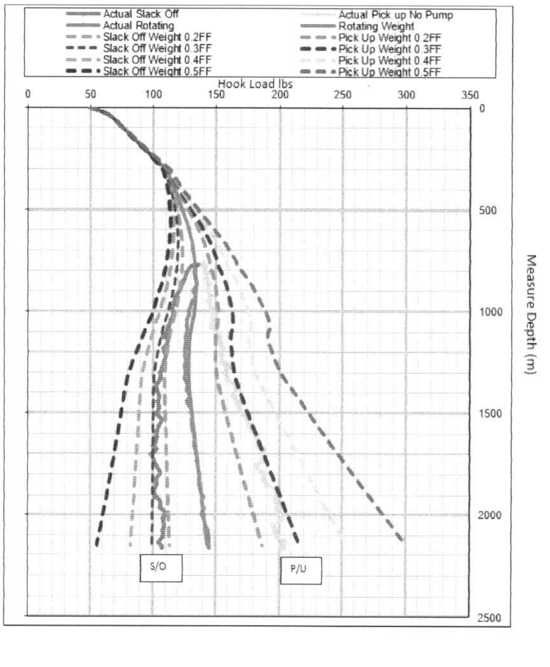

Figure 12: Hook Load effect vs Depth (Old Design)

5.1.6. Performance of old design of solid control system on Hooks load.

Drag force is known as difference between free rotating weight and the force required to move the string up and down in the hole. Pick-up (P/U) drag force usually higher than the free rotating weight while slack-off (S/O) force is lower than the free rotating weight. String weight or free rotating weight on this analysis represented by the green curve. Data was taken while the string was off bottom to compare the modelling string weight and actual weight. As designed expectation, the plots should tabulate across and along the modelling string weight at any well depth. Drilling the Well A using old design system, more solids were not properly separated from the mud. This increased the mud viscosity and altered the mud properties. In this chart, the hook load readings when P/U the drill string was tabulated between 0.2FF and 0.3FF as shown in figure 12. Occasionally, the plot exceeded that friction factor line as observed at depth around 1550m MDRT and 1900m MDRT. Pile up of drilled cuttings bed and accumulation increased the drill string contact with the wellbore. Mud circulation for hole cleaning was performed to remove the cuttings from the well. S/O weight of the drill string in this chart was observed between 0.2FF to 0.3FF. The plots were occasionally tabulated exceeding the 0.3FF. This condition indicated that more drilled cutting were still sticking and restricting the drill string to move down. It required extra cumulative axial force to free the drill string. Lower S/O FF represents high hook load and vice versa. Failure to effectively transport the cuttings to surface result in a number of drilling problems including: excessive overpull on trips, high rotary torque, stuck pipe, hole-pack off, excessive ECDs and cutting accumulation.

5.1.7. Performance of new design of solid control system on hook load.

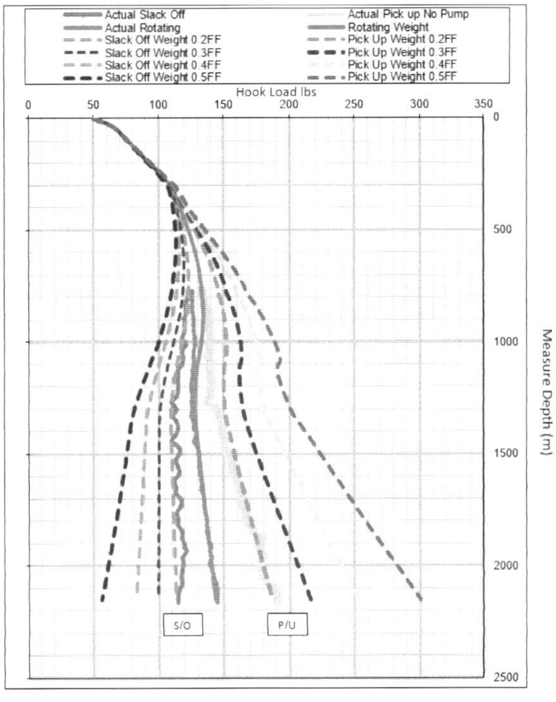

Figure 13: Hook Load effect vs Depth (New Design)

The new design system effectively separated the drilled cuttings from the mud and allowed each of the equipment to work at peak performance. Mud properties were within the design specification with no severe solid contamination result improvement on PV, YP and LGS reading. The hook load chart indicated that P/U weight was concentrated at 0.2FF but occasionally exceeded 0.2FF. Good control of mud return and consistent distribution flow to solid control system allowed effective solid particles separation. S/O weight of the drill string was observed tabulated below the 0.2FF but occasionally exceeded 0.2FF. This indication represented that the hole was not piled up with drilled cuttings. Less axial force was required to move the drill string as there was no severe contact of drilled cuttings with the string.

IOP Conf. Series: Materials Science and Engineering **267** (2017) 012038 doi:10.1088/1757-899X/267/1/012038

6. Conclusion

A new solid control system was developed with detailed studies on the effect of drilled solids to mud properties and drilling performance. The following conclusion can be drawn from this investigation and as an improvement of the rig solid control system.

1. The installation of mud distributor tank offers flow stability and control of mud return from the well. This effectively removes the solids from the mud, minimize the tendency of frequent mud flow and screened mesh plugging
2. Comparing the old design while drilling 12.25-inch hole section, the new design improved the mud properties of PV by 14%, YP by 17% and LGS by 25%.
3. New design improved the ROP by 20% while drilling the 12.25-inch hole section
4. New design improved average ECD margin by reducing additional pressure exerted using original mud from 4.9% to 2.9% at 12.25 inch. High ECD margin is not recommended because it can break the weak formation.
5. Drill string drag effect to the hook load for P/U and S/O at Old design were tabulated between 0.2FF to 0.3FF while drilling the section with New design were mostly tabulated at 0.2FF and less.

7. Acknowledgment

The authors wish to thank UTP for providing the resources and opportunity to conduct this research.

8. Nomenclature

PV	Plastic Viscosity
YP	Yield Point
ROP	Rate of Penetration
SCE	Solid Control Equipment
LGS	Low Gravity Solid
ECD	Equivalent Circulating Density
NPT	Non-Productive Time
CCC	Cutting carrying capacity
P/U	Pick up
S/O	Slack off
FF	Friction Factor

IOP Conf. Series: Materials Science and Engineering **267** (2017) 012038 doi:10.1088/1757-899X/267/1/012038

9. References

[1] B. Dahl, T.H. Omland and A. Saasen, "Optimised Solids Control In Artic Environments", SPE Russian Oil & Gas Exploration & Production Technical Conference and Exhibition, Moscow, Russia, October 2012.

[2] J.U. Akpbio, P.B Inyang and I.C Iheaka, The effect of Mud Density on Rate of Penetration, International Research Journal or Engineering and Technlogy (IRJET), 9 Dec 2015

[3] L. Robinson, "Drilling Fluid Processing Handbook", ASME – Elsevier, 2005.

[4] B.Dahl, A.Saasen and T.H. Omland, "Successful Drilling Oil and Gas by Optimal Drilling – Fluid Solid Control – A practical and Theoretical Evaluation", SPE Drilling & Completion, pp. 409-414, December 2008.

[5] A.O.A. Moses, and F. Egbon, "Semi-Analytical Models of the Effect of Drilling Fluid Properties on Rate of Penetration (ROP)", SPE Nigeria Annual International Conference and Exhibition, Abuja, Nigeria, August 2011.

[6] H.C.H Darley and G.R. Gray, "Composition and Properties of Drilling and Completion Fluids", Fifth Edition, 1988.

[7] G.V. Chilingarian, E. Alp, M. Al-Salem, S. Uslu, S. Gonzales and R. Dorovi, "Drilling Fluid Evaluation using Yield point-Plastic Viscosity Correlation" SPE General, 1983.

[8] L. Robinson, "Economic Consequences of Poor Solids Control", American Association of Drilling Engineer (AADE), 2006.

[9] Edition Technip, "Drilling Mud and Cement Slurry Rheology Manual", 1982.

[10] M.E. Baumert, E.N. Allouche and I.D. Moore, "Drilling Fluid Consideration in Design of Engineered Horizontal Directional Drilling Installation", International Journal of Geomechanic, vol. 5, issue 4, pp. 339-349, December 2005

[11] J.C. Machado and P.F. Castilho, "Solid control and Low solids: Important Binary to Drill Deep wells", Second Latin American Petroleum Engineering Conference, Caracas, Venezuela, March 1992.

[12] A.M. Paiaman, M.K. Ghassem and B. Salamani, B.D. Al-Anazi and M. Masihi, "Effect of fluid properties on Rate of Penetration", NAFTA 60 (3) pp. 129-134, 2009.

[13] A. Saasen and G. Lokingholm, "The Effect Of Drilling Fluid Rheological Properties On Hole Cleaning", IADC/SPE Drilling Conference, Dallas, U.S.A, February 2002.

[14] API recommended Practice 13D – Rheology and Hydraulics of oil-well drilling fluids, API2009.